Lecture Notes in Mathematics 1895

Editors:
J.-M. Morel, Cachan
F. Takens, Groningen
B. Teissier, Paris

L. Molnár

Selected Preserver Problems on Algebraic Structures of Linear Operators and on Function Spaces

 Springer

Author

Lajos Molnár
Institute of Mathematics
University of Debrecen
4010 Debrecen, P.O. Box 12
Hungary
e-mail: molnarl@math.klte.hu

Library of Congress Control Number: 2006931934

Mathematics Subject Classification (2000): Primary: 47B49, 81R15

ISSN print edition: 0075-8434
ISSN electronic edition: 1617-9692
ISBN-10   3-540-39944-5 Springer Berlin Heidelberg New York
ISBN-13   978-3-540-39944-5 Springer Berlin Heidelberg New York

DOI 10.1007/3-540-39944-5

Springer is a part of Springer Science+Business Media
springer.com
© Springer-Verlag Berlin Heidelberg 2007

Typesetting by the author and SPi using a Springer LaTeX package
Cover design: WMXDesign GmbH, Heidelberg

Printed on acid-free paper   SPIN: 11857136   VA41/3100/SPi   5 4 3 2 1 0

To my family

Preface

Originally this work served as the author's dissertation for the scientific title "Doctor of the Hungarian Academy of Sciences". To publish it in the form of a book has been initiated by the referees who formulated this suggestion in their reports.

In what follows we present a cross-section of recent research concerning preserver problems (both linear and non-linear) and local transformations (namely, local automorphisms and local isometries) defined on algebraic structures of linear operators and on function spaces.

Generally speaking, preserver problems concern the question of determining or describing the general form of all transformations of a given structure $\mathcal{X}$ which preserve

- a quantity attached to the elements of $\mathcal{X}$, or
- a distinguished set of elements of $\mathcal{X}$, or
- a given relation among the elements of $\mathcal{X}$,

etc. Such problems arise in most parts of mathematics. In fact, it turns out that in many cases the corresponding results provide important information on the automorphisms of the underlying structures. However, preserver problems are systematically studied only within the scope of matrix theory and, recently, within that of operator theory.

In this work we give a picture about some parts of that area of research by presenting several of our related results and referring to corresponding results of other authors.

The first group of questions we study here consists of preserver problems in the 'classical' sense, i.e., of problems which can be classified into that systematic study mentioned above. The second group is about preserver problems on certain fundamental structures of linear operators which appear in the mathematical foundations of quantum mechanics. Those results can also be viewed as the descriptions of certain automorphisms of the underlying quantum structures or, in other words, quantum mechanical symmetries.

The vague formulation of the basic problem concerning local transformations reads as follows. We are given a structure and a distinguished collection of transformations on it. Is it true that any transformation which locally belongs to that collection (i.e., which has the property that at every point its value coincides with the value of a transformation from our collection at that point) belongs to it also globally? If the answer to this question turns to be affirmative, then one can say that the collection under consideration is completely determined by its local actions. In this work we present some corresponding results which concern the collections (in fact, groups) of automorphisms and surjective isometries of operator algebras and function algebras. Although the ground problem of local transformations seems to have nothing to do with preserver problems, this is still an area of applications of the theory of preservers at least in part. In fact, as one can see in the proofs of the presented results we often use 'preserver arguments' and several particular preserver results.

Some words about the structure of the book which is very fixed and, hopefully, rather clear and logical.

The chapter 'Introduction' describes the considered research areas and presents the basic problems and important results obtained by other researchers. This is done in Sections 0.1, 0.3 and 0.5. These sections are followed by short surveys of our corresponding results which are included in the book. This is the content of Sections 0.2, 0.4 and 0.6. Section 0.7 collects the most frequently used notation and definitions.

Our results are presented in detail in the sections of Chapters 1, 2 and 3. Basically, they are taken from our corresponding papers but with a large number of modifications and revision, of course. These sections begin with short summaries describing the main results therein. After that more detailed introductions to the treated problems are presented and the precise formulations of the results are given. These subsections are followed by the complete proofs. At the end of each section some remarks are made on further results of ours or other authors, or on possible further directions of research.

In Appendix some basic results are collected that we use several times in Chapters 1, 2 and 3.

This was about the content of the book. Let me close the preface with some acknowledgements. In fact, it would have been much much harder to write this work without the support I have been lucky to get from the following three sources: my family, the Alexander von Humboldt Foundation and my friend Werner Timmermann. I am very grateful to all of them. To my family for the continuous love and for the support in the critical periods of my life. To the Alexander von Humboldt Foundation whose research fellowship I held during my stay at the Technical University of Dresden in the academic year 2002-2003 for the great possibility that for one complete year I could concentrate all of my energy on research and on the composition of this material. And to Werner who was my academic host in Dresden for the exceptionally warm hospitality including many illuminating discussions not only about mathematics but also about the matters of life.

Finally, special thanks are due to László Kérchy, Béla Nagy, Zoltán Sebestyén (referees of the original dissertation) and Werner Timmermann for the encouragement to publish this book, to the anonymous referees at Springer for their helpful comments and suggestions, and to Marina Reizakis from Springer for her kind assistance during the process of publication.

Hajdúszoboszló,   *Lajos Molnár*
May 2006

Contents

Introduction

0.1 Linear Preserver Problems

Particular preserver problems appear in many parts of mathematics. This is not surprising since it is a natural question in many contexts to ask that what are the transformations on a given structure which preserve 'something' (the meaning of this 'something' is of course well-defined in the problem under consideration and is in connection with the underlying structure). However, it seems that the only field in mathematics where preserver problems are studied systematically are matrix theory and, recently, its infinite dimensional variant, i.e., operator theory.

In fact, the so-called linear preserver problems (abbreviated as LPPs) represent one of the most active research areas in matrix theory. For surveys of the topic we refer to the papers [208, 136, 137]. According to the linear character of matrix theory, preserver problems mean here the characterizations of all linear transformations on a given linear space of matrices that leave certain functions, subsets, relations, etc. invariant. In the last decades there have been remarkable interest in similar problems also in the infinite dimensional case, i.e., when the underlying space consists of bounded linear operators (acting on a Hilbert space or on a Banach space) rather than matrices. For a corresponding survey see [29]. As in our work we are interested in problems of this latter kind, we are going to present the classification of linear preserver problems due to Li and Tsing [136] in such a setting.

Let $\mathcal{M}$ be a linear space of bounded linear operators on a Hilbert space. The first group of LPPs is concerned with the study of those linear transformations on $\mathcal{M}$ which preserve certain functions.

PROBLEM I. Let F be a (scalar-valued, vector-valued, or set-valued) given function on $\mathcal{M}$. Characterize those linear transformations ϕ on $\mathcal{M}$ which satisfy

$$F(\phi(A)) = F(A) \qquad (A \in \mathcal{M}). \qquad (0.1.1)$$

To give an example for such a problem, we recall a well-known result of Frobenius from 1897 describing the general form of all determinant preserving linear maps on matrix algebras which is commonly considered as the first result on LPPs. Namely, Frobenius proved in [76] that if $\mathcal{M} = M_n(\mathbb{C})$ (the algebra of all $n \times n$ complex matrices) and $F(A) = \det A$, then every linear transformation ϕ which satisfies (0.1.1) is either of the form

$$\phi(A) = MAN \qquad (A \in \mathcal{M})$$

or of the form

$$\phi(A) = MA^{tr}N \qquad (A \in \mathcal{M})$$

for some nonsingular matrices $M, N \in M_n(\mathbb{C})$ with $\det MN = 1$. (A^{tr} stands for the transpose of A.)

Next, observe that the problem of describing the surjective linear isometries of operator algebras can also be classified into this group of problems. In fact, we define F to be the norm function. (In relation with this, we remark that one would be right when saying that the characterization of all isometries of a given metric space is also a preserver problem in a more general but similar sense. See the discussion after the classification of LPPs.) In his famous paper [117], Kadison described the surjective linear isometries of C^*-algebras. He proved that every such transformation can be written as a so-called Jordan *-isomorphism multiplied by a fixed unitary element. As we see from Theorem A.9 in Appendix, this general result implies that the surjective linear isometries of the Banach space $B(H)$ of all bounded linear operators acting on the Hilbert space H can be described in the following way. The surjective linear map $\phi : B(H) \to B(H)$ is an isometry if and only if there are unitary operators U, V on H such that ϕ is either of the form

$$\phi(A) = UAV \qquad (A \in B(H))$$

or of the form

$$\phi(A) = UA^{tr}V \qquad (A \in B(H))$$

(here and in what follows ^{tr} denotes the transpose of operators with respect to an arbitrary but fixed complete orthonormal system).

As another important preserver problem of the same type, we mention the problem of spectrum preserving transformations. Clearly, here $F(A)$ equals $\sigma(A)$, the spectrum of the operator A. In [107], Jafarian and Sourour proved that if $\phi : B(H) \to B(H)$ is a surjective linear transformation with the property that $\sigma(\phi(A)) = \sigma(A)$ holds for every $A \in B(H)$, then there is an invertible operator $T \in B(H)$ such that ϕ is either of the form

$$\phi(A) = TAT^{-1} \qquad (A \in B(H))$$

or of the form

$$\phi(A) = TA^{tr}T^{-1} \qquad (A \in B(H)).$$

(In fact, in [107] the authors presented a similar result concerning Banach space operators not only Hilbert space operators.) As for the much more general problem of spectrum preserving surjective linear maps between semi-simple Banach algebras we refer to [9] and some of the references therein as well.

Now we turn to the second group of LPPs which concern linear transformations that preserve certain subsets.

PROBLEM II. Let $\mathcal{S}$ be a given subset of $\mathcal{M}$. Characterize those linear transformations ϕ on $\mathcal{M}$ which satisfy

$$A \in \mathcal{S} \Longrightarrow \phi(A) \in \mathcal{S} \qquad (A \in \mathcal{M})$$

or satisfy

$$A \in \mathcal{S} \Longleftrightarrow \phi(A) \in \mathcal{S} \qquad (A \in \mathcal{M}).$$

In the first case we say that ϕ preserves the elements of $\mathcal{S}$ in one direction (or, more simply, that ϕ preserves the elements of $\mathcal{S}$) while in the second case we say that ϕ preserves the elements of $\mathcal{S}$ in both directions.

When giving examples for this type of LPPs, one should not forget to mention the famous Kaplansky's problem on invertibility preservers although it concerns general Banach algebras not merely operator algebras. Motivated by the results in [65, 151] on invertibility preserving linear maps of matrix algebras and by the famous Gleason-Kahane-Żelazko theorem on the characterization of multiplicative linear functionals of commutative Banach algebras, Kaplansky [122] asked when must invertibility preserving linear transformations on algebras be Jordan homomorphisms (for the definition see the section Notation). For general Banach algebras the answer turned to be negative. In fact, borrowing a simple example from [234], consider the algebra of 3×3 upper triangular complex matrices and the transformation

$$\phi : \begin{bmatrix} a_{11} & a_{12} & a_{13} \\ 0 & a_{22} & a_{23} \\ 0 & 0 & a_{33} \end{bmatrix} \longmapsto \begin{bmatrix} a_{11} & a_{13} & a_{12} \\ 0 & a_{22} & a_{23} \\ 0 & 0 & a_{33} \end{bmatrix}.$$

Obviously, this is a surjective unital linear map from a Banach algebra onto itself which preserves invertibility in both directions but it is not a Jordan homomorphism since for the matrix

$$C = \begin{bmatrix} 0 & 0 & 1 \\ 0 & 0 & 1 \\ 0 & 0 & 0 \end{bmatrix}$$

we have $\phi(C^2) = 0$, $\phi(C)^2 \neq 0$. Therefore, the question was modified and Kaplansky's Problem now reads as follows.

KAPLANSKY'S PROBLEM. Let $\mathcal{A}, \mathcal{B}$ be semi-simple Banach algebras and let $\phi : \mathcal{A} \to \mathcal{B}$ be a surjective linear transformation with the properties that $\phi(1) = 1$ (i.e., ϕ is unital) and $\phi(A) \in \mathcal{B}$ is invertible whenever so is $A \in \mathcal{A}$ (i.e., ϕ preserves invertibility in one direction). Is it true that ϕ is necessarily a Jordan homomorphism?

Observe that this problem is in an intimate relation with spectrum preservers since if a linear transformation $\phi : \mathcal{A} \to \mathcal{B}$ is unital, then it preserves invertibility in both directions if and only if $\sigma(\phi(A)) = \sigma(A)$ $(A \in \mathcal{A})$ while it preserves invertibility in one direction if and only if $\sigma(\phi(A)) \subset \sigma(A)$ $(A \in \mathcal{A})$. Although a lot of work has been devoted to the solution of Kaplansky's Problem and several important results have been obtained, the problem is still open in its full generality. We mention only two breakthrough steps on the way to a possible solution. The first one is due to Sourour who proved in [234] that the conclusion in Kaplansky's Problem is true if $\mathcal{A}, \mathcal{B}$ are full operator algebras on Banach spaces. The second one is due to Aupetit who showed in [10] that every surjective unital linear map between von Neumann algebras which preserves invertibility in both directions is a Jordan isomorphism.[1]

Another important problem from the same group of LPPs is the one of unitary preservers. In [149], Marcus described the form of all linear transformations on $M_n(\mathbb{C})$ which preserve the unitary matrices in one direction. In [219], Russo and Dye generalized this result for arbitrary C^*-algebras showing that every linear transformation $\phi : \mathcal{A} \to \mathcal{B}$ between C^*-algebras which maps the unitary elements into unitary elements can be written as a unital Jordan *-homomorphism multiplied by a fixed unitary element. (We also refer to the paper [213] of Rais where the particular case $\mathcal{A} = \mathcal{B} = B(H)$ was treated.)

Our last example here is about linear transformations preserving operators with fixed rank. LPPs concerning rank are among the most important preserver problems. The reason is that it turns out in many cases that, with some effort, the original LPP under consideration can be reduced to such a problem. Among a number of relevant examples we only refer to the arguments in [107, 234] where such reductions were the clues of the proofs. The probably most fundamental result concerning rank preservers on operator algebras is due to Hou. In the paper [104] he described the general form of all weakly continuous linear maps on the whole operator algebra of a Banach space which preserve the rank-one operators. (We also refer to [203] for some results on additive maps preserving rank-one operators.)

The third group of LPPs concerns linear transformations preserving certain relations.

---

[1] Added in revision: We remark that one of the referees of the book pointed out that the same conclusion holds also for bijective unital maps preserving invertibility only in one direction. This follows from the discussion in [10, Remark 2.7]. Next, we remark that in the paper [60] Cui and Hou could strengthen the results of Aupetit. Namely, they proved that every surjective unital invertibility preserving linear map from a von Neumann algebra onto a semi-simple Banach algebra is a Jordan homomorphism.

PROBLEM III. Let $\sim$ be a relation on $\mathcal{M}$. Characterize those linear transformations ϕ on $\mathcal{M}$ which satisfy

$$A \sim B \Longrightarrow \phi(A) \sim \phi(B) \qquad (A, B \in \mathcal{M})$$

or satisfy

$$A \sim B \Longleftrightarrow \phi(A) \sim \phi(B) \qquad (A, B \in \mathcal{M})$$

In the first case we say that ϕ preserves the relation $\sim$ in one direction (or, more simply, that ϕ preserves the relation $\sim$) while in the second case we say that ϕ preserves $\sim$ in both directions.

A problem which should be certainly mentioned here is the problem of preserving commutativity. Although there are many results on the problem, this is still an active research topic. (For a recent remarkable achievement of the recent years see [202].) The basic result concerning commutativity preserving linear transformations on operator algebras is due to Omladič. In the paper [201] he described the structure of all bijective linear transformations on the whole operator algebra of a Banach space which preserve the commutativity in both directions. In the Hilbert space setting his result reads as follows. Let H be a Hilbert space of dimension at least 3 and let $\phi : B(H) \to B(H)$ be a bijective linear transformation which preserves commutativity in both directions. Then there exist a nonzero scalar λ, an invertible operator $T \in B(H)$ and a linear functional f on $B(H)$ such that ϕ is either of the form

$$\phi(A) = \lambda T A T^{-1} + f(A)I \qquad (A \in B(H))$$

or of the form

$$\phi(A) = \lambda T A^{tr} T^{-1} + f(A)I \qquad (A \in B(H)).$$

(We remark that the key step in Omladič's proof is to show that the transformation under consideration has a certain rank preserving property; see the last paragraph in the discussion of Problem II.)

To give another example, we mention the problem of zero product preserving mappings. These are the linear transformations $\phi : \mathcal{M} \to \mathcal{M}$ which satisfy

$$AB = 0 \Longleftrightarrow \phi(A)\phi(B) = 0 \qquad (A, B \in \mathcal{M}).$$

(We remark that similar transformations on function algebras are usually considered under the name 'disjointness preserving maps' or 'separating maps'.) For recent results on such transformations defined on operator algebras, we refer to [6], [58] and [105] (also see [50] for some general algebraic results). One of the statements in [105] asserts that if a surjective linear map $\phi : B(H) \to B(H)$ preserves zero product in both directions, then it is of the form

$$\phi(A) = \lambda T A T^{-1} \qquad (A \in B(H))$$

where λ is a nonzero scalar and $T \in B(H)$ is an invertible operator. For an interesting non-linear extension of this result we refer to last section of the paper [229].

Finally, the fourth group of LPPs concerns linear transformations which commute with certain maps on $\mathcal{M}$.

PROBLEM IV. Given a map $F : \mathcal{M} \to \mathcal{M}$, characterize those linear transformations ϕ on $\mathcal{M}$ which satisfy

$$F(\phi(A)) = \phi(F(A)) \qquad (A \in \mathcal{M}).$$

Although in this case we sometimes say that ϕ preserves F, this is not to be confused with Problem I.

To mention an interesting problem of this kind, we refer to the paper [48], where all the linear transformations on a full matrix algebra were described which preserve the kth power (here $F(A) = A^k$, k being a fixed integer not less than 2). The importance of that problem lies in the fact that when $k = 2$, the corresponding preservers are exactly the Jordan homomorphisms of the underlying algebra. Using deep algebraic techniques, an analogous problem was solved in [24] concerning additive maps of general prime rings. As a very particular case of the result presented there, we obtain that if $\phi : B(H) \to B(H)$ is a surjective linear transformation with the property $\phi(A^k) = \phi(A)^k$ $(A \in B(H))$, then there exist an $(k-1)$th root of unity λ and an invertible operator $T \in B(H)$ such that ϕ is either of the form

$$\phi(A) = \lambda T A T^{-1} \qquad (A \in B(H))$$

or of the form

$$\phi(A) = \lambda T A^{tr} T^{-1} \qquad (A \in B(H)).$$

The problem of linear transformations on operator algebras preserving the absolute value (here $F(A) = |A|$) also belongs to this group of LPPs. For corresponding results we refer to [159, 211, 210] where the problem concerning merely additive (not necessarily linear) transformations were considered and we also mention the paper [18].

This is the classification scheme of linear preserver problems á la Li and Tsing [136]. It should be clear that in the formulations of the problems above it is not really essential to assume that the underlying structures are operator algebras or that the transformations what we consider are all linear. We mean that analogous problems can be raised on arbitrary structures concerning transformations of any kind. In fact, in that way the territory of preserver problems can be enormously enlarged and many problems from very different parts of mathematics can be regarded as preserver problems.

For a very broad class of transformations that appear in every part of mathematics and can be viewed as preservers in some sense, we refer to morphisms as transformations on a given structure which preserve 'something' that is peculiar to their domain. Let us mention some specific examples, for instance, from geometry.

First we recall Felix Klein and his Erlanger Programm. In his famous talk in 1872 on the occasion that he was appointed to a professorship in Erlangen, Klein presented a new unified approach to geometry. In his view, geometry is the study of the properties of a space that are invariant under a particular group of transformations. In that way, according to the inclusions among those transformation groups, he could make a hierarchical order among the geometries of his times. This was an outstanding achievement because from the late 18th century until the end of the 19th century approximately, geometry literally exploded and flourished into a complex and apparently disconnected tree. Klein's approach profoundly influenced the mathematical development of geometry and it is now the standard accepted view.

Now, we turn to our examples.

- In Klein's approach Euclidean geometry is the study of properties of figures in the plane $\mathbb{R}^2$ which are invariant under the rigid motions, that is, the isometries of $\mathbb{R}^2$ when it is equipped with the usual Euclidean metric d_2. Since, according to Klein's view, the geometrical content is captured by the corresponding transformation group, it is an important problem to explore the structure of all transformations belonging to that group. In the present case, the problem is that how one can describe the structure of all isometries of $\mathbb{R}^2$. It is a remarkable result that every such transformation can be obtained from rotation, reflection and translation. Formulating the result in a more precise way, we have the following. If $\phi : \mathbb{R}^2 \to \mathbb{R}^2$ is a map which satisfies

$$d_2(\phi(x), \phi(y)) = d_2(x, y) \qquad (x, y \in \mathbb{R}^2)$$

(i.e., it preserves the Euclidean distance), then ϕ is of the form

$$\phi(x) = Ux + a \qquad (x \in \mathbb{R}^2)$$

where U is a 2×2 orthogonal matrix and $a \in \mathbb{R}^2$ is a fixed vector. This result is valid in higher dimensions as well and what concerns infinite dimensions, we refer to the famous Mazur-Ulam theorem [157] stating that every surjective isometry between real normed spaces is affine. (According to a result of Charzyński, in finite dimensions the assumption of surjectivity can be omitted; see [49] or [20].)

One should observe the analogy (howsoever weak it is) between the present problem and the ones which belong to the class 'Problem I'.

- Our second example concerns affine plane geometry. As the underlying space we take $\mathbb{R}^2$ as above. In Euclidean geometry, the notion of distance is fundamental: circles with different radii are different. There are, however, geometric properties which are independent of metric information. Circles have properties as circles, not as circles of a certain radius. The same is true for all other geometric figures. Developing geometry with these requirements in mind leads to affine geometry. The relevant transformations of the plane in this case are the collinearities: a collinearity is a bijective map $\phi : \mathbb{R}^2 \to \mathbb{R}^2$ satisfying the condition that for all triples x, z, y of distinct points, x, y, z are collinear if and only if $\phi(x), \phi(y), \phi(z)$ are collinear. It is a natural question that how one can describe those transformations in a more explicit way. The answer is given in a famous theorem due to Darboux. It states that the bijective map $\phi : \mathbb{R}^2 \to \mathbb{R}^2$ is a collinearity if and only if it is of the form

$$\phi(x) = Ax + b \qquad (x \in \mathbb{R}^2),$$

where A is an invertible 2×2 matrix and b is a fixed vector in $\mathbb{R}^2$. Sometimes, Darboux theorem is formulated in a different way. Namely, one can say that the bijective map $\phi : \mathbb{R}^2 \to \mathbb{R}^2$ maps lines to lines if and only if it is of the form above.

One should remark that the first formulation of the theorem is in a quite close relation with Problem III while the second one is in some looser connection with Problem II.

- Our next example comes from projective geometry. One of the several forms of the fundamental theorem of projective geometry reads as follows. Let X be a linear space over a field and denote by $\mathcal{L}$ the set of all one-dimensional subspaces of X. A projectivity is a bijective map $\phi : \mathcal{L} \to \mathcal{L}$ which maps any three coplanar elements of $\mathcal{L}$ to coplanar elements. In other words ϕ has the property that whenever $e, f, g \in \mathcal{L}$ are such that $e \subset f + g$, we have $\phi(e) \subset \phi(f) + \phi(g)$. Now, the theorem states that if $\dim X \geq 3$ then a bijective map $\phi : \mathcal{L} \to \mathcal{L}$ is a projectivity if and only if it is of the form

$$\phi(e) = Se \qquad (e \in \mathcal{L})$$

where $S : X \to X$ is a bijective semilinear operator.

It is clear that the projectivities are defined as maps which preserve the relation 'coplanarity' and hence they can be considered as preservers in a sense similar to what appears in Problem III.

- Observe that the group automorphisms can be also viewed as preservers in some general sense. In fact, roughly speaking, these transformations are the bijective maps on a given group which preserve the group operation in a sense similar to what we have seen in Problem IV. We mean that the group automorphisms are precisely the bijective maps which commute (in a certain sense) with the function of 'taking products'. Of course, this problem can not be classified directly into the group 'Problem IV', but we feel unquestionable that it is of similar spirit. If this argument is accepted, then going further on the way one can say that the isomorphisms between rings, algebras, etc. are also preservers in certain sense.

Regarding the automorphisms as preservers, in the particular cases we would like to know whether they are all of some nice form. To show one such case, we recall a beautiful result of Schreier and van der Waerden [222] concerning the form of all automorphisms of the general linear group. This group, that is the group $GL_n(\mathbb{F})$ of all nonsingular $n \times n$ matrices over the field $\mathbb{F}$ is a distinguished algebraic structure because of many reasons. Here we are interested in it because in the case when $n = 2$ and the underlying field is $\mathbb{R}$, it is the transformation group which represents 'normalized' affine plane geometry. Namely, it is the transformation group with collinearity and origin position as invariants. In Klein's approach the algebraic study of transformation groups is of fundamental importance. Therefore, it is important to explore the structure of the automorphisms of those groups. As for the present case, the corresponding result of Schreier and van der Waerden reads as follows. Let ϕ be an automorphism of $GL_n(\mathbb{F})$. Then there is a homomorphism $\gamma : GL_n(\mathbb{F}) \to \mathbb{F} \setminus \{0\}$, a field automorphism $\tau : \mathbb{F} \to \mathbb{F}$ and an element $T \in GL_n(\mathbb{F})$ such that ϕ is either of the form

$$\phi(A) = \gamma(A)T\tau(A)T^{-1} \qquad (A \in GL_n(\mathbb{F}))$$

or of the form

$$\phi(A) = \gamma(A)T\tau(A^{-1^{tr}})T^{-1} \qquad (A \in GL_n(\mathbb{F})).$$

(Here $\tau(A)$ is the matrix which one obtains when applying τ for every entry of A.)

With the above list of examples we have hopefully given some evidence that preserver problems in the general sense appear in many parts of mathematics and hence the study of them certainly deserves attention.

In this book we present a number of our results concerning preserver problems in both the original strict sense and the extended sense as well. Moreover, we collect several of our results on the so-called local automorphisms of operator algebras in the study of which the 'preserver approach' has proved to be very useful.

0.2 Survey of the Results of Chapter 1

In Chapter 1 we collect some of our results concerning linear and multiplicative preserver problems on operator algebras and on function algebras.

In Section 1.1 we consider linear transformations on operator algebras which preserve operators having infinite rank and infinite corank. As we have already mentioned, preserver problems which concern rank play distinguished role among LPPs and a lot of work has been done on them both in the finite and in the infinite dimensional cases. To mention only some of the corresponding papers, we refer to [17, 21, 66, 140, 246] for the finite dimensional case and to [58, 104, 203] for the infinite dimensional case. In our paper [90] we considered the problem of linear maps on operator algebras which preserve operators of a fixed corank. This problem is in some sense complementary to the problem of rank preservers. In Section 1.1 we study linear maps which are 'between' (finite) rank preservers and (finite) corank preservers, i.e., linear maps which preserve operators of infinite rank and infinite corank in both directions. It turns out immediately that, in general, transformations of this kind can be quite 'irregular': most of them can not be written in any of the nice compact forms that we have got used to when dealing with LPPs. Therefore, we restrict our attention to certain important subsets of operators with infinite rank and infinite corank such as the corresponding sets of idempotents (Theorem 1.1.1), projections (Theorem 1.1.2), and partial isometries (Theorem 1.1.3). We shall see that in those cases the corresponding preservers are all of one of the forms $X \mapsto AXB$, $X \mapsto AX^{tr}B$ with suitable operators A, B. In the last result of that section (Theorem 1.1.4) we describe the structure of the linear bijections of $B(H)$ which preserve left ideals in both directions. We prove that every such transformation is a two-sided multiplication corresponding to invertible operators.

In Section 1.2 we determine the structure of all linear maps on a von Neumann factor which preserve the extreme points of the unit ball. The study of this problem is motivated among others by a celebrated result of Kadison characterizing the surjective linear isometries of C^*-algebras. Clearly, every such isometry preserves the extreme points of the unit ball. In the main results of this section (Theorem 1.2.1 and Theorem 1.2.4) we see that on von Neumann factors the structure of all linear maps (not only the surjective ones!) which preserve the extreme points of the unit ball can be characterized in a way very similar to what Kadison obtained for the surjective isometries. For example, Theorem 1.2.1 shows that on an infinite factor each of our preservers can be written as a unitary multiple of a unital *-homomorphism or *-antihomomorphism.

In Section 1.3 we consider a preserver problem on the function algebra $C(X)$ of all complex valued continuous functions on the compact Hausdorff space X. The famous Banach-Stone theorem (see Theorem A.10) describes the general form of all surjective linear isometries of $C(X)$ with respect to the sup-norm. But beside the sup-norm, one can find it also natural to 'measure' a function with the diameter of its range. In view of the Banach-Stone theorem it is an immediate question that what are the linear bijections of $C(X)$ which preserve this quantity. In Theorem 1.3.1 we obtain that if X is first countable then our preservers are of a nice form. In fact, we prove that every linear bijection of $C(X)$ which preserves the diameter of the range of the functions $f \in C(X)$ is induced by a fixed homeomorphism of the underlying space X, a fixed rotation on $\mathbb{C}$ and translations in the vector space $C(X)$ with constant functions which depend linearly on f.

In Section 1.4 we present a result which can be viewed as a solution of a so-called multiplicative preserver problem. Such problems, that is, preserver problems concerning (not linear but) multiplicative maps on matrix algebras or, more generally, on operator algebras were first considered by Hochwald in [101]. (As recent papers, we refer to [52, 83] and [4] concerning the finite and infinite dimensional case, respectively.) In this section we study the *-semigroup endomorphisms of the operator algebra $B(H)$, that is, the multiplicative maps on $B(H)$ which preserve the adjoint in the sense as in Problem IV above. In Theorem 1.4.1 we show that if H is separable and infinite dimensional, then every such transformation satisfying a certain continuity condition (namely, having the Lipschitz property on commutative C^*-subalgebras) can be written in a nice form: it is a direct sum of the constant 0 map, the constant I map, some (unspecified number of) copies of the identity $A \mapsto A$, and some copies of the map $A \mapsto A^{*tr}$. We emphasize that we do not assume linearity. In fact, if the *-semigroup endomorphism under consideration is linear, then the problem can be reformulated asking that what are the *-representations of the operator algebra $B(H)$ on a separable Hilbert space. The answer to this question is well-known. Namely, by some deep results in the representation theory of $B(H)$ culminating in [121, 10.4.14 Corollary] we have that every such representation is the direct sum of some copies of the identity $A \mapsto A$.

0.3 Some Structures of Linear Operators in Quantum Mechanics

In the Hilbert space formulation of quantum mechanics, which is mainly due to von Neumann, several mathematical objects appear whose physical meaning is connected with the probabilistic aspects of the theory. The corresponding objects in which we are interested in the present work are the following. Let H be a (preferably complex) Hilbert space attached to a given quantum system.

(i) $S(H)$ denotes the set of all positive trace-class operators on H with trace 1. The elements of $S(H)$ are called the (normal) states of the system.

(ii) The extreme points of $S(H)$ as a convex set in $B(H)$ are called pure states. It is easy to see that they are exactly the rank-one projections on H. Their set is denoted by $P_1(H)$.

(iii) $P(H)$ stands for the set of all projections on H. This is an orthomodular lattice which describes the logical structure of quantum theory. In fact, the elements of $P(H)$ represent quantum events.

(iv) $B_s(H)$ denotes the set of all self-adjoint bounded linear operators on H. The elements of this set correspond to observable physical quantities.

(v) $E(H)$ stands for the set of all positive bounded linear operators on H which are majorized by the identity. The elements of $E(H)$ are called (Hilbert space) effects. They describe yes-no measurements which can be unsharp.

These sets are equipped with certain scalar valued functions and/or algebraic operations and/or binary relations which all have important physical content. The corresponding automorphisms, that is, the bijective maps on those sets which preserve the relevant structure represent different kinds of quantum mechanical symmetries. To describe those symmetries and to study the relations among them are important problems which were considered by a number of mathematicians and theoretical physicists. In what follows we briefly discuss the most well-known results obtained so far (see Chapter 2 of the recent book [47]).

We begin with Wigner's classical theorem on symmetry transformations [250, pp. 251-254] which concerns the set of all pure states equipped with the notion of transition probability. As written above, the set of pure states coincides with the set $P_1(H)$ of all rank-one projections on the Hilbert space H. A rank-one projection can be trivially identified with its range (which is a one-dimensional subspace) or with any unit vector which generates its range. Hence, one can regard pure states in (at least) three different ways: rank-one projections, one-dimensional subspaces, unit vectors (in this latter case the identification is one-to-one only up to multiplication by a scalar of modulus 1). We turn to the concept of transition probability. If $P, Q \in P_1(H)$ are pure states, then the transition probability between them is defined by $\operatorname{tr} PQ$, where tr stands for the usual trace functional. The mathematical meaning of this quantity can be better expressed by means of one-dimensional subspaces or unit vectors. In fact, it is easy to see that $\operatorname{tr} PQ$ is equal to $\cos^2 \theta$, where θ is the angle between the ranges of P and Q as one-dimensional subspaces. Moreover, $\operatorname{tr} PQ$ is also equal to $|\langle \varphi, \psi \rangle|^2$, the square of the absolute value of the inner product of φ and ψ which are arbitrary representing unit vectors from the ranges of P and Q, respectively.

According to the above three different representations (we mean (a) rank-one projections and the trace of their product, (b) one-dimensional subspaces and the angle between them, (c) unit vectors and the absolute value of their inner product), we have three different but essentially equivalent concepts of symmetry transformations. Here, we mention only that one which concerns maps on $P_1(H)$. A bijective map $\phi : P_1(H) \to P_1(H)$ is called a symmetry transformation if it preserves the transition probability, i.e., if it satisfies

$$\operatorname{tr} \phi(P)\phi(Q) = \operatorname{tr} PQ \qquad (P, Q \in P_1(H)).$$

In this language, Wigner classical result reads as follows. (We shall meet other formulations of Wigner's theorem later.)

Wigner's theorem. *The bijective map $\phi : P_1(H) \to P_1(H)$ is a symmetry transformation if and only if there is an either unitary or antiunitary operator U on H such that*

$$\phi(P) = UPU^* \qquad (P \in P_1(H)). \tag{0.3.1}$$

The operator U is unique up to multiplication by a scalar of modulus 1.

In other words, the result says that any automorphism of $P_1(H)$ with respect to the structure induced by the transition probability arises from a semi-automorphism of the underlying Hilbert space (here we use the word 'semi' to designate that U can also be conjugate-linear). Moreover, the operator algebraic reformulation of this classical result tells us that any symmetry transformation of $P_1(H)$ extends to a *-automorphism or to a *-antiautomorphism of the C^*-algebra $B(H)$.

There is a nice and important generalization of Wigner's theorem due to Uhlhorn [237]. His result in the language what we have used to formulate Wigner's theorem reads as follows.

Uhlhorn's theorem. *Suppose that* $\dim H \geq 3$. *Let* $\phi : P_1(H) \rightarrow P_1(H)$ *be a bijective transformation which preserves the orthogonality between the elements of* $P_1(H)$ *in both directions, i.e., assume that* ϕ *has the property*

$$PQ = 0 \Longleftrightarrow \phi(P)\phi(Q) = 0 \qquad (P, Q \in P_1(H)).$$

Then ϕ *is of the form* (0.3.1) *with some unitary or antiunitary operator* U *on* H.

It is easy to see that $\operatorname{tr} PQ = 0$ is equivalent to $PQ = 0$. Hence, this result is really stronger than Wigner's original theorem as this one requires merely the preservation of zero transition probability. One can say that Uhlhorn's transformations preserve only the logical structure of a quantum mechanical system while Wigner's transformations preserve its complete probabilistic structure. The point is that in the case when $\dim H \geq 3$, Uhlhorn was able to obtain the same conclusion as Wigner.

These are the most fundamental results concerning the symmetries of the set of pure states.

As for the set $S(H)$ of all states, we have already mentioned that it is a convex set. The operation of convex combination on $S(H)$ is sometimes called mixture. The corresponding automorphisms are called mixture automorphisms (or affine automorphisms or S-automorphisms in the terminology of [45] or Kadison automorphisms in the terminology of [232]). Clearly, the mixture automorphisms are simply the bijective affine maps of $S(H)$. These are the most common automorphisms of the set of states. It is well-known (see e.g. [45, 232]) that the map $\phi : S(H) \rightarrow S(H)$ is a mixture automorphism if and only if it is of the form

$$\phi(A) = UAU^* \qquad (A \in S(H))$$

for some unitary or antiunitary operator U on H.

By an automorphism of the set $P(H)$ of all projections one usually means an ortho-order automorphism which is a bijective map $\phi : P(H) \rightarrow P(H)$ that preserves the order $\leq$ in both directions (this is just the usual order among self-adjoint operators) and the operation $\perp: P \mapsto I - P$ of orthocomplementation (which means that $\phi(P^\perp) = \phi(P)^\perp$ holds for every P). It follows from the fundamental theorem of projective geometry that, in case $\dim H \geq 3$, any ortho-order automorphism ϕ of $P(H)$ is of the form

$$\phi(P) = UPU^* \qquad (P \in P(H))$$

for some unitary or antiunitary operator U on H (see [45, 243]; a stronger result can be found in [71]).

Next consider the set $B_s(H)$ of all self-adjoint bounded linear operators (or, in other words, bounded observables) on H. This set is usually equipped with the operations of addition, multiplication by real scalars, and the so-called Jordan product $(A, B) \mapsto (AB + BA)/2$. The obtained structure is a real Jordan algebra. The most important automorphisms of $B_s(H)$ are the bijective maps $\phi : B_s(H) \rightarrow B_s(H)$ which preserve these operations. They are usually called Jordan automorphisms (or Segal automorphisms in the terminology of [232]). It is well-known that every such automorphism ϕ is implemented by an either unitary or antiunitary operator U on H which means that ϕ is of the form

$$\phi(A) = UAU^* \qquad (A \in B_s(H))$$

(see [45, 232]).

The notion of quantum effects is also a basic concept in the foundational studies of quantum physics [38, 63, 69, 130, 146]. These objects correspond to yes-no measurements that can be unsharp. In the Hilbert space formalism, quantum effects are represented by operators A on a Hilbert space H which satisfy $0 \leq A \leq I$. The set of all such operators on H (which are also called Hilbert space effects) is denoted by $E(H)$.

There are several relations and operations usually considered on $E(H)$ which all have serious physical meaning. We briefly summarize the most important ones.

(E1) There is a partial addition $\oplus$ on $E(H)$ which is defined in the following way. If $A, B \in E(H)$ are such that $A + B$ (the usual sum of operators) belongs to $E(H)$, then $A \oplus B := A + B$.

(E2) The usual ordering among self-adjoint operators gives rise to a partial order on $E(H)$. This is denoted by $\leq$. The map $\perp \colon A \mapsto A^\perp = I - A$ defines a kind of orthocomplementation on $E(H)$.

(E3) The set $E(H)$ is a convex subset of the linear space $B_s(H)$. Any convex combination $\lambda A + (1 - \lambda)B$ of effects A, B is called a mixture.

(E4) There is a kind of product called sequential product on $E(H)$ which is defined in the following way. If $A, B \in E(H)$, then their sequential product is $A \circ B := A^{1/2}BA^{1/2}$ which is easily seen to be an effect again.

All of the mentioned operations and relation are important because of the role they play in different aspects of quantum theory. Clearly, according to the above four items we obtain four different algebraic structures on the same set $E(H)$. Hence, we have four different kinds of automorphisms of $E(H)$.

(EA1) The automorphisms of $E(H)$ which correspond to the partial addition appearing in (E1) are called E-automorphisms [45] (or effect automorphisms [146]). Therefore, a bijective map $\phi : E(H) \rightarrow E(H)$ is an

E-automorphism if for any $A, B \in E(H)$ it satisfies

$$A + B \in E(H) \Longleftrightarrow \phi(A) + \phi(B) \in E(H)$$

and, in this case, we have

$$\phi(A + B) = \phi(A) + \phi(B).$$

(EA2) The automorphisms of $E(H)$ with respect to the partial order $\leq$ and the orthocomplementation $\perp$ are nowadays called ortho-order automorphisms (they were called *-automorphism in [146]). So, a bijective map $\phi : E(H) \to E(H)$ is an ortho-order automorphism if for any $A, B \in E(H)$ we

$$A \leq B \Longleftrightarrow \phi(A) \leq \phi(B)$$

and

$$\phi(A^{\perp}) = \phi(I - A) = I - \phi(A) = \phi(A)^{\perp}.$$

(EA3) The automorphisms of $E(H)$ corresponding to the operation of mixtures are called mixture automorphisms. Therefore, a bijective map $\phi : E(H) \to E(H)$ is a mixture automorphism if it is affine, i.e., satisfies

$$\phi(\lambda A + (1 - \lambda)B) = \lambda\phi(A) + (1 - \lambda)\phi(B)$$

for all $A, B \in E(H)$ and $\lambda \in [0, 1]$.

(EA4) The automorphisms of $E(H)$ with respect to the sequential product $\circ$ are called sequential automorphisms [79]. Trivially, a bijective map $\phi : E(H) \to E(H)$ is a sequential automorphism if

$$\phi(A \circ B) = \phi(A) \circ \phi(B)$$

holds for all $A, B \in E(H)$.

Just as with any algebraic structure in mathematics in general, it is an important question to explore and determine (if possible) the form of those automorphisms. Moreover, as they represent different kinds of quantum mechanical symmetries, to know their explicit forms is important also because of the applications.

Fortunately, we do know the structure of those automorphisms very well. Namely, any E-automorphism ϕ of $E(H)$ is of the form

$$\phi(A) = UAU^* \qquad (A \in E(H)) \tag{0.3.2}$$

where U is an either unitary or antiunitary operator [45, 146]. The ortho-order automorphisms are of the same form in the case when $\dim H \geq 2$. In fact, if $\dim H \geq 3$, this was proved by Ludwig in [146, Section V.5] (although the proof presented there contained some gaps which were recently clarified in [46]). In our paper [190] we showed that the same form applies also in the two-dimensional case and this solved a problem that was open for several years. As for the mixture automorphisms of $E(H)$, they were determined in our paper [170]. It turned out that any such automorphism ϕ is either of the form

$$\phi(A) = UAU^* \qquad (A \in E(H)),$$

or of the form

$$\phi(A) = U(I - A)U^* \qquad (A \in E(H)),$$

where U is a unitary or antiunitary operator on H. Finally, what concerns the sequential automorphisms of $E(H)$, as an easy corollary of one of our results in [170], in their paper [79] Gudder and Greechie obtained that if $\dim H \geq 3$, then every sequential automorphism of $E(H)$ is of the form (0.3.2).

In Chapter 2 we collect some of our results on the above mentioned automorphisms and on certain preservers defined on different quantum structures.

0.4 Survey of the Results of Chapter 2

In Sections 2.1 and 2.2 we present generalizations of Wigner's and Uhlhorn's theorems. In order to formulate our results we have to continue the discussion about symmetry transformations that has been begun in the previous section. As we have mentioned there, there are several ways to define symmetry transformations which are essentially equivalent to each other.

Above we have said that a symmetry transformation is a bijective map on the set $P_1(H)$ of all pure states which preserves the transition probability (i.e., the trace of the product of rank-one projections). However, historically pure states were first viewed as rays in the Hilbert space H. If we have a unit vector $x \in H$, then the ray $\underline{x}$ corresponding to x is the set of all nonzero scalar multiples of x. If $x, y \in H$ are unit vectors, then we define

$$|\langle \underline{x}, \underline{y} \rangle| = |\langle x, y \rangle|.$$

(Recall that if P, Q are the rank-one projections onto the one-dimensional subspaces generated by x and y, respectively, then the above quantity equals the square-root of $\operatorname{tr} PQ$.) In this representation a symmetry transformation is a bijective map T on the set of all rays in H which satisfies

$$|\langle T\underline{x}, T\underline{y} \rangle| = |\langle \underline{x}, \underline{y} \rangle|$$

for every $\underline{x}, \underline{y}$. In this language, Wigner's theorem states that every symmetry transformation is induced by an either unitary or antiunitary operator on H in the following way. If T is a symmetry transformation, then there is a unitary or antiunitary operator U on H such that

$$T\underline{x} = \underline{Ux} \tag{0.4.1}$$

holds for every unit vector $x \in H$. It belongs to the history of Wigner's theorem that in the book [250] which appeared in 1931 Wigner did not give a rigorous mathematical proof of his result. In fact, the first such proofs were published only in the 60's in the papers [14, 135]. The common feature of those proofs and most of the further ones (see, for example, [230, 212, 45] ordered chronologically) is that they manipulate in the underlying Hilbert space. This is not surprising since the problem is to 'linearize' the ray transformation T somehow. Therefore, such approaches are very natural on the one hand, but unfortunately the so-obtained proofs are quite lengthy and sometimes hard to follow. Just to imagine: in the language of vectors, the result says that if T is a symmetry transformation, then for every nonzero vector x we can choose a vector x' from the ray $T(x/\|x\|)$ with norm $\|x'\| = \|x\|$ such that the correspondence $x \mapsto x'$ is either linear or conjugate-linear. Obviously, we have the freedom to change the phase (i.e., we can multiply the vectors by any complex number of modulus 1), but this is all what we have. So, we start with some vector and move from one vector to another probably changing phases in every step in such a way that at the end the obtained transformation must be linear or conjugate-linear.

Instead of this and similar quite constructive but, in our eyes, very complicated arguments, in our paper [160] we invented a completely different approach to attack the problem. In fact, our idea proved to be very fruitful since it allowed us not only to give a short and easily understandable proof of Wigner's original result but also to generalize Wigner's theorem for different structures where the first mentioned approach would certainly break down. Our corresponding papers are [164, 166, 169, 171, 172, 174, 183].

The idea of our proof which is algebraic in character can be described in few sentences as follows. We consider a symmetry transformation ϕ on $P_1(H)$. For any finite system $P_1, P_2, \ldots, P_n$ of (not necessarily orthogonal) rank-one projections and any corresponding system $\lambda_1, \lambda_2, \ldots, \lambda_n$ of real numbers we define

$$\Phi\Big(\sum_k \lambda_k P_k\Big) = \sum_k \lambda_k \phi(P_k).$$

As ϕ preserves the transition probability, we can easily show that Φ is a well-defined real-linear transformation on the set $F_s(H)$ of all finite rank self-adjoint bounded linear operators on H. Using the properties of ϕ we deduce that Φ preserves the rank-one projections and the orthogonality between them. By the spectral theorem it follows readily that Φ has the property that

$$\Phi(A^2) = \Phi(A)^2 \qquad (A \in F_s(H)).$$

Next, we extend Φ to whole set $F(H)$ of finite rank (bounded linear) operators on H by the formula

$$\tilde{\Phi}(A + iB) = \Phi(A) + i\Phi(B) \qquad (A, B \in F_s(H)).$$

It is easy to check that $\tilde{\Phi}$ is a linear transformation on $F(H)$ which preserves squares, i.e., it has the property that

$$\tilde{\Phi}(T^2) = \tilde{\Phi}(T)^2 \qquad (T \in F(H)).$$

This means that $\tilde{\Phi}$ is a Jordan homomorphism of $F(H)$. As ϕ is bijective, it can be seen that $\tilde{\Phi}$ is also bijective. Now, it follows from the well-known result Theorem A.7 of Herstein [100] that $\tilde{\Phi}$ is either an automorphism or an antiautomorphism of $F(H)$. But the form of those transformations is described in Theorem A.8. The proof can now be completed very easily. For details see [160] or [172, Remark] and we also refer to the proof of Theorem 2.1.1 in the present work.

The above described operator algebraic approach made possible for us to generalize Wigner's original theorem in several directions. In Section 2.1 we present such a result which has interesting geometrical content. Namely, we prove a Wigner-type result for transformations defined on the set of all n-dimensional subspaces of H (n is a fixed positive integer) which preserve the so-called principal angles. To formulate the result observe that one can state Wigner's original theorem also in the following way: Any bijective transformation T on the set of all one-dimensional subspaces of a Hilbert space H which preserves the angles between those subspaces is induced by either a unitary or an antiunitary operator on H. More precisely, for any such T there is an either unitary or antiunitary operator U on H for which

$$T(L) = U[L] = \{Ux : x \in L\} \qquad (0.4.2)$$

holds for every one-dimensional subspace L. The main result Theorem 2.1.1 in Section 2.1 states that every transformation (not necessarily bijective) on the set of all n-dimensional subspaces of a real or complex Hilbert space which preserves the principal angles is induced by a linear or conjugate-linear isometry of H in the same way as in (0.4.2).

In Section 2.2 we present a generalization of Uhlhorn's version of Wigner's theorem concerning ray transformations on indefinite inner product spaces. Our corresponding result significantly generalizes a result of Van den Broek. In fact, in [32] he obtained an Uhlhorn-type result for indefinite inner product spaces which are induced by nonsingular self-adjoint operators on a finite (at least 3-)dimensional complex Hilbert space. Our result is a far-reaching extension of Van den Broek's theorem since it holds for both real and complex Hilbert spaces of any dimension (not less than 3) and indefinite inner products induced by arbitrary nonsingular operators.

The proof of this result of ours is again operator algebraic in character. In fact, the main result Theorem 2.2.1 of that section (from which we can deduce our Uhlhorn-type result) gives a complete description of the bijective transformations ϕ on the set of all rank-one idempotents on a given real or complex Banach space X with $\dim X \geq 3$ which preserve zero product in both directions. For example, in the infinite dimensional case the result asserts that every such transformation ϕ extends to a linear or conjugate-linear algebra automorphism of $B(X)$, i.e., there is an invertible bounded linear or conjugate-linear operator $A \in B(X)$ such that ϕ is of the form

$$\phi(P) = APA^{-1}$$

for every rank-one idempotent P on X. Using this theorem we can prove our Uhlhorn-type result Corollary 2.2.3 concerning ray transformations on indefinite inner product spaces. It turns out that under the conditions what we have mentioned in the previous paragraph, i.e., given a real or complex Hilbert space H with dim $H \geq 3$ and an arbitrary nonsingular operator on H, for any ray transformation T on the induced indefinite inner product space which preserves orthogonality in both directions there exists an invertible either linear or conjugate-linear operator U on H such that

$$T\underline{x} = \underline{Ux}$$

for every nonzero $x \in H$.

In Sections 2.3 and 2.4 we present the descriptions of all bijective transformations of the set $S(H)$ of all states which preserve a certain given function operating on pairs of states.

The results in Section 2.3 concern those bijective transformations of $S(H)$ which preserve the so-called fidelity. If A, B are states (or, more generally, positive trace-class operators), then the fidelity between them is defined by

$$F(A, B) = \text{tr}(A^{1/2}BA^{1/2})^{1/2}.$$

The concept in this form was introduced by Uhlmann in [241] but one should consult also the papers [116] and [238]. Nowadays this notion plays very important role in several extensive research areas in quantum theory like quantum computation and quantum information theory. Fidelity can be viewed as the natural extension of the notion of transition probability from the case of pure states to the case of all states. In fact, if P, Q are rank-one projections, then it is easy to see that we have

$$F(P, Q) = (\text{tr } PQ)^{1/2},$$

i.e., $F(P, Q)$ is equal to the square-root of the transition probability between P and Q. Keeping Wigner's fundamental result in mind, it is a natural question that what are the bijective transformations on $S(H)$ which preserve the fidelity. This problem was raised by Uhlmann. The answer is presented in Theorem 2.3.2 which tells us that every such transformation is implemented by an either unitary or antiunitary operator. This is the main result in Section 2.3.

Motivated by certain problems in physics, in many cases the set of all states is considered as a metric space. In fact, there are a number of metrics on $S(H)$ which fit to different physical problems. However, as one can learn from [93], the most important such metrics can be derived from two fundamental ones, namely, from the Bures metric and from the one which is induced by the trace-norm. (We note that the Bures metric is in an intimate connection with

Uhlmann's fidelity as we shall see.) Since the corresponding isometries represent certain kinds of symmetries, Uhlmann raised the problem to determine their forms. We give the solution in Section 2.4. Namely, in Theorems 2.4.2 and 2.4.4 we shall see that all those isometries are implemented by unitary or antiunitary operators.

In Sections 2.5 and 2.6 we present some of our results concerning preservers on the set $B_s(H)$ of all bounded observables.

The main result of Section 2.5, namely Theorem 2.5.2 describes the general form of all bijective maps (no linearity is assumed) of $B_s(H)$ which preserve the usual order $\leq$ in both directions. Surprisingly enough, it turns out that in the case when $\dim H > 1$, every such transformation is automatically affine and of the form

$$A \mapsto TAT^* + X$$

with some invertible bounded either linear or conjugate-linear operator T on H and a fixed element $X \in B_s(H)$. Some corollaries of this result concerning bijective maps of $B_s(H)$ which preserve the order and other physically relevant relations (like compatibility or complementarity) are also presented.

In Section 2.6 we consider linear transformations on $B_s(H)$ which preserve important probabilistic characteristics of observables such as moments or variance. We show that each bijective linear transformation which preserves any of those quantities is 'almost' a Jordan automorphism of $B_s(H)$. For example, Theorem 2.6.2 (which is the main result of that section) states that if a bijective linear transformation $\phi : B_s(H) \to B_s(H)$ preserves the so-called maximal deviation of observables (that is, the supremum of the set of square-roots of the variances in all possible pure states; see (2.6.3)), then it is of the form

$$\phi(A) = \pm UAU^* + f(A)I,$$

where U is a unitary or antiunitary operator on H and f is a linear functional on $B_s(H)$. The solution of a non-linear version of the problem is also presented.

In Sections 2.7 and 2.8 we consider certain algebraic structures on the set of Hilbert space effects and study their transformations.

In Section 2.7 we deal with a certain class of preservers on $E(H)$. They are the bijective maps on $E(H)$ which preserve the order and zero product in both directions. In the main result of the section we determine the general form of all those transformations. Namely, Theorem 2.7.1 tells us that if $\dim H \geq 2$ and the bijective map $\phi : E(H) \to E(H)$ preserves the order and zero product in both directions, then there is an either unitary or antiunitary operator U on H and a real number $p < 1$ such that with the real function $f_p(x) = \frac{x}{xp+(1-p)}$ ($x \in [0,1]$) we have

$$\phi(A) = Uf_p(A)U^* \qquad (A \in E(H)).$$

(Here $f_p(A)$ denotes the image of the function f_p under the continuous function calculus belonging to the self-adjoint operator A.) In Corollary 2.7.3 we obtain that the same nice form holds for any bijection of $E(H)$ which preserves the order in both directions and sends one single nontrivial scalar operator to an operator of the same type. In Corollary 2.7.5 we easily deduce that every bijective map $\phi : E(H) \rightarrow E(H)$ which preserves the order and the so-called coexistency in both directions is of the form

$$\phi(A) = UAU^* \qquad (A \in E(H)) \qquad (0.4.3)$$

for some unitary or antiunitary operator U on H. In the last two results Corollary 2.7.6 and Corollary 2.7.7 we obtain that the ortho-order automorphisms and the sequential automorphisms of $E(H)$ are implemented by unitary or antiunitary operators just as in (0.4.3). In fact, we shall see that both of those two types of automorphisms preserve the order and zero product in both directions when they are defined on $E(H)$ and, moreover, also when their domain is the set all effects which belong to a given von Neumann algebra. Therefore, the bijective maps which preserve the order and zero product in both directions provide a common frame to study the ortho-order automorphisms and sequential automorphisms. It should also be emphasized that the results of the section hold for Hilbert spaces of dimension not less than 2. In the two-dimensional case we obtain extensions of several former results which were known before only for the case dim $H \geq 3$.

After introducing the concept of effects in an arbitrary C^*-algebra, in Section 2.8 we present a description of the sequential isomorphisms between the sets of von Neumann algebra effects. The main result Theorem 2.8.1 of this section is a far-reaching generalization of a result of Gudder and Greechie [79] which gives the general form of sequential automorphisms of $E(H)$ under the condition that dim $H \geq 3$. We shall see that in the particular case when the underlying von Neumann algebras $\mathcal{A}$ and $\mathcal{B}$ have no commutative direct summands, every sequential isomorphism $\phi : E(\mathcal{A}) \rightarrow E(\mathcal{B})$ between the sets of their effects extends to the direct sum of a linear *-isomorphism and a linear *-antiisomorphism which maps $\mathcal{A}$ onto $\mathcal{B}$.

0.5 Local Derivations, Local Automorphisms and Local Isometries of Operator Algebras and Function Algebras

In the last decades considerable work has been done on certain local maps of operator algebras. The main problem in this area of research is to answer the question whether the local actions of some important classes of transformations (like derivations, automorphisms, isometries) on a given operator algebra determine the class under consideration completely.

The originators of investigations of this kind are Kadison, Larson and Sourour. In [119] Kadison studied the local derivations of a von Neumann algebra $\mathcal{R}$. He called a continuous linear map $\delta : \mathcal{R} \rightarrow \mathcal{R}$ a local derivation if it coincides with some derivation at each point in the algebra (the derivations possibly differing from point to point). More precisely, it is supposed that for every $a \in \mathcal{R}$ there exists a derivation δ_a on $\mathcal{R}$ such that $\delta(a) = \delta_a(a)$. (In fact, in [119] Kadison considered the more general case of dual $\mathcal{R}$-bimodule valued local derivations.) Kadison's investigation was motivated by some problems concerning the Hochschild cohomology of operator algebras. He proved in [119] that in the above setting, every local derivation is a (global) derivation. Independently and approximately at the same time, Larson and Sourour proved in [134] that similar conclusion holds true for the local derivations of the full operator algebra $B(X)$ on a Banach space X. (We note that in [134] local derivation means any linear map on $B(X)$ which pointwise equals a derivation, so continuity is not assumed there.)

Beside derivations, there are at least two additional very important classes of transformations on operator algebras which certainly deserve attention from the point of view described above. These are the group of automorphisms and the group of surjective linear isometries. The automorphism group reflects the algebraic properties of the underlying algebra while the isometry group reflects its geometrical structure. In [133, Some concluding remarks (5), p. 298], motivated by the extensive research on reflexive linear subspaces of $B(H)$, Larson initiated the study of local automorphisms of Banach algebras. The definition is straightforward: a local automorphism is a linear map ϕ on a given Banach algebra $\mathcal{A}$ with the property that for every $x \in \mathcal{A}$ there exists an automorphism ϕ_x of $\mathcal{A}$ such that $\phi(x) = \phi_x(x)$. In other words, ϕ pointwise equals an automorphism that may vary from point to point. In the paper [134], Larson and Sourour proved that if X is an infinite dimensional Banach space, then every surjective local automorphism of $B(X)$ is an automorphism (also see [25]). Afterwards, in their important paper [27] Brešar and Šemrl showed that in the case of a separable infinite dimensional Hilbert space H the above conclusion holds true without the assumption of surjectivity, i.e., every local automorphism of $B(H)$ is an automorphism.

In relation with the name of Larson, above we have mentioned the concept of reflexivity. In fact, one can regard local maps also from this perspective. We recall the notion of reflexivity in the sense of Loginov and Shulman [141]. If $\mathcal{S}$ is a linear subspace of the algebra $B(H)$ of all bounded linear operators on the Hilbert space H, then set

$$\text{ref}\,\mathcal{S} = \{T \in B(H) \,:\, Tx \in \overline{\mathcal{S}x} \text{ for all } x \in H\}$$

(bar stands for the closure operation). We say that $\mathcal{S}$ is reflexive if $\text{ref}\,\mathcal{S} = \mathcal{S}$. The study of reflexive linear subspaces is one of the main research areas in operator theory as it is intimately connected with the fundamental problem of invariant subspaces. A nice introduction to the subject can be found, for example, in the recent book [55] of Conway. (For an interesting general view of reflexivity we refer to the paper [91] of Hadwin.)

Obviously, in the above definition of reflexivity the assumptions that the underlying space is a Hilbert space and $\mathcal{S}$ is a linear subspace are in fact not essential. Therefore, we can introduce the following notions. Let $\mathcal{X}$ be a Banach space and $\mathcal{E}$ an arbitrary subset of $B(\mathcal{X})$. Define

$$\mathrm{ref}_{al}\,\mathcal{E} = \{T \in B(\mathcal{X}) : Tx \in \mathcal{E}x \text{ for all } x \in \mathcal{X}\}$$

and

$$\mathrm{ref}_{to}\,\mathcal{E} = \{T \in B(\mathcal{X}) : Tx \in \overline{\mathcal{E}x} \text{ for all } x \in \mathcal{X}\}.$$

The sets $\mathrm{ref}_{al}\,\mathcal{E}$ and $\mathrm{ref}_{to}\,\mathcal{E}$ are called the algebraic reflexive closure of $\mathcal{E}$ and the topological reflexive closure of $\mathcal{E}$, respectively. The collection $\mathcal{E}$ of transformations is called algebraically reflexive if $\mathrm{ref}_{al}\,\mathcal{E} = \mathcal{E}$. Similarly, $\mathcal{E}$ is said to be topologically reflexive if $\mathrm{ref}_{to}\,\mathcal{E} = \mathcal{E}$.

With this terminology we can reformulate Kadison's result in [119] mentioned above saying that the set (or, more precisely, Lie algebra) of all derivations of a von Neumann algebra $\mathcal{R}$ is algebraically reflexive (in this case the Banach space $\mathcal{X}$ is equal to $\mathcal{R}$). Similarly, it follows from [134] that the same conclusion holds for the derivation algebra of $B(X)$, X being any Banach space (in this case we have $\mathcal{X} = B(X)$).

It is undoubtedly a remarkable fact on an operator algebra (or, more generally, on a Banach algebra) if the Lie algebra of all of its derivations or the group of all of its automorphisms is algebraically or topologically reflexive. In fact, this means that those collections of transformations are completely determined by their local actions (in the case of algebraic reflexivity) or by their approximate local actions (in the case of topological reflexivity). Roughly speaking, the algebraic reflexivity of a given collection of transformations means that if a continuous linear map pointwise belongs to the collection, then it globally belongs to it. Similarly, the topological reflexivity of a given collection of transformations means that if a continuous linear map pointwise approximately belongs to the collection, then it globally belongs to it.

The study of such questions issued in several important and interesting results. Beside the above mentioned papers we also refer

- to [23, 56, 92, 114, 200, 221, 249, 251, 252, 254] for results on local derivations and on the algebraic reflexivity of the derivation algebra,
- to [111, 112, 231] for results on the topological reflexivity of the derivation algebra,
- to [41, 88, 92, 109, 214, 216, 221] for results on local automorphisms and local isometries, and on the algebraic or topological reflexivity of the automorphism and isometry groups.

Our results on the algebraic or topological reflexivity of the automorphism and isometry groups of operator algebras and function algebras appeared in the papers [16, 161, 162, 163, 168, 188, 191, 196, 197] (also see [173, 192]).

Above we have considered local maps in the following sense. Given a linear algebraic structure $\mathcal{X}$ (mainly an operator algebra) and a collection $\mathcal{E}$ of its linear transformations, the corresponding local maps are the functions ϕ which have the following properties:

(a) ϕ is linear and

(b) for every $x \in \mathcal{X}$ there exists a transformation $\phi_x \in \mathcal{E}$ such that $\phi(x) = \phi_x(x)$.

The results we have referred to show that in many important cases our local maps are all 'global', i.e., they belong to the given class of transformations. All those results concern linear algebraic structures. Since they have proved to be nice and important achievements, it is a natural idea to extend the territory of such investigations from linear structures to more general ones. In order to do so, we clearly have to omit the condition (a) requiring linearity. But in that case it is quite apparent that the second property (b) alone is too weak to give reasonable results even on operator algebras. For example, if we consider any function ϕ on the operator algebra $B(H)$ such that for every $A \in B(H)$ the operator $\phi(A)$ belongs to the similarity orbit of A, then ϕ pointwise belongs to the automorphism group of $B(H)$ but in general ϕ fails to be an automorphism. So, if we would like to omit the condition (a), we have to strengthen the condition (b).

One simple idea leads to the concept of 2-locality. In fact, in his paper [224] Šemrl introduced the following definition. Let $\mathcal{A}$ be an algebra. The transformation $\phi : \mathcal{A} \to \mathcal{A}$ (linearity is not assumed) is called a 2-local automorphism if for every $x, y \in \mathcal{A}$, there is an automorphism $\phi_{x,y}$ of $\mathcal{A}$ for which $\phi(x) = \phi_{x,y}(x)$ and $\phi(y) = \phi_{x,y}(y)$. The definition of 2-local derivations is similar. The main results of [224] show that if H is an infinite dimensional separable Hilbert space, then every 2-local automorphism of $B(H)$ is an automorphism and similar assertion holds concerning the 2-local derivations. These nice and highly nontrivial results justify the usefulness of Šemrl's definition. Clearly, the above notion of 2-local automorphisms has the great advantage that it can be defined on arbitrary algebraic structures not only on algebras and this is what we have been looking for. Now one can define 2-local maps of different kinds and study them on different structures. Obviously, the main problem to answer should be the following question: Given an algebraic structure and a class of transformations on it, is it true that the corresponding 2-local maps are all global?

This general concept of 2-locality is rather new and relatively few results have been obtained so far concerning it. We mention the papers [13, 57, 85, 126, 127, 253] and our corresponding works [179, 182, 184, 193]. Nevertheless, we believe that the study of 2-local automorphisms is an important problem because it can give valuable new information on the automorphism groups appearing in different parts of mathematics. Therefore, we expect that the attention paid to this kind of investigations and the intensity of research in this direction will increase in the near future.

Finally, we note that the basic problem of local transformations seems to have nothing to do with preserver problems. However, as it will turn out in the proofs of our results on local maps, we often use 'preserver arguments' as well as several particular preserver results. In this respect the area of local automorphisms and isometries can be considered as a field of applications of preservers.

0.6 Survey of the Results of Chapter 3

In order to present the results of Chapter 3 we have to continue the discussion about former results concerning the reflexivity of certain classes of transformations on operator algebras.

There is an important and beautiful result of Shulman on the reflexivity of the derivation algebra. He proved in [231] that it is topologically reflexive in the case of any C^*-algebra. If such a general result holds for derivations, one can ask that what is the situation with the automorphisms. Unfortunately, this is a completely different story. In fact, there exists a (unital and commutative) C^*-algebra whose automorphism group is not reflexive even algebraically. An example is given in Remark 3.2.2. The absence of such a general result motivates us to consider important particular C^*-algebras. The most simple and fundamental such algebras are $B(H)$, the algebra of all bounded linear operators on a Hilbert space H and $C(X)$, the algebra of all continuous complex valued functions on a compact Hausdorff space X.

As for $B(H)$, we have already learnt from [27] that if H is infinite dimensional separable then the automorphism group of $B(H)$ is algebraically reflexive. (In the finite dimensional case the result is no longer valid, see Remark 3.1.7.) In Section 3.1 we show that even more is true. Namely, Theorems 3.1.2 and 3.1.3 tell us that the automorphism and isometry groups of $B(H)$ (H being infinite dimensional separable) are topologically reflexive. The proofs of these statements rest heavily on the rather surprising result Theorem 3.1.1 which asserts that if a Jordan homomorphism ϕ of $B(H)$ has the property that its range contains two extreme operators, one with rank one and one with dense range, then ϕ is automatically surjective and, furthermore, it is either an automorphism or an antiautomorphism of $B(H)$.

In Section 3.2 we consider the reflexivity problem of the automorphism and isometry groups for function algebras. There we deal only with the case of the most fundamental such algebra which is $C(X)$, the algebra of all continuous complex valued functions on a compact Hausdorff space X. As it follows from Remark 3.2.2, for general X we do not have even the algebraic reflexivity of the considered groups. Nevertheless, in the simple result Theorem 3.2.1 we obtain that if X is a first countable compact Hausdorff space then the automorphism and isometry groups of $C(X)$ are algebraically reflexive. (As for topological reflexivity, we note that it does not hold even when X is a compact interval of the real numbers. See Subsection 3.1.2.)

So, we have results concerning the topological reflexivity of the automorphism and isometry groups of $B(H)$ and a result concerning the algebraic reflexivity of the automorphism and isometry groups of $C(X)$. It is a natural question that what happens to reflexivity if we compose a new algebra from two given ones with nice reflexivity properties. For example, one should be interested in taking tensor product. Unfortunately, as we shall see in Section 3.3, the problem in full generality seems to be completely hopeless. But in the case of the function algebra $C_0(X)$ of all continuous complex valued functions on a locally compact Hausdorff space vanishing at infinity and $B(H)$ (H being infinite dimensional separable) we have positive results. In fact, in Theorem 3.3.2 we prove that if the automorphism group of $C_0(X)$ is algebraically reflexive, then so is the automorphism group of the tensor product $C_0(X) \otimes B(H)$. As for the isometry group, we have a similar result in the case when the underlying topological space X is σ-compact. This is the content of Theorem 3.3.3. In Theorem 3.3.4 we show that the automorphism and isometry groups of $C_0(\mathbb{R}^n)$ are algebraically reflexive. As a consequence, in Corollary 3.3.5 we obtain that the same holds for the tensor product $C_0(\mathbb{R}) \otimes B(H)$ as well. The reason to emphasize this corollary is that tensor products of the form $C_0(\mathbb{R}) \otimes \mathcal{A}$ ($\mathcal{A}$ being an arbitrary C^*-algebra) are particularly important. In fact, $C_0(\mathbb{R}) \otimes \mathcal{A}$ is called the suspension of the C^*-algebra $\mathcal{A}$ and is usually denoted by $S\mathcal{A}$. This concept plays very important role in the K-theory of C^*-algebras since the K_1-group of $\mathcal{A}$ is well-known to be isomorphic to the K_0-group of $S\mathcal{A}$.

In Sections 3.4 and 3.5 we present some results of ours on 2-local automorphisms. As mentioned above, in his paper [224] Šemrl proved that for an infinite dimensional separable Hilbert space H, every 2-local automorphism of $B(H)$ is an automorphism. Moreover, he admitted that the same is true in the finite dimensional case and, as he could give only a long proof of it involving tedious computations, he raised the problem to present a shorter argument. In Section 3.4 we give such a short proof. In fact, in Corollary 3.4.2, respectively in Proposition 3.4.3 we prove that every 2-local automorphism of the algebra of all $n \times n$ matrices over an algebraically closed field of characteristic 0 or over the field of real numbers is an automorphism. Our approach to attack the problem also leads to a remarkable generalization of Šemrl's original result. Namely, in Theorem 3.4.4 we obtain the following result. If X is a real or complex Banach space with a Schauder basis and $\mathcal{A}$ is a subalgebra of $B(X)$ which contains all compact operators, then every 2-local automorphism of $\mathcal{A}$ is an automorphism. As most classical separable Banach spaces have Schauder bases, this is a substantial generalization of the result in [224].

In Section 3.5 we consider the 2-local automorphisms of some non-linear quantum structures. In Theorem 3.5.1 we prove that every continuous 2-local automorphism of the poset (partially ordered set) of all idempotents on an infinite dimensional separable Hilbert space is an automorphism. Theorem 3.5.2 tells us that similar assertion holds for the 2-local automorphisms of the orthomodular lattice of all projections of H even without assuming continuity. Finally, we obtain in Theorem 3.5.3 that every 2-local automorphism of the Jordan algebra $B_s(H)$ of all self-adjoint operators is an automorphism.

0.7 Notation

In this section we fix the notation and some definitions that we shall use (or have already used) frequently in our work.

Firstly, unless explicitly stated otherwise, every linear space which appears in the book is considered over the complex field $\mathbb{C}$.

For Chapters 1 and 3 we fix the following.

For an arbitrary positive integer n, $M_n(\mathbb{C})$ denotes the algebra of all $n \times n$ complex matrices. The transpose of a matrix $A \in M_n(\mathbb{C})$ is denoted by A^{tr}.

If X is a Banach space, then $B(X)$ stands for the algebra of all bounded linear operators on X. The ideals of all finite rank operators and all compact operators in $B(X)$ are denoted by $F(X)$ and $K(X)$, respectively. Any subalgebra of $B(X)$ which contains $F(X)$ is called a standard operator algebra on X.

If $A \in B(X)$, then $\ker A$ denotes the kernel of A while $\operatorname{rng} A$ stands for the range of A. The spectrum of A is denoted by $\sigma(A)$.

The operator $P \in B(X)$ is called an idempotent if $P^2 = P$.

Let H be a Hilbert space. The inner product on H is denoted by $\langle .,. \rangle$. If $x, y \in H$, then $x \otimes y$ stands for the operator defined by $(x \otimes y)z = \langle z, y \rangle x$ ($z \in H$). If $A \in B(H)$, then its adjoint (in the Hilbert space sense) is denoted by A^*. Fixing an arbitrary complete orthonormal system in H and considering the corresponding matrix representation of operators, one can define the transpose A^{tr} of an arbitrary operator $A \in B(H)$ in the obvious way.

The operator $P \in B(H)$ is called a projection if it is a self-adjoint idempotent. If $U \in B(H)$ is a surjective isometry, then it is called unitary. By an antiunitary operator we mean a surjective conjugate-linear isometry on H. The operator $W \in B(H)$ is called a partial isometry if it is an isometry on a closed subspace M of H and is 0 on $M^\perp$, the orthogonal complement of M.

If X is a compact Hausdorff space, then $C(X)$ denotes the algebra of all continuous complex valued functions on X.

Let $\mathcal{A}$ be a C^*-algebra. If not explicitly stated otherwise, we automatically assume that $\mathcal{A}$ is unital.

Let $\mathcal{A}, \mathcal{B}$ be algebras. The linear transformation $\phi : \mathcal{A} \to \mathcal{B}$ is called a Jordan homomorphism if it satisfies

$$\phi(A^2) = \phi(A)^2 \qquad (A \in \mathcal{A}). \tag{0.7.1}$$

It is easy to see that this equality is equivalent to

$$\phi(AB + BA) = \phi(A)\phi(B) + \phi(B)\phi(A) \qquad (A, B \in \mathcal{A})$$

(just write $A + B$ in the place of A in (0.7.1) and compute). A linear transformation $\psi : \mathcal{A} \to \mathcal{B}$ is called a homomorphism if it satisfies

$$\psi(AB) = \psi(A)\psi(B) \qquad (A, B \in \mathcal{A})$$

while ψ is said to be an antihomomorphism if it satisfies

$$\psi(AB) = \psi(B)\psi(A) \qquad (A, B \in \mathcal{A}).$$

Obviously, homomorphisms and antihomomorphisms between algebras are examples for Jordan homomorphisms.

The meaning of the corresponding concepts like Jordan automorphism, Jordan isomorphism, etc. are supposed to be clear.

Let $\mathcal{A}, \mathcal{B}$ be *-algebras. A linear transformation $\phi : \mathcal{A} \to \mathcal{B}$ is said to be a Jordan *-homomorphism if it is a Jordan homomorphism which preserves the adjoint operation, i.e., which satisfies

$$\phi(A^*) = \phi(A)^* \qquad (A \in \mathcal{A}).$$

The definition of a *-homomorphism or a *-antihomomorphism is analogous.

In addition to the above, in Chapter 2 we use the following notation and definitions.

The ideal of all trace-class operators in $B(H)$ is denoted by $C_1(H)$ and tr stands for the usual trace functional on it. The set of all positive elements in $C_1(H)$ which we call density operators is denoted by $C_1^+(H)$. The normalized elements of $C_1^+(H)$, i.e., the ones with trace 1 are called (normal) states and they form the set $S(H)$.

The set of all projections on H is denoted by $P(H)$ and $P_1(H)$ stands for the set of all rank-one elements in $P(H)$. The operators in $P_1(H)$ are also called pure states.

The set of all self-adjoint bounded linear operators on H is denoted by $B_s(H)$. The elements of this set are also called (bounded) observables as they represent observable physical quantities.

The set of all positive bounded linear operators on H which are bounded by the identity I (i.e., the set of all operators $A \in B(H)$ for which $0 \leq A \leq I$) is denoted by $E(H)$. The elements of $E(H)$ are called Hilbert space effects. They represent yes-no quantum measurements which can be unsharp.

1

Some Linear and Multiplicative Preserver Problems on Operator Algebras and Function Algebras

1.1 Some Linear Preserver Problems on $B(H)$ Concerning Rank and Corank

1.1.1 Summary

As a natural continuation of the work concerning linear maps on operator algebras which preserve certain subsets of operators with finite rank, or finite corank, here we consider the problem 'between', that is, we treat the question of preserving operators with infinite rank and infinite corank. Since, as it turns out, in this generality our preservers cannot be written in any regular form what we have got used to when dealing with linear preserver problems, hence we restrict our attention to certain important classes of operators like idempotents, or projections, or partial isometries and describe the structures of the corresponding preservers. Moreover, we obtain the general form of all linear bijections of $B(H)$ which preserve the left ideals in both directions. The results of this section appeared in the paper [165].

1.1.2 Formulation of the Results

As mentioned in Section 0.1 of the Introduction it is an important class of preserver problems on matrix algebras and on operator algebras which concerns the rank (see PROBLEM II there). This is because in many cases the studied preserver problem can be reduced to a rank preserver problem. Therefore, it is not surprising that a lot of work has been done on rank preservers. See, for example, [17, 21, 66, 140, 246] for the finite dimensional case and [104, 203] for the infinite dimensional case as well as the references therein.

The first results concerning rank preservers on operator algebras appeared in the papers [104, 201] (we mention that such maps were considered implicitly also in [107]). In [104] Hou proved among others the following. (In fact, he formulated his results for Banach spaces that we recall here for Hilbert spaces.) Let H be a (complex) Hilbert space and $\phi : B(H) \to B(H)$ be a weakly continuous linear map which preserves the rank-one operators (i.e., it sends rank-one operators to rank-one operators). If the range of ϕ contains an operator with rank greater than 1, then there is an injective operator $A \in B(H)$ and an operator $B \in B(H)$ with dense range such that ϕ is either of the form

$$\phi(T) = ATB \qquad (T \in B(H))$$

or of the form

$$\phi(T) = AT^{tr}B \qquad (T \in B(H)).$$

In the paper [90] we considered the similar problem of corank preservers which deserves attention, of course, only in the infinite dimensional case. Our result [90, Theorem 3] reads as follows. Let H be an infinite dimensional Hilbert space and $\phi : B(H) \to B(H)$ a bijective linear map which is weakly continuous on norm bounded sets. If ϕ preserves the corank-k operators in both directions, then there exist invertible operators $A, B \in B(H)$ such that ϕ is of the form $\phi(T) = ATB$ $(T \in B(H))$.

Having these in mind, it is a natural question to consider the problem of such preservers which are "between" (finite) rank preservers and (finite) corank preservers, that is, to determine those linear maps which preserve the operators with infinite rank and infinite corank. We say that an operator $A \in B(H)$ has infinite rank and infinite corank if the (Hilbert space) dimensions of $\overline{\operatorname{rng} A}$ and $\overline{\operatorname{rng} A}^{\perp}$ are both infinite. We consider separable Hilbert spaces since in this case there is only one sort of infinite dimension. Unfortunately, in general these preservers do not have such a nice form which we have got used to when dealing with linear preserver problems. Namely, there exist preservers of the above kind which cannot be expressed in terms of multiplications by fixed operators and, possibly, by transposition. To see this, let $\psi : B(H) \to B(H)$ be a linear map with norm less than 1 whose range consists of finite rank operators. Then it follows from a basic Banach algebra fact that the linear map ϕ defined by

$$\phi(T) = T - \psi(T) \qquad (T \in B(H))$$

is a bijection of $B(H)$ onto itself, and it is easy to check that ϕ preserves the operators with infinite rank and infinite corank in both directions. Therefore, in order to obtain a more "regular" form for our preservers we should somehow modify the problem by, for example, restricting the set of operators with infinite rank and infinite corank which we want to preserve. This is exactly what we are doing here considering the important sets of idempotents, projections and partial isometries, respectively.

Turning to the formulation of our results, in the first statement below we describe the structure of all linear bijections of $B(H)$ which preserve the idempotents with infinite rank and infinite corank in both directions.

Theorem 1.1.1. *Let H be a separable infinite dimensional Hilbert space. Suppose that $\phi : B(H) \to B(H)$ is a linear bijection which preserves the idempotents of infinite rank and infinite corank in both directions. Then there is an invertible operator $A \in B(H)$ such that ϕ is either of the form*

$$\phi(T) = ATA^{-1} \qquad (T \in B(H))$$

or of the form

$$\phi(T) = AT^{tr}A^{-1} \qquad (T \in B(H)).$$

The next result treats projections in the place of idempotents above and reads as follows.

Theorem 1.1.2. *Let H be a separable infinite dimensional Hilbert space. Suppose that $\phi : B(H) \to B(H)$ is a linear bijection which preserves the projections of infinite rank and infinite corank in both directions. Then there is a unitary operator $U \in B(H)$ such that ϕ is either of the form*

$$\phi(T) = UTU^* \qquad (T \in B(H))$$

or of the form

$$\phi(T) = UT^{tr}U^* \qquad (T \in B(H)).$$

The following result provides the description of all linear bijections ϕ of $B(H)$ which preserve the partial isometries of infinite rank and infinite corank in both directions.

Theorem 1.1.3. *Let H be a separable infinite dimensional Hilbert space. Let $\phi : B(H) \to B(H)$ be a linear bijection which preserves the partial isometries of infinite rank and infinite corank in both directions. Then there exist unitary operators $U, V \in B(H)$ such that ϕ is either of the form*

$$\phi(T) = UTV \qquad (T \in B(H))$$

or of the form

$$\phi(T) = UT^{tr}V \qquad (T \in B(H)).$$

In the last result of this section we describe the linear bijections ϕ of $B(H)$ which preserve the left ideals in both directions (this means that $\mathcal{L} \subset B(H)$ is a left ideal if and only if $\phi(\mathcal{L})$ is a left ideal). As it will be clear from the proof, this problem is closely related to the problem of rank preservers.

Theorem 1.1.4. *Let H be a Hilbert space. Suppose that $\phi : B(H) \to B(H)$ is a linear bijection preserving the left ideals of $B(H)$ in both directions. Then there are invertible operators $A, B \in B(H)$ such that ϕ is of the form*

$$\phi(T) = ATB \qquad (T \in B(H)).$$

1.1.3 Proofs

In the proof of Theorem 1.1.1 we shall use the following lemma which is certainly well-known and is included here only for the sake of completeness.

Lemma 1.1.5. *If $P, Q \in B(H)$ are idempotents, then*

(i) $P + Q$ is an idempotent if and only if $PQ = QP = 0$;
(ii) $P - Q$ is an idempotent if and only if $PQ = QP = Q$.

Proof. It follows from elementary algebraic computations.   □

Proof of Theorem 1.1.1. If $P, Q \in B(H)$ are idempotents, then we write $P \leq Q$ if $PQ = QP = P$. Clearly, this is equivalent to the condition that $\operatorname{rng} P \subset \operatorname{rng} Q$ and $\ker Q \subset \ker P$. For temporary use, we say that an idempotent $P \in B(H)$ is regular if it has infinite rank and infinite corank. We prove that for any two regular idempotents P, Q we have $P \leq Q$ if and only if for every regular idempotent $R \in B(H)$, if $Q + R$ is a regular idempotent, then so is $P + R$. The necessity is evident. To the sufficiency suppose first that $\operatorname{rng} P \subset \operatorname{rng} Q$ does not hold. Let $x \in H$ be such that $Px = x$ and $Qx \neq x$. Choose a regular idempotent $R \leq I - Q$ for which $Q + R$ is a regular idempotent and $Rx \neq 0$ (observe that $(I - Q)x \neq 0$). Since $P + R$ is an idempotent, we have $PR = RP = 0$. It follows that $0 = RPx = Rx$ which is a contradiction. Hence, we have $\operatorname{rng} P \subset \operatorname{rng} Q$. The relation $\ker Q \subset \ker P$ can be proved in a similar manner. Using the above characterization and the preserving property of ϕ, we obtain that ϕ preserves the relation $\leq$ between regular idempotents. Now, if R is a finite rank idempotent, then R can be written in the form $R = Q - P$ with some regular idempotents $P \leq Q$. Since $\phi(P) \leq \phi(Q)$, it follows that $\phi(R) = \phi(Q) - \phi(P)$ is also an idempotent. We prove that $\phi(R)$ is of finite rank. Choosing a regular idempotent P with $R \leq P$, it follows that $P - R$ is a regular idempotent and hence $\phi(P) - \phi(R)$ is also an idempotent. By Lemma 1.1.5 (ii) this implies that $\phi(R) \leq \phi(P)$. If $\phi(R)$ is not of finite rank, then it is regular which implies that R is also regular and this is a contradiction. Therefore, using the preserving properties of ϕ and ϕ^{-1} we obtain that ϕ preserves the finite rank idempotents in both directions. It is now easy to see that ϕ is a linear bijection of $F(H)$ onto itself. By Lemma 1.1.5 (i), for any idempotents $R, R' \in F(H)$ we have $RR' = R'R = 0$ if and only if $\phi(R)\phi(R') = \phi(R')\phi(R) = 0$. Using this property it is easy to verify that ϕ preserves the rank-one idempotents in both directions. In [203, Theorem 4.4] Omladič and Šemrl described the form of all additive surjective mappings of the ideal of all finite rank operators on an infinite dimensional complex Hilbert space which preserve the rank-one idempotents and their linear spans. Applying that result and taking into consideration that our map ϕ is linear, we obtain that there is an invertible bounded linear operator $A \in B(H)$ such that ϕ is either of the form

$$\phi(T) = ATA^{-1} \qquad (T \in F(H))$$

or of the form

$$\phi(T) = AT^{tr}A^{-1} \qquad (T \in F(H)).$$

Without loss of generality we may assume that ϕ is of the first form and then that $A = I$. We intend to show that $\phi(T) = T$ ($T \in B(H)$). Let $P \in B(H)$ be a regular idempotent. If R is any finite rank idempotent with $R \leq P$, then just as above, we obtain $R = \phi(R) \leq \phi(P)$. As $R \leq P$ was arbitrary, it now follows that $P \leq \phi(P)$. Since ϕ^{-1} has the same preserving property as ϕ, it follows that $P \leq \phi^{-1}(P)$. But ϕ preserves the order between the regular idempotents. Hence, we have $\phi(P) \leq P$. Therefore, $\phi(P) = P$ for every regular idempotent P. It is clear that every idempotent of finite corank is the sum of two regular idempotents. Thus we obtain that $\phi(P) = P$ holds for every idempotent $P \in B(H)$. According to the result [70, Theorem 2] of Fillmore, every element of $B(H)$ is a finite linear combination of projections. This gives us that $\phi(T) = T$ is valid for every $T \in B(H)$ completing the proof. $\square$

Since the proof of Theorem 1.1.2 is quite similar in spirit to what we have presented above, hence we omit it.

In the proof of Theorem 1.1.3 we shall use the following two auxiliary results.

Lemma 1.1.6. *Let $T, S \in B(H)$ be partial isometries with $S = ST^*S$. Then we have $TT^*S = S$ and $ST^*T = S$.*

Proof. Denote $Q = TS^*$. Since SS^* and T^*T are projections, we compute

$$SS^* = ST^*SS^*TS^* = Q^*(SS^*)Q \leq Q^*Q = S(T^*T)S^* \leq SS^*.$$

This implies $Q^*Q = SS^*$. In particular, we obtain $\|Q\| \leq 1$ (in fact, the norm of Q is either 0 or 1). But Q is an idempotent. Indeed, we have

$$Q^2 = TS^*TS^* = T(ST^*S)^* = TS^* = Q.$$

So, Q is a contractive idempotent. It is easy to see that this implies that Q is a self-adjoint idempotent, that is, a projection. To verify this, pick arbitrary elements $x \in \ker Q$ and $y \in \operatorname{rng} Q$. Then we have

$$\|y\|^2 = \|Q(\mu x + y)\|^2 \leq \|\mu x + y\|^2 \qquad (\mu \in \mathbb{C}).$$

An elementary argument shows that this implies that $x \perp y$. Hence the kernel and the range of Q are orthogonal to each other and this verifies that Q is a projection. Now, from $Q^*Q = SS^*$ we obtain $Q = SS^*$. Therefore, $TS^* = SS^*$ and, as SS^* is the projection onto $\operatorname{rng} S$, it follows that the range of S is included in that of T. Since TT^* is the projection onto $\operatorname{rng} T$, we have $TT^*S = S$. So, we have proved that $S = ST^*S$ implies $TT^*S = S$. Considering S^* and T^* in the place of S and T, respectively, we obtain from this implication that

$T^*TS^* = S^*$ also holds. But this is equivalent to $ST^*T = S$. This completes the proof. $\qquad\square$

In what follows we need the concept of orthogonality between arbitrary operators in $B(H)$. We say that the operators $A, B \in B(H)$ are orthogonal to each other if $A^*B = AB^* = 0$. This means that the ranges of A and B as well as the orthogonal complements of their kernels are orthogonal to each other.

Lemma 1.1.7. *Suppose that $T, S \in B(H)$ are partial isometries. The operator $T + \lambda S$ is a partial isometry for every $\lambda \in \mathbb{C}$ with $|\lambda| = 1$ if and only if T and S are orthogonal to each other.*

Proof. Suppose first that

$$(T + \lambda S)(T + \lambda S)^*(T + \lambda S) = T + \lambda S$$

holds for every $\lambda \in \mathbb{C}$ with $|\lambda| = 1$. Using the fact that T, S are partial isometries, one can conclude that

$$0 = \lambda^2 ST^*S + \lambda(TT^*S + ST^*T) + \bar{\lambda}TS^*T + SS^*T + TS^*S.$$

Since this is valid for every $\lambda \in \mathbb{C}$ of modulus 1, choosing the particular values $\lambda = 1, -1, i, -i$, it is easy to deduce that

$$ST^*S = 0 \tag{1.1.1}$$

$$TT^*S + ST^*T = 0 \tag{1.1.2}$$

$$TS^*T = 0 \tag{1.1.3}$$

$$SS^*T + TS^*S = 0. \tag{1.1.4}$$

Multiplying (1.1.2) by T^* from the left and taking (1.1.3) into account, we obtain $T^*S = 0$. Similarly, multiplying (1.1.4) by S^* from the right and taking (1.1.1) into account, we have $TS^* = 0$. So, T and S are orthogonal. As for the reverse implication, if T, S are mutually orthogonal partial isometries, then it is just a simple calculation that $T + \lambda S$ is a partial isometry for every $\lambda \in \mathbb{C}$ of modulus 1. $\qquad\square$

Proof of Theorem 1.1.3. Let $\{x_1, \ldots, x_k\} \subset H$ and $\{y_1, \ldots, y_k\} \subset H$ be two systems of pairwise orthogonal unit vectors. We claim that the image of the finite rank partial isometry $R = \sum_{j=1}^k x_j \otimes y_j$ under ϕ is also a finite rank partial isometry. Let (e_n) be an orthonormal sequence in the orthogonal complement of $\{x_1, \ldots, x_k\}$ which generates a closed subspace of infinite codimension. Similarly, let (f_n) be an orthonormal sequence in $\{y_1, \ldots, y_k\}^\perp$. Denote $U = \sum_n e_n \otimes f_n$ and let $V = U + R$. Clearly, U and V are partial isometries of infinite rank and infinite corank. Moreover, for every $\lambda \in \mathbb{C}$ of modulus 1, the operator $R + \lambda U = (V - U) + \lambda U$ is also a partial isometry of infinite rank and infinite corank. Therefore, $\phi(V) + (\lambda - 1)\phi(U)$ is a partial isometry for every

$\lambda \in \mathbb{C}$ with $|\lambda| = 1$. This means that with the notation $V' = \phi(V), U' = \phi(U)$ we have

$$(V' + (\lambda - 1)U')(V' + (\lambda - 1)U')^*(V' + (\lambda - 1)U') = (V' + (\lambda - 1)U')$$

for every $\lambda \in \mathbb{C}$ of modulus 1. Performing the above operations and arranging the terms to one side of the equality, we obtain a polynomial in $\lambda, \bar{\lambda}$ with operator coefficients which equals 0 on the perimeter of the unit disc in the complex plane. Just as in the proof of Lemma 1.1.7, choosing the particular values $\lambda = 1, -1, i, -i$ we find that the coefficients of the polynomial in question are all 0. Therefore, we have

$$-U' + U'V'^*U' = 0, \tag{1.1.5}$$

$$2U' + V'V'^*U' + U'V'^*V' - U'U'^*V' - 2U'V'^*U' - V'U'^*U' = 0, \tag{1.1.6}$$

$$U' + V'U'^*V' - U'U'^*V' - V'U'^*U' = 0, \tag{1.1.7}$$

and

$$-2U' - V'V'^*U' - V'U'^*V' - U'V'^*V' + \\ 2U'U'^*V' + U'V'^*U' + 2V'U'^*U' = 0, \tag{1.1.8}$$

where the left hand sides of (1.1.5), (1.1.6), (1.1.7), (1.1.8) are the coefficients of λ^2, λ, $\bar{\lambda}$ and 1, respectively. From (1.1.5) and (1.1.6) we deduce

$$V'V'^*U' + U'V'^*V' = U'U'^*V' + V'U'^*U'. \tag{1.1.9}$$

We prove that $\phi(R) = V' - U'$ is a partial isometry. Indeed, we compute

$$(V' - U')(V' - U')^*(V' - U') =$$

$$V' - U' - V'V'^*U' - V'U'^*V' - U'V'^*V' + U'U'^*V' + U'V'^*U' + V'U'^*U'.$$

From (1.1.9) we know that

$$-V'V'^*U' - U'V'^*V' + U'U'^*V' + V'U'^*U' = 0.$$

So, we have to show that $V' - U' - V'U'^*V' + U'V'^*U' = V' - U'$. By (1.1.5) we have $U'V'^*U' = U'$. It remains to verify that $V'U'^*V' = U'$. From (1.1.7) and (1.1.9) we infer that

$$U' + V'U'^*V' - V'V'^*U' - U'V'^*V' = 0. \tag{1.1.10}$$

As $U'V'^*U' = U'$, by Lemma 1.1.6 it follows that $V'V'^*U' = U'$ and $U'V'^*V' = U'$. Now, (1.1.10) gives $V'U'^*V' = U'$ and this was to be verified. Hence, $\phi(R)$ is a partial isometry.

Now we prove that $\phi(R)$ has finite rank. By the preserving property of ϕ it is sufficient to prove that $\phi(R)$ has infinite corank. We have seen that for every $\lambda \in \mathbb{C}$ of modulus 1, the operator $R + \lambda U$ is a partial isometry of infinite rank and infinite corank. This implies that for $R' = \phi(R)$, the operator $R' + \lambda U'$

is a partial isometry for every $\lambda \in \mathbb{C}$ with $|\lambda| = 1$. According to Lemma 1.1.7, we obtain that R' and U' are orthogonal to each other. Since the range of U' is infinite dimensional, it follows that R' is of infinite corank which implies that R' is a finite rank partial isometry.

We next prove that ϕ preserves the partial isometries in general. To see this, let W be a partial isometry. If it is of finite rank, then there is now nothing to prove. So, let W be of infinite rank. In that case we have an orthogonal sequence (W_n) of partial isometries of infinite rank and infinite corank whose sum is W. By the preserving property of ϕ it follows that the operators $A_n = \phi(W) - \sum_{k=1}^{n+1} \phi(W_k) = \phi(W - \sum_{k=1}^{n+1} W_k)$ and $B_n = \sum_{k=1}^{n} \phi(W_k) = \phi(\sum_{k=1}^{n} W_k)$ are partial isometries. Because of the same reason, $A_n + \lambda B_n$ is a partial isometry for every $\lambda \in \mathbb{C}$ of modulus 1. By Lemma 1.1.7 this implies that A_n and B_n are orthogonal to each other. It is known that the series of pairwise orthogonal partial isometries is convergent in the strong operator topology and its sum is also a partial isometry (cf. [196, Lemma 1.3]). Consider the operators $A = \phi(W) - \sum_n \phi(W_n)$ and $B = \sum_n \phi(W_n)$. By the just mentioned result $\sum \phi(W_n)^*$ is strongly convergent as well, and $\sum_n \phi(W_n)^* = B^*$. We then also have $\phi(W)^* - \sum_n \phi(W_n)^* = A^*$. Since A_n and B_n are mutually orthogonal for every $n \in \mathbb{N}$, it is now easy to verify that A is orthogonal to B. The operator B is a partial isometry. As for A, we know that (A_n) strongly converges to A and, as we have seen, (A_n^*) strongly converges to A^*. It is well-known that the multiplication is strongly continuous on the norm-bounded subsets of $B(H)$. Consequently, we infer that $(A_n A_n^*)$ strongly converges to AA^* and then that $(A_n A_n^* A_n)$ strongly converges to AA^*A. Since A_n is a partial isometry for every n, we obtain that A is also a partial isometry. Now, since $\phi(W)$ is the sum of the mutually orthogonal partial isometries A and B, it follows that $\phi(W)$ is a partial isometry as well. We have assumed that ϕ^{-1} has the same preserving properties as ϕ. Therefore, ϕ preserves the partial isometries in both directions.

It is well-known that every operator A with norm less than 1 is the arithmetic mean of finitely many unitaries (see, for example, [8, Theorem 6.2.13]). As the image of a unitary operator under ϕ is a partial isometry, we infer that $\|\phi(A)\| \leq 1$. It is obvious that ϕ is contractive. Since ϕ^{-1} has the same properties as ϕ, it follows that ϕ is in fact an isometry of $B(H)$. The structure of those maps is described in Theorem A.9. In fact, they are of one of the forms appearing in the formulation of our theorem. $\square$

Now we turn to the proof of the last result of the section.

Proof of Theorem 1.1.4. The minimal left ideals of $B(H)$ are precisely the sets $\{x \otimes y : x \in H\}$ for nonzero $y \in H$. Since ϕ clearly preserves the minimal left ideals of $B(H)$ in both directions, we easily deduce that ϕ is a linear bijection of $F(H)$ onto itself which preserves the rank-one operators. The general form of such transformations can be described. It follows from Theorem A.1 that either there are linear bijections $A, B : H \to H$ such that

$$\phi(x \otimes y) = Ax \otimes By \qquad (x, y \in H) \qquad (1.1.11)$$

or there are conjugate-linear bijections $A', B' : H \to H$ such that

$$\phi(x \otimes y) = A'y \otimes B'x \qquad (x, y \in H).$$

Since ϕ is left ideal preserving, the second possibility above obviously cannot occur.

We prove that $\phi(I)$ is invertible. First we note the following. It is true in any algebra with unit that an element fails to have a left inverse if and only if this element can be included in a maximal left ideal (recall that every proper left ideal is included in a maximal left ideal). Therefore, ϕ preserves the left invertible elements of $B(H)$ in both directions. We recall that an operator S in $B(H)$ is left invertible if and only if S is injective and S has closed range. Now, let $x, y \in H$ be arbitrary nonzero vectors. Let $\lambda \in \mathbb{C}$. By Fredholm alternative, the operator $x \otimes y - \lambda I$ is injective if and only if it is surjective. This gives us that $x \otimes y - \lambda I$ is left invertible if and only if it is invertible. Since the spectrum of any element in $B(H)$ is nonempty, we infer that there is a $\lambda \in \mathbb{C}$ for which $x \otimes y - \lambda I$ is not left invertible. Suppose that $x \otimes y$ is not quasinilpotent, that is, $\langle x, y \rangle \neq 0$. Then the scalar λ above can be chosen to be nonzero. It follows that $Ax \otimes By - \lambda \phi(I)$ is not left invertible. On the other hand, $\phi(I)$ is left invertible and hence it is a left Fredholm operator (see [54, 2.3. Definition, p. 356]). But any compact perturbation of a left Fredholm operator has closed range [54, 2.5. Theorem, p. 356]. So, the operator $Ax \otimes By - \lambda \phi(I)$ is not left invertible but it has closed range. Therefore, this operator is not injective, that is, there exists a nonzero vector $z \in H$ such that $\lambda \phi(I)z = \langle z, By \rangle Ax$. Clearly, this implies that $Ax \in \operatorname{rng} \phi(I)$. Since $x \in H$ was arbitrary, we conclude that $H = \operatorname{rng} A \subset \operatorname{rng} \phi(I)$ which means that $\phi(I)$ is surjective. This gives us that $\phi(I)$ is invertible.

We next show that the linear operators A, B in (1.1.11) are bounded. Let $x, y \in H$. We have seen above that $x \otimes y - \lambda I$ is not left invertible if and only if $\lambda \in \sigma(x \otimes y)$. Similarly, $\phi(I)^{-1}(Ax \otimes By - \lambda \phi(I))$ is not left invertible if and only if $\lambda \in \sigma(\phi(I)^{-1}Ax \otimes By)$. Since ϕ preserves the left invertible operators in both directions, we obtain

$$\sigma(x \otimes y) = \sigma(\phi(I)^{-1}Ax \otimes By).$$

By the spectral radius formula we have

$$|\langle x, y \rangle| = |\langle \phi(I)^{-1}Ax, By \rangle| \qquad (x, y \in H).$$

Now, an easy application of the closed graph theorem shows that A, B are continuous.

Evidently, we may suppose without any loss of generality that $A = B = I$. Let $S \in B(H)$ be invertible and write $C = \phi(S)$. We claim that $C = S$. Let $x \in H$ be an arbitrary unit vector. Then $S(I - \lambda x \otimes x)$ has a left inverse if and only if $\lambda \neq 1$. Consequently, the operator $C - \lambda Sx \otimes x$ is injective for every $\lambda \neq 1$ and for every unit vector $x \in H$. Let $z \in H$ be a nonzero vector. Let $y = S^{-1}Cz$ which is also nonzero since $C = \phi(S)$ is left invertible. If $\langle z, y \rangle \neq 0$, then choosing $\lambda = \|y\|^2/\langle z, y \rangle$ we see that

$$Cz - \lambda(1/\|y\|^2 Sy \otimes y)(z) = Sy - Sy = 0.$$

Since z is nonzero, we deduce $\lambda = 1$ which means $\|y\|^2 = \langle z, y \rangle$. Therefore, for every nonzero vector $z \in H$ we have two possibilities. Either $\langle z, S^{-1}Cz \rangle = 0$ or $\langle S^{-1}Cz, S^{-1}Cz \rangle = \langle z, S^{-1}Cz \rangle$. Clearly, the set of all nonzero vectors satisfying the first equality as well as the set of those ones which satisfy the second equality are both closed in $H \setminus \{0\}$. Since $H \setminus \{0\}$ is a connected set, we infer that either

$$\langle z, S^{-1}Cz \rangle = 0 \qquad (z \in H)$$

or

$$\langle S^{-1}Cz, S^{-1}Cz \rangle = \langle z, S^{-1}Cz \rangle \qquad (z \in H).$$

The first possibility would imply that $S^{-1}C = 0$. Therefore, we have the second one which can be reformulated as $(S^{-1}C)^*(S^{-1}C) = S^{-1}C$. This shows that $S^{-1}C$ is a projection. But, on the other hand, it is left invertible, and hence we have $S^{-1}C = I$ which results in $\phi(S) = C = S$. Since $B(H)$ is linearly generated by the set of all invertible operators, it follows that ϕ is the identity on $B(H)$. This completes the proof.   □

1.1.4 Remarks

In the paper [165] we emphasized that in the proof of Theorem 1.1.4 we used the preservation of only two extreme kinds of left ideals, namely, that of the minimal ones and that of the maximal ones. As we have seen above, preserving minimal left ideals is connected with the problem of rank-one preservers. On the other hand, preserving maximal left ideals is connected with the problem of left invertibility preservers. Since there is so much interest in preserver problems concerning rank or invertibility, this remark motivated Kim [124] and later Cui and Hou [59] to consider the problem of preserving minimal or, rather, maximal one-sided ideals. They obtained some nice results in a quite general setting, but under the restrictive assumption that the transformations in question are unital, i.e., they map the unit into the unit, or that they send the unit at least to an invertible element. (Recall that it was an important and considerable part of the proof of Theorem 1.1.4 to show that $\phi(I)$ is invertible.)

We emphasize the probably main result of the paper [59] which states that a unital linear bijection from a semi-simple Banach algebra onto a C^*-algebra of real rank zero which preserves the maximal left ideals is necessarily a Jordan homomorphism.

Finally, we also refer to the paper [105] of Hou and Cui concerning additive left invertibility preservers.

1.2 Linear Maps on Factors Which Preserve the Extreme Points of the Unit Ball

1.2.1 Summary

In this section we characterize those linear maps from a von Neumann factor $\mathcal{A}$ into itself which preserve the extreme points of the unit ball. We show that if $\mathcal{A}$ is infinite, then every such preserver can be written as an either unital *-homomorphism or a unital *-antihomomorphism multiplied by a fixed unitary element. The results of this section appeared in the paper [155].

1.2.2 Formulation of the Results

The linear preserver problem we investigate below is in an intimate connection with the problem of unitary group preservers on matrix algebras which are the linear maps leaving the set of unitaries in $M_n(\mathbb{C})$ invariant. This problem was treated by Marcus in [149], while its infinite dimensional analogue concerning the operator algebra $B(H)$ or, more generally, a general C^*-algebra were solved by Rais in [213] and by Russo and Dye in [219] (see Theorem A.2), respectively.

In finite dimension the extreme points of the unit ball of $B(H)$ are exactly the unitaries. This is no longer true in infinite dimension. Accordingly, in [213, Section 4] Rais raised the problem of characterizing those linear maps on $B(H)$ which preserve the extreme points of the unit ball of $B(H)$ and concerning bijective linear transformations of $B(H)$ which preserve the extreme points of the unit ball in both directions he obtained a complete description. The precise connection between unitary group preservers on $B(H)$ and linear maps preserving the extreme points of the unit ball of $B(H)$ is that in the first case our maps preserve the set of all bijective partial isometries while in the second case they preserve the set of all injective or surjective partial isometries (see [97, Sections 98, 99], cf. Lemma 1.2.6, Lemma 1.2.7 below).

We mention two other motivations which have led us to consider the present problem. The first one is the following. When studying the structure of the surjective linear isometries of operator algebras, in the arguments it is very usual to use the easy fact that every such map preserves the extreme points of the unit ball (we refer only to two of the most famous papers on the topic [117, 7]). This property turns to include important information and, together with the norm preserving property, is of great help in the proofs of the corresponding results. Having this in mind, it is a natural problem to study the extreme point preserving property alone.

The other motivation to our present investigations is the following. In the paper [132] linear maps between C^*-algebras whose adjoints preserve the extreme points of the dual ball were studied and they turned out to give a valuable clue as to what objects may be regarded as 'non-commutative composition operators'. It seems to be a natural problem to consider linear maps in general (i.e. without the assumption of being the adjoints of linear maps on a C^*-algebra) which preserve the extreme points of the unit ball. For example, since $B(H)$ is the dual space of the Banach algebra of all trace-class operators on H (which is a highly non-C^*-algebra), in that particular but undoubtedly very important case the problem has nothing to do with the one treated in [132]. In fact, as one can see below, we follow a completely different approach to attack the problem.

Now we formulate the results of this section. The first one of the main results reads as follows.

Theorem 1.2.1. *Let $\mathcal{A}$ be an infinite factor. The linear map $\phi : \mathcal{A} \to \mathcal{A}$ preserves the extreme points of the unit ball of $\mathcal{A}$ if and only if either there are a unitary operator $U \in \mathcal{A}$ and a unital *-homomorphism $\psi : \mathcal{A} \to \mathcal{A}$ such that ϕ is of the form*

$$\phi(A) = U\psi(A) \qquad (A \in \mathcal{A})$$

*or there are a unitary operator $U' \in \mathcal{A}$ and a unital *-antihomomorphism $\psi' : \mathcal{A} \to \mathcal{A}$ such that ϕ is of the form*

$$\phi(A) = U'\psi'(A) \qquad (A \in \mathcal{A}).$$

As a consequence of this result, we immediately have the structure of all surjective linear transformations of $B(H)$ which preserve the extreme points of the unit ball. In fact, Corollary 1 below is a significant improvement of a result of Rais [213, Lemma 3 and Corollary 1] who obtained a similar conclusion but under the much more restrictive assumption that the transformations under consideration are bijective and preserve the extreme points of the unit ball in both directions.

Corollary 1.2.2. *Let H be an infinite dimensional Hilbert space. Then the surjective linear map $\phi : B(H) \to B(H)$ preserves the extreme points of the unit ball of $B(H)$ if and only if either there are unitaries $U, V \in B(H)$ such that ϕ is of the form*

$$\phi(A) = UAV \qquad (A \in B(H))$$

or there are antiunitaries $U', V' \in B(H)$ such that ϕ is of the form

$$\phi(A) = U'A^*V' \qquad (A \in B(H)).$$

As a small but interesting observation we note that, as one can easily see from the proof of Corollary 1.2.2, it would have been sufficient to assume that the range of ϕ contains merely one rank-one operator instead of supposing that ϕ is surjective.

If the underlying Hilbert space is separable, then we can write our linear preservers on $B(H)$ in a form which is more explicit than what has appeared in Theorem 1.2.1.

Corollary 1.2.3. *Let H be a separable infinite dimensional Hilbert space. The linear map $\phi : B(H) \to B(H)$ preserves the extreme points of the unit ball of $B(H)$ if and only if either there are a unitary operator V and a collection $\{U_\alpha\}$ of isometries with pairwise orthogonal ranges which generate H such that ϕ is of the form*

$$\phi(A) = V(\sum_\alpha U_\alpha A U_\alpha^*) \qquad (A \in B(H))$$

or there are a unitary operator V and a family of antiisometries (i.e., conjugate-linear isometries) $\{V_\alpha\}$ with pairwise orthogonal ranges which generate H such that ϕ is of the form

$$\phi(A) = V(\sum_\beta V_\beta A^* V_\beta^*) \qquad (A \in B(H)).$$

Our second main result describes the linear preservers under consideration on arbitrary finite von Neumann algebras (not only on factors).

Theorem 1.2.4. *Let $\mathcal{A}$ be a finite von Neumann algebra. The linear map $\phi : \mathcal{A} \to \mathcal{A}$ preserves the extreme points of the unit ball of $\mathcal{A}$ if and only if there exist a unitary operator $U \in \mathcal{A}$ and a unital Jordan *-homomorphism ψ such that*

$$\phi(A) = U\psi(A) \qquad (A \in \mathcal{A}).$$

Concerning matrix algebras, we immediately have the following assertion which is the last one in this section (cf. [149]). Here, $M_n(\mathbb{C})$ is equipped with the operator norm, of course.

Corollary 1.2.5. *The linear map $\phi : M_n(\mathbb{C}) \to M_n(\mathbb{C})$ preserves the extreme points of the unit ball of $M_n(\mathbb{C})$ if and only if there are unitary matrices $U, V \in M_n(\mathbb{C})$ such that ϕ is either of the form*

$$\phi(A) = UAV \qquad (A \in M_n(\mathbb{C}))$$

or of the form

$$\phi(A) = UA^{tr}V \qquad (A \in M_n(\mathbb{C})).$$

1.2.3 Proofs

The statements of Lemma 1.2.6 and Lemma 1.2.7 below are guessed to be well-known. However, since we have not found any trace of them in the monograph [120, 121] of Kadison and Ringrose or in its bibliography, for the sake of completeness we present them with proofs. Recall that an element W in a C^*-algebra $\mathcal{B}$ is called a partial isometry if it satisfies $WW^*W = W$.

Lemma 1.2.6. *Let $\mathcal{A}$ be a factor. The operator $A \in \mathcal{A}$ is an extreme point of the unit ball of $\mathcal{A}$ if and only if A is either an isometry or a coisometry (i.e., the adjoint of an isometry).*

Proof. In an arbitrary C^*-algebra $\mathcal{B}$, the extreme points of the unit ball can be characterized algebraically in the following way. By [121, 7.3.1. Theorem], these points are exactly those partial isometries $W \in \mathcal{B}$ for which we have

$$(I - W^*W)\mathcal{B}(I - WW^*) = \{0\}. \tag{1.2.1}$$

It is known that in a factor every two projections are comparable (see [121, 6.2.6. Proposition]). So, for example, let $I - W^*W \precsim I - WW^*$. Then there is a partial isometry $V \in \mathcal{A}$ such that $I - W^*W = V^*V$ and VV^* is a subprojection of $I - WW^*$. By (1.2.1) we have $(V^*V)(V^*)(VV^*) = 0$. But V is a partial isometry and hence $VV^*V = V$. Consequently, we obtain that $0 = (V^*V)(V^*VV^*) = V^*VV^* = V^*$ which implies $V = 0$. This gives us that W is an isometry.

In the remaining case when $I - WW^* \precsim I - W^*W$ we obtain that W^* is an isometry, i.e., W is a coisometry.   □

Proof of Theorem 1.2.1. By the characterization of the extreme points of the unit ball of $\mathcal{A}$ given in Lemma 1.2.6, the sufficiency is trivial to check. So, let us assume that ϕ is a linear map on $\mathcal{A}$ which preserves the extreme points of the unit ball. First observe that ϕ is necessarily norm-continuous. Indeed, similarly to the last part of the proof of Theorem 1.1.3 we obtain that ϕ is a contraction.

Consider the operator $V = \phi(I)$. Since V is either an isometry or a coisometry, without loss of generality we may and do suppose that $V^*V = I$. Since the unitary group is arcwise connected in $\mathcal{A}$, it follows from the uniform continuity of ϕ that $\phi(U)$ is an isometry for every unitary $U \in \mathcal{A}$.

Let us define a linear map $\psi : \mathcal{A} \to \mathcal{A}$ by $\psi(A) = V^*\phi(A)$ $(A \in \mathcal{A})$. In what follows we prove that ψ is a Jordan *-homomorphism. To verify this, first observe that by the fact that ϕ sends unitaries to isometries we have

$$\phi(e^{itS})^*\phi(e^{itS}) = I \qquad (t \in \mathbb{R})$$

for every self-adjoint operator $S \in \mathcal{A}$. Using the power series expansion of the exponential function as well as the uniqueness of the coefficients, it is easy to conclude that

$$\phi(I)^*\phi(S) - \phi(S)^*\phi(I) = 0$$
$$-\frac{1}{2}\phi(I)^*\phi(S^2) + \phi(S)^*\phi(S) - \frac{1}{2}\phi(S^2)^*\phi(I) = 0.$$

These identities imply that

$$\phi(S)^*\phi(S) = \phi(I)^*\phi(S^2). \tag{1.2.2}$$

Since $\psi(S^2) = \phi(I)^*\phi(S^2)$, this shows that ψ preserves the positive operators and then we obtain that ψ preserves the self-adjoint operators as well. If we replace S by $S + T$ in (1.2.2) where $S, T \in \mathcal{A}$ are self-adjoint, then we obtain

$$\phi(S)^*\phi(T) + \phi(T)^*\phi(S) = \phi(I)^*\phi(ST + TS)$$

and one can check that this results in

$$\phi(A^*)^*\phi(A) = \phi(I)^*\phi(A^2) \qquad (A \in \mathcal{A}).$$

If we linearize this equation, i.e. replace A by $A + B$, we get

$$\phi(A^*)^*\phi(B) + \phi(B^*)^*\phi(A) = \phi(I)^*\phi(AB + BA) \qquad (A, B \in \mathcal{A}). \quad (1.2.3)$$

It is known that in an infinite factor every projection has the property that either itself or its orthogonal complement is equivalent to the identity (see [236, E.4.11, p. 105]). Let $P \in \mathcal{A}$ be an arbitrary projection. It follows that either $P \sim I$ or $I - P \sim I$. Suppose that this latter possibility is the case. Then we have an isometry $U \in \mathcal{A}$ for which $UU^* = I - P$. By (1.2.3) we can compute

$$\begin{aligned}\phi(U^*)^*\phi(U^*) + \phi(U)^*\phi(U) &= \phi(I)^*\phi(UU^* + U^*U) = \\ &= \phi(I)^*\phi(2I - P) = 2 - \psi(P).\end{aligned} \qquad (1.2.4)$$

Since, by our assumption on ϕ, the operators $\phi(U), \phi(U^*)$ are partial isometries, it follows that $\phi(U^*)^*\phi(U^*)$ and $\phi(U)^*\phi(U)$ are projections. It follows from (1.2.4) that there are projections $Q_1, Q_2 \in \mathcal{A}$ such that

$$\psi(P) = Q_1 + Q_2.$$

On the other hand, using the property of ψ that it preserves the self-adjoint operators and considering the equation (1.2.2) we have

$$\psi(P)^2 = \psi(P)^*\psi(P) = \phi(P)^*VV^*\phi(P) \le \phi(P)^*\phi(P) = \psi(P).$$

This yields

$$Q_1Q_2 + Q_2Q_1 \le 0.$$

Since Q_1, Q_2 are projections, it is not hard to verify that this inequality implies $Q_1Q_2 = Q_2Q_1 = 0$. This gives us that $\psi(P)$ is a projection.

So, ψ is a continuous linear transformation which preserves the projections. By Theorem A.4 we obtain that ψ is a Jordan *-homomorphism.

Our next claim is that $\phi(A) = V\psi(A)$ $(A \in \mathcal{A})$. To this end, let $U \in \mathcal{A}$ be an arbitrary unitary operator. Then $W = \phi(U)$ is an isometry and we have

$$\psi(U)^*\psi(U) + \psi(U)\psi(U)^* = \psi(U^*U + UU^*) = \psi(2I) = 2I,$$

i.e.,

$$W^*VV^*W + V^*WW^*V = 2I. \qquad (1.2.5)$$

Since $W^*VV^*W \le I$ and $V^*WW^*V \le I$, from (1.2.5) it follows that

$$W^*VV^*W = I, \quad V^*WW^*V = I.$$

Since VV^* and WW^* are projections onto the ranges of V and W, respectively, it is not hard to see that the above equalities imply $\operatorname{rng} W \subset \operatorname{rng} V$ and $\operatorname{rng} V \subset \operatorname{rng} W$. Consequently, $\operatorname{rng} \phi(U) = \operatorname{rng} V$ for every unitary operator $U \in \mathcal{A}$. Since the linear span of the unitaries in $\mathcal{A}$ is $\mathcal{A}$, it follows that $\operatorname{rng} \phi(A) \subset \operatorname{rng} V$ for every $A \in \mathcal{A}$, and this gives us that $V\psi(A) = VV^*\phi(A) = \phi(A)$ $(A \in \mathcal{A})$.

We now prove that V is unitary. Since ψ is a unital Jordan *-homomorphism, by the result Theorem A.6 of Størmer, there exists a projection E in the center of the von Neumann algebra generated by the image of ψ such that the maps ψ_1 and ψ_2 on $\mathcal{A}$ defined by $\psi_1(A) = \psi(A)E$ and $\psi_2(A) = \psi(A)(I - E)$ are a *-homomorphism and a *-antihomomorphism, respectively. If we suppose on the contrary that $\operatorname{rng} V \neq H$, then by the extreme point preserving property of ϕ and the equality $\phi(A) = V\psi(A)$ it follows that ϕ sends every isometry and coisometry to a proper isometry. Let $W \in \mathcal{A}$ be an arbitrary isometry or coisometry, that is, suppose that $W^*W = I$ or $WW^* = I$. In both cases we have

$$I = \psi(W)^*V^*V\psi(W) = \psi(W)^*\psi(W) = \psi_1(W^*W) + \psi_2(WW^*).$$

This readily implies that

$$\psi_1(W^*W) = E \quad \text{and} \quad \psi_2(WW^*) = I - E. \tag{1.2.6}$$

This shows that ψ_1 and ψ_2 map every projection equivalent to I into E and $I - E$, respectively. In an infinite factor there is a projection such that both itself and its orthogonal complement are equivalent to the identity. This is a particular case of the so-called Halving Lemma (see [121, 6.3.3. Lemma]). So, we can pick a projection $P \in \mathcal{A}$ such that $P \sim I$ and $I - P \sim I$. Then we have

$$\psi_1(P) = \psi_1(I - P) = E, \quad \psi_2(P) = \psi_2(I - P) = I - E.$$

As $\psi_1(I) = E$ and $\psi_2(I) = I - E$, we deduce $E = 0$, $I - E = 0$ which is an obvious contradiction. Therefore, we obtain that V is unitary.

It only remains to prove that either $\psi_1 = 0$ or $\psi_2 = 0$. Just as above, let $P \in \mathcal{A}$ be a projection for which $P \sim I$ and $I - P \sim I$. Let $W \in \mathcal{A}$ be such that $W^*W = I$ and $WW^* = P$. Then $\psi(W)$ is either an isometry or a coisometry. Suppose that this latter one is the case. Then we have

$$I = \psi(W)\psi(W)^* = \psi_1(WW^*) + \psi_2(W^*W)$$

which gives us that $\psi_1(P) = E = \psi_1(I)$. If W' is an isometry for which $I - P = W'W'^*$, then we have $\psi_1(W')\psi_1(W')^* = \psi_1(I - P) = 0$. Therefore, $\psi_1(W') = 0$ and hence $0 = \psi_1(W')^*\psi_1(W') = \psi_1(I)$. It follows that $\psi_1 = 0$ and thus we obtain that ψ is a *-antihomomorphism. In case $\psi(W)$ is an isometry, one can argue in a very similar way to obtain that $\psi_2 = 0$ which means that ψ is a *-homomorphism. $\qquad\square$

Proof of Corollary 1.2.2. Using Theorem 1.2.1, without a serious loss of generality we may and do suppose that our map ϕ is a surjective unital *-homomorphism of $B(H)$. We assert that ϕ is injective. Let P be a rank-one projection. Then there is an operator $A \in B(H)$ such that $\phi(A) = P$. Since $\phi(A^*A) = \phi(A)^*\phi(A) = P$, we can assume that our operator A is positive. Let us consider its spectral resolution. By the monotonicity and continuity of ϕ it follows that there is a Borel subset B of the spectrum of A having a positive distance from 0 for which $E(B)$ (E is the spectral measure corresponding to A) has nonzero image under ϕ. Since there is a positive constant c for which $cE(B) \leq A$, we obtain $0 \neq c\phi(E(B)) \leq P$. As $\phi(E(B))$ is a nonzero projection and P is of rank one, we have $\phi(E(B)) = P$. Now, if ϕ is not injective, then its kernel, being a nontrivial ideal of $B(H)$, contains the ideal of all finite rank operators. Hence, the projection $E(B)$ is infinite dimensional. But in this case $E(B)$ is the sum of two orthogonal projections Q_1, Q_2 which are both equivalent to $E(B)$. As P is of rank one, we see that either $\phi(Q_1) = 0$ or $\phi(Q_2) = 0$. Since Q_1 and Q_2 are equivalent to each other and ϕ is a *-homomorphism, thus if any of Q_1, Q_2 is in the kernel of ϕ, then so is the other one. Hence, we deduce $0 = \phi(Q_1) + \phi(Q_2) = P$ which is a contradiction. Consequently, we obtain the injectivity of ϕ and so ϕ is a *-automorphism of $B(H)$. It is a folk result (see Theorem A.8) that every *-automorphism of $B(H)$ is inner, i.e., there is a unitary operator $U \in B(H)$ such that $\phi(A) = UAU^*$ ($A \in B(H)$). This completes the proof.   □

Proof of Corollary 1.2.3. By Theorem 1.2.1 we can suppose without loss of generality that $\phi : B(H) \to B(H)$ is a unital *-homomorphism. Now, we can apply Theorem A.11 on the form of the *-representations of $B(H)$ and obtain the assertion.   □

Lemma 1.2.7. *Let $\mathcal{A}$ be a finite von Neumann algebra. The extreme points of the unit ball of $\mathcal{A}$ are exactly the unitaries in $\mathcal{A}$.*

Proof. We use the algebraic characterization of the extreme points of the unit ball of a C^*-algebra that was mentioned in the beginning of the proof of Lemma 1.2.6. First it follows that every unitary operator in $\mathcal{A}$ is an extreme point of the unit ball of $\mathcal{A}$. Next, suppose that $V \in \mathcal{A}$ is an extremal point of the unit ball which means that it is a partial isometry such that $(I - V^*V)\mathcal{A}(I - VV^*) = \{0\}$. Let $V^*V = E, VV^* = F$. In a finite von Neumann algebra equivalent projections have equivalent orthogonal complements. This is the content of [121, 6.9.6. Exercise]. So, we have $I - E \sim I - F$. Let $W \in \mathcal{A}$ be a partial isometry for which $I - E = W^*W$ and $I - F = WW^*$. Since $W^*WAWW^* = \{0\}$, it follows that $0 = W^*WW^*WW^* = W^*WW^* = W^*$. Consequently, we have $E = F = I$ which means that V is unitary.   □

Proof of Theorem 1.2.4. Let ψ be a unital Jordan *-homomorphism on $\mathcal{A}$. Let $U \in \mathcal{A}$ be unitary. Then we have

$$\psi(U)^*\psi(U) + \psi(U)\psi(U)^* = \psi(2I) = 2I.$$

On the other hand, taking into consideration that ψ is necessarily a contraction (see the first paragraph in the proof of Theorem 1.2.1), we get $\psi(U)^*\psi(U), \psi(U)\psi(U)^* \leq I$. Therefore, we have

$$\psi(U)^*\psi(U) = \psi(U)\psi(U)^* = I$$

which means that $\psi(U)$ is unitary. Consequently, we obtain that ψ preserves the unitary elements of $\mathcal{A}$, and by Lemma 1.2.7 we conclude that ψ preserves the extreme points of the unit ball.

Conversely, if ϕ preserves the extreme points of $\mathcal{A}$, i.e. the unitary group of the C^*-algebra $\mathcal{A}$, then by the result Theorem A.2 of Russo and Dye we obtain that ϕ is of the desired form. □

Proof of Corollary 1.2.5. The sufficiency is obvious. To the necessity we may suppose that ϕ is a unital Jordan *-homomorphism. It is well-known that every Jordan ideal of $M_n(\mathbb{C})$ is an associative ideal (see, for example, [106, Theorem 11]). Since the algebra $M_n(\mathbb{C})$ is simple, we obtain that ϕ is an injective and hence surjective Jordan *-homomorphism, i.e. a Jordan *-automorphism. By the well-known result Theorem A.7 of Herstein, it follows that ϕ is either a *-automorphism or a *-antiautomorphism and, similarly to the proof of Corollary 1.2.2, we are done. □

1.2.4 Remarks

As it can be seen from the proofs above, our arguments used heavily the fact that the underlying structures are von Neumann algebras. Nevertheless, as there is so much interest in the geometrical properties of C^*-algebras, we believe that it would be a prosperous (and probably difficult) problem to study the extreme point preservers in that context.

1.3 Diameter Preserving Linear Bijections of $C(X)$

1.3.1 Summary

Let X be a first countable compact Hausdorff space. We show that every linear bijection $\phi : C(X) \to C(X)$ which preserves the diameter of the range, i.e., which has the property that

$$\mathrm{diam}(\phi(f)(X)) = \mathrm{diam}(f(X)) \qquad (f \in C(X))$$

is of the form

$$\phi(f) = \tau \cdot f \circ \varphi + t(f)1 \qquad (f \in C(X)),$$

where $\tau \in \mathbb{C}$, $|\tau| = 1$, $\varphi : X \to X$ is a homeomorphism and $t : C(X) \to \mathbb{C}$ is a linear functional. The result of this section appeared in the paper [89].

1.3.2 Formulation of the Result

With regard to function algebras, the main linear preserver problems studied so far have been the characterizations of linear bijections preserving some given norm, or preserving the disjointness of the support (these latter maps are also called separating). To mention some papers from the not so far past, we refer to [110, 245, 248], and [5, 73, 99, 108] (for a recent result also see [123]).

One way of measuring a function is to consider its sup-norm. The linear bijections of $C(X)$ which preserve this quantity are determined in the famous Banach-Stone theorem. According to Theorem A.10, any surjective linear isometry $\phi : C(X) \to C(X)$ is of the form

$$\phi(f) = \tau \cdot f \circ \varphi \qquad (f \in C(X)),$$

where $\tau \in C(X)$ is a function of modulus 1 and $\varphi : X \to X$ is a homeomorphism.

Beside the sup-norm, it is another sensible possibility to measure the functions by some data which reflect how large their range is. For example, this can be done by considering the diameter of the range. The aim of this section is to determine all linear bijections of $C(X)$ which preserve the seminorm $f \mapsto \mathrm{diam}(f(X))$. For brevity, we call these maps diameter preserving. Obviously, every automorphism $f \mapsto f \circ \varphi$ ($\varphi : X \to X$ is a homeomorphism) of $C(X)$ is diameter preserving. The diameter of sets in $\mathbb{C}$ is clearly invariant under rotation and translation. This implies that every linear map of the form

$$\phi(f) = \tau \cdot f \circ \varphi + t(f)1 \qquad (f \in C(X)),$$

where $\tau \in \mathbb{C}$ is of modulus 1 and $t : C(X) \to \mathbb{C}$ is a linear functional, preserves the diameter of the ranges of functions in $C(X)$. The result which follows shows that, under a mild condition on the topological space X, every bijective diameter preserving linear map on $C(X)$ is of the above form.

Theorem 1.3.1. Let X be a first countable compact Hausdorff space. A bijective linear map $\phi : C(X) \to C(X)$ is diameter preserving if and only if there exists a complex number τ of modulus 1, a homeomorphism $\varphi : X \to X$ and a linear functional $t : C(X) \to \mathbb{C}$ with $t(1) \neq -\tau$ such that ϕ is of the form

$$\phi(f) = \tau \cdot f \circ \varphi + t(f)1 \qquad (f \in C(X)). \tag{1.3.1}$$

We remark that the statement of our theorem also holds for the algebra of all continuous real valued functions on X. Of course, in this situation, we have $\tau = \pm 1$ and $t : C(X) \to \mathbb{R}$. In fact, in this case the proof can be accomplished more simply.

We also note that, as a particular case, this result gives the form of all nonsingular linear transformations on $\mathbb{K}^n$ ($\mathbb{K}$ stands for $\mathbb{C}$ or $\mathbb{R}$) which preserve the maximal distance between the coordinates. Clearly, this gives us the solution of a classical LPP.

1.3.3 Proof

Proof of Theorem 1.3.1. It is straightforward to check that every linear map ϕ of the form (1.3.1) with τ, φ, t being as in the statement above is a diameter preserving linear bijection of $C(X)$.

Now, suppose that $\phi : C(X) \to C(X)$ is a linear bijection which preserves the diameter of the ranges of functions in $C(X)$. Since the theorem is easy to verify in the cases in which X has one or two points, for the proof of necessity, we suppose that X has at least three points.

We introduce the following notation. Let $\tilde{X}$ stand for the collection of all subsets of X having exactly two elements. For any $f \in C(X)$, let

$$S(f) = \{\{x, y\} \in \tilde{X} \; : \; |f(x) - f(y)| = \operatorname{diam}(f(X))\},$$
$$P(f) = \{(x, y) \in X \times X \; : \; |f(x) - f(y)| = \operatorname{diam}(f(X))\},$$
$$T(f) = \{(x, y, u) \in X \times X \times \mathbb{C} \; : \; |f(x) - f(y)| = \operatorname{diam}(f(X)),$$
$$u = f(x) - f(y)\},$$

and, for every $\{x, y\} \in \tilde{X}$ and $u \in \mathbb{C}$, let

$$\mathcal{S}(\{x, y\}) = \{f \in C(X) \; : \; \{x, y\} \in S(f)\},$$
$$\mathcal{S}_s(\{x, y\}) = \{f \in C(X) \; : \; \{\{x, y\}\} = S(f)\},$$
$$\mathcal{T}(x, y, u) = \{f \in C(X) \; : \; (x, y, u) \in T(f)\},$$
$$\mathcal{T}_s(x, y, u) = \{f \in C(X) \; : \; \{(x, y, u), (y, x, -u)\} = T(f)\}.$$

Finally, define

$$G(\{x, y\}) = \bigcap\{S(\phi(f)) \; : \; f \in C(X), \{x, y\} \in S(f)\}$$
$$H(x, y, u) = \bigcap\{T(\phi(f)) \; : \; f \in C(X), (x, y, u) \in T(f)\}.$$

Obviously, for every nonconstant function $f \in C(X)$, the sets $S(f)$, $P(f)$ and $T(f)$ are nonempty. Since X is first countable, using Uryson's lemma we see easily that, for every pair $x, y \in X, x \neq y$ of points, there exists a continuous real valued function f from X into $[0, 1]$ such that $f(x) = 1, f(y) = 0$ and $0 < f(z) < 1$ ($z \in X, z \neq x, z \neq y$). This shows that for any element $\{x, y\} \in \tilde{X}$ and $0 \neq u \in \mathbb{C}$, the sets $\mathcal{S}_s(\{x, y\})$, $\mathcal{T}_s(x, y, u)$ are also nonempty.

After this preparation we can begin the proof of the necessity of our statement which will be carried out through a series of steps. The following observation will be used repeatedly in what follows.

Lemma 1.3.2. *For an arbitrary integer $n \in \mathbb{N}$ and functions $f_1, \dots, f_n \in C(X)$ we have*

$$\operatorname{diam}((f_1 + \dots + f_n)(X)) = \operatorname{diam}(f_1(X)) + \dots + \operatorname{diam}(f_n(X))$$

if and only if there exists an $\{x, y\} \in \tilde{X}$ and a complex number v of modulus 1 such that $f_i \in \mathcal{T}(x, y, \lambda_i v)$ holds for every $i = 1, \ldots, n$, where $\lambda_i = \mathrm{diam}(f_i(X))$ $(i = 1, \ldots, n)$.

Proof. Assume that

$$\mathrm{diam}((f_1 + \ldots + f_n)(X)) = \mathrm{diam}(f_1(X)) + \ldots + \mathrm{diam}(f_n(X)).$$

Then there exists $\{x, y\} \in \tilde{X}$ and a complex number v of modulus 1 such that

$$f_1 + \ldots + f_n \in \mathcal{T}(x, y, (\lambda_1 + \ldots + \lambda_n)v).$$

Now, we compute

$$\lambda_1 + \ldots + \lambda_n = |(f_1 + \ldots + f_n)(x) - (f_1 + \ldots + f_n)(y)| \leq$$
$$\leq |f_1(x) - f_1(y)| + \ldots + |f_n(x) - f_n(y)| \leq \lambda_1 + \ldots + \lambda_n.$$

It readily follows that $f_i \in \mathcal{T}(x, y, \lambda_i v)$ $(i = 1, \ldots, n)$. The converse statement of the lemma is trivial. □

Step 1. *For arbitrary $\{x, y\} \in \tilde{X}$ and $0 \neq u \in \mathbb{C}$, we have $G(\{x, y\}) \neq \emptyset$ and $H(x, y, u) \neq \emptyset$.*

Let $\{x, y\} \in \tilde{X}$. We first show that

$$\bigcap \{P(\phi(f)) : f \in C(X), \{x, y\} \in S(f)\} \neq \emptyset. \tag{1.3.2}$$

Since the collection of sets $P(\phi(f))$ in (1.3.2) consists of closed subsets of the compact Hausdorff space $X \times X$, in order to verify (1.3.2), it is sufficient to show that this collection has the finite intersection property. Accordingly, let $f_1, \ldots, f_n \in C(X)$ be such that $\{x, y\} \in S(f_1), \ldots, S(f_n)$. Define $u_i = f_i(x) - f_i(y)$ $(i = 1, \ldots, n)$. Then there exist complex numbers μ_i with $|\mu_i| = 1$ for which $\mu_i u_i \geq 0$. Since the diameter of the range of $\mu_i f_i$ is $\mu_i u_i$ and

$$|(\mu_1 f_1 + \ldots + \mu_n f_n)(x) - (\mu_1 f_1 + \ldots + \mu_n f_n)(y)| =$$
$$|\mu_1(f_1(x) - f_1(y)) + \ldots + \mu_n(f_n(x) - f_n(y))| =$$
$$|\mu_1 u_1 + \ldots + \mu_n u_n| = \mu_1 u_1 + \ldots + \mu_n u_n = |u_1| + \ldots + |u_n|,$$

we deduce that the diameter of the range of $\mu_1 f_1 + \ldots + \mu_n f_n$ is $|u_1| + \ldots + |u_n|$. From the diameter preserving property of ϕ it follows that the diameter of the range of $\phi(\mu_1 f_1 + \ldots + \mu_n f_n)$ is $|u_1| + \ldots + |u_n|$ which equals the sum of the diameters of the ranges of $\phi(\mu_1 f_1), \ldots, \phi(\mu_n f_n)$. By the lemma above we conclude that there exists $\{x_0, y_0\} \in \tilde{X}$ and a complex number v of modulus 1 for which $\phi(\mu_i f_i) \in \mathcal{T}(x_0, y_0, |u_i|v)$ $(i = 1, \ldots, n)$. Obviously, we have $(x_0, y_0) \in P(\phi(\mu_i f_i)) = P(\phi(f_i))$. This shows the desired finite intersection property and hence we have (1.3.2). Since there is a nonconstant function in $S(\{x, y\})$, every element of the intersection appearing in (1.3.2) has different coordinates. This implies that $G(\{x, y\}) \neq \emptyset$.

We now prove the remaining assertion

$$H(x, y, u) = \bigcap \{T(\phi(f)) : f \in C(X), (x, y, u) \in T(f)\} \neq \emptyset. \qquad (1.3.3)$$

It is easy to see that the sets $T(\phi(f))$ are compact subsets of $X \times X \times \mathbb{C}$. Therefore, just as above, in order to verify (1.3.3), it is sufficient to check that the system $\{T(\phi(f)) : f \in C(X), (x, y, u) \in T(f)\}$ has the finite intersection property. Let $f_1, \ldots, f_n \in C(X)$ be such that $(x, y, u) \in T(f_1), \ldots, T(f_n)$. We evidently have $f_1 + \ldots + f_n \in \mathcal{T}(x, y, nu)$ and

$$\mathrm{diam}((f_1 + \ldots + f_n)(X)) = n|u|.$$

From the diameter preserving property of ϕ we deduce that

$$\mathrm{diam}(\phi(f_1 + \ldots + f_n)(X)) = \mathrm{diam}(\phi(f_1)(X)) + \ldots + \mathrm{diam}(\phi(f_n)(X)).$$

By Lemma 1.3.2, there exists $\{x_0, y_0\} \in \tilde{X}$ and a complex number v of modulus 1 such that $\phi(f_i) \in \mathcal{T}(x_0, y_0, |u|v)$ $(i = 1, \ldots, n)$. Plainly, this can be reformulated as $(x_0, y_0, |u|v) \in \cap_{i=1}^{n} T(\phi(f_i))$, thus verifying the claimed finite intersection property. Hence, we obtain (1.3.3).

Step 2. *If $\{x_1, y_1\}, \{x_2, y_2\} \in \tilde{X}$ and $\{x_1, y_1\} \neq \{x_2, y_2\}$, then we have $G(\{x_1, y_1\}) \cap G(\{x_2, y_2\}) = \emptyset$.*

Suppose on the contrary that $G(\{x_1, y_1\}) \cap G(\{x_2, y_2\}) \neq \emptyset$. Clearly, we may assume that $x_1 \neq x_2$ and $y_1, y_2 \notin \{x_1, x_2\}$. Let $\{x, y\} \in G(\{x_1, y_1\}) \cap G(\{x_2, y_2\})$. By Uryson's lemma there are functions f_1, f_2 with disjoint supports and with ranges in $[0, 1]$ such that

$$f_1(x_1) = 1, \ f_1(y_1) = 0, \ f_1(x_2) = 0, \ f_1(y_2) = 0$$
$$f_2(x_1) = 0, \ f_2(y_1) = 0, \ f_2(x_2) = 1, \ f_2(y_2) = 0.$$

Clearly, we have $f_1 \in \mathcal{T}(x_1, y_1, 1)$ and $f_2 \in \mathcal{T}(x_2, y_2, 1)$. Let

$$u_1 = \phi(f_1)(x) - \phi(f_1)(y) \quad \text{and} \quad u_2 = \phi(f_2)(x) - \phi(f_2)(y).$$

Then, by the definition of G, we have $|u_1| = |u_2| = 1$ and $(x, y, u_1) \in T(\phi(f_1))$, $(x, y, u_2) \in T(\phi(f_2))$. Let $t \in [-\pi/3, \pi/3]$ be arbitrary and set $\mu_t = e^{it}$. This yields $f_1 + \mu_t f_2 \in \mathcal{T}(x_1, y_1, 1)$ and hence $\{x, y\} \in S(\phi(f_1 + \mu_t f_2))$. We compute

$$|u_1 + \mu_t u_2| = |(\phi(f_1)(x) - \phi(f_1)(y)) + (\phi(\mu_t f_2)(x) - \phi(\mu_t f_2)(y))| =$$
$$|\phi(f_1 + \mu_t f_2)(x) - \phi(f_1 + \mu_t f_2)(y)| = 1. \qquad (1.3.4)$$

Since $|u_1| = |u_2| = 1$ and (1.3.4) holds for every $t \in [-\pi/3, \pi/3]$, we easily arrive at a contradiction.

Step 3. *For every $\{x, y\} \in \tilde{X}$, the set $G(\{x, y\})$ has exactly one element in $\tilde{X}$. The function $G' : \tilde{X} \to \tilde{X}$ defined by $\{G'(\{x, y\})\} = G(\{x, y\})$ is a bijection.*

Let $\{x, y\} \in \tilde{X}$. Since $G(\{x, y\})$ is nonempty, we can choose $\{x_0, y_0\} \in \tilde{X}$ such that

$$\{x_0, y_0\} \in G(\{x, y\}). \tag{1.3.5}$$

Because of the surjectivity of ϕ, there exists a function $f \in C(X)$ for which $\phi(f) \in \mathcal{T}_s(x_0, y_0, 1)$. Let $\{x_1, y_1\} \in S(f)$. We have $G(\{x_1, y_1\}) \subset S(\phi(f)) = \{\{x_0, y_0\}\}$. Therefore, by Step 1 we obtain $G(\{x_1, y_1\}) = \{\{x_0, y_0\}\}$. Using (1.3.5) we deduce that $G(\{x_1, y_1\}) \subset G(\{x, y\})$. By Step 2 this implies that $\{x, y\} = \{x_1, y_1\}$ and we conclude that $G(\{x, y\}) = G(\{x_1, y_1\}) = \{\{x_0, y_0\}\}$.

We prove now that the function G' is bijective. In view of Step 2 the injectivity is obvious. To show the surjectivity, let $\{x, y\} \in \tilde{X}$ and take $f \in C(X)$ for which $\phi(f) \in \mathcal{T}_s(x, y, 1)$. If $\{x_0, y_0\} \in S(f)$, then we have $G'(\{x_0, y_0\}) \in S(\phi(f))$. It follows that $G'(\{x_0, y_0\}) = \{x, y\}$, thus verifying our claim.

Step 4. *Let $\{x, y\} \in \tilde{X}$ and pick any $f \in C(X)$. If $\phi(f) \in \mathcal{S}_s(G'(\{x, y\}))$, then $f \in \mathcal{S}_s(\{x, y\})$.*

If $\{x_0, y_0\} \in S(f)$ is arbitrary, then $G(\{x_0, y_0\}) \subset S(\phi(f)) = G(\{x, y\})$. From Step 2 it follows that $\{x_0, y_0\} = \{x, y\}$. Consequently, $S(f) = \{\{x, y\}\}$.

Step 5. *If $\{x_1, y_1\}, \{x_2, y_2\} \in \tilde{X}$ and $\{x_1, y_1\} \cap \{x_2, y_2\} \neq \emptyset$, then we have*

$$G'(\{x_1, y_1\}) \cap G'(\{x_2, y_2\}) \neq \emptyset.$$

From Step 3 it follows that, if $\{x_1, y_1\}, \{x_2, y_2\} \in \tilde{X}$ have exactly one element in common, then the same holds true for $G'(\{x_1, y_1\})$ and $G'(\{x_2, y_2\})$.

Let $\{x, y_1\}, \{x, y_2\} \in \tilde{X}$ with $y_1 \neq y_2$. Assume that

$$G'(\{x, y_1\}) \cap G'(\{x, y_2\}) = \emptyset.$$

The surjectivity of ϕ implies that there exist functions $\phi(f_1), \phi(f_2) \in C(X)$ with the following properties. The supports of $\phi(f_1)$ and $\phi(f_2)$ are disjoint,

$$\phi(f_1) \in \mathcal{S}_s(G'(\{x, y_1\})) \quad \text{and} \quad \phi(f_2) \in \mathcal{S}_s(G'(\{x, y_2\})),$$

the ranges of $\phi(f_1), \phi(f_2)$ are included in $[-1/2, 1/2]$ and, finally, we have

$$\mathrm{diam}(\phi(f_1)(X)) = \mathrm{diam}(\phi(f_2)(X)) = 1.$$

By Step 4, we infer that $f_1 \in \mathcal{S}_s(\{x, y_1\})$, $f_2 \in \mathcal{S}_s(\{x, y_2\})$. For every complex number μ with $|\mu| = 1$, the diameter of the range of $\phi(f_1) + \mu\phi(f_2)$ is 1 and hence the same must hold true for $f_1 + \mu f_2$.

Since $f_2 \in \mathcal{S}_s(\{x, y_2\})$, we have $f_2(y_1) \neq f_2(x)$. Define

$$\mu = \frac{f_1(x) - f_1(y_1)}{f_2(x) - f_2(y_1)} |f_2(x) - f_2(y_1)|.$$

It follows that $|\mu| = 1$ and

$$|(f_1 + \mu f_2)(x) - (f_1 + \mu f_2)(y_1)| = |(f_1(x) - f_1(y_1)) + \mu(f_2(x) - f_2(y_1))| =$$
$$|(f_1(x) - f_1(y_1))(1 + |f_2(x) - f_2(y_1)|)| = 1 + |f_2(x) - f_2(y_1)| > 1,$$

which is untenable, since the diameter of the range of $f_1 + \mu f_2$ is 1. This justifies our assertion.

Step 6. *Let* $\{x_1, y_1\}, \{x_2, y_2\} \in \tilde{X}$. *Then* $\{x_1, y_1\} \cap \{x_2, y_2\} = \emptyset$ *if and only if* $G'(\{x_1, y_1\}) \cap G'(\{x_2, y_2\}) = \emptyset$.

The sufficiency follows from Step 5. As for the necessity, observe that, if G'_{-1} denotes the function corresponding to ϕ^{-1} as G' corresponds to ϕ in Step 3, then we have $G'_{-1} = (G')^{-1}$. Indeed, pick an $\{x, y\} \in \tilde{X}$. Let $G'_{-1}(\{x, y\}) = \{a, b\}$ and $G'(\{a, b\}) = \{x', y'\}$. Applying Step 4 for ϕ and then for ϕ^{-1}, we find that for a function $f \in C(X)$ with $\phi(f) \in \mathcal{S}_s(\{x', y'\})$ we have $\phi^{-1}(\phi(f)) = f \in \mathcal{S}_s(\{a, b\})$ and then $\phi(f) \in \mathcal{S}_s(\{x, y\})$. Consequently, $\{x, y\} = \{x', y'\} = G'(G'_{-1}(\{x, y\}))$. The assertion is now obvious.

Step 7. *Let* $x \in X$. *There exists a unique element* $g(x) \in X$ *such that* $g(x) \in G'(\{x, y\})$ *for every* $x \neq y \in X$. *The function* $g : X \to X$ *is bijective and* $\{g(x), g(y)\} = G'(\{x, y\})$ $(\{x, y\} \in \tilde{X})$.

Let $y_1, y_2 \in X$ be such that $x \neq y_1, x \neq y_2$ and $y_1 \neq y_2$. Let $g(x)$ denote the unique element of the set $G'(\{x, y_1\}) \cap G'(\{x, y_2\})$ (see Step 5). We shall show that $g(x) \in G'(\{x, y\})$ holds for every $x \neq y \in X$. If X has only three elements, this is obvious. Otherwise, pick $x \neq y \in X$ for which $y \neq y_1, y \neq y_2$ and suppose on the contrary that $g(x) \notin G'(\{x, y\})$. If $a_1, a_2 \in X$ are such that $G'(\{x, y_1\}) = \{g(x), a_1\}$ and $G'(\{x, y_2\}) = \{g(x), a_2\}$, then, taking the fact into account that by Step 6 the sets $G'(\{x, y\}) \cap G'(\{x, y_1\})$ and $G'(\{x, y\}) \cap G'(\{x, y_2\})$ are nonempty, we deduce that $G'(\{x, y\}) = \{a_1, a_2\}$. Similarly, we infer that $G'(\{y, y_1\})$ contains an element from $G'(\{x, y\}) = \{a_1, a_2\}$ in addition to one from $G'(\{x, y_1\}) = \{g(x), a_1\}$. If we had $g(x), a_2 \in G'(\{y, y_1\})$, then we would obtain $G'(\{x, y_2\}) = \{g(x), a_2\} = G'(\{y, y_1\})$ which would further imply that $\{x, y_2\} = \{y, y_1\}$, an obvious contradiction. Therefore, we have $a_1 \in G'(\{y, y_1\})$ and a similar argument can be applied to show that $a_2 \in G'(\{y, y_2\})$. Hence, by Step 6 again, we can choose a point $g(x) \neq b \in X$ such that $G'(\{y, y_1\}) = \{a_1, b\}$ and $G'(\{y, y_2\}) = \{a_2, b\}$. In the same fashion as above, one can check that $G'(\{y_1, y_2\}) = \{g(x), b\}$. To sum up, we have the following relations

$$G'(\{x, y_1\}) = \{g(x), a_1\}, \; G'(\{x, y_2\}) = \{g(x), a_2\},$$
$$G'(\{x, y\}) = \{a_1, a_2\}, \; G'(\{y, y_1\}) = \{a_1, b\}, \tag{1.3.6}$$
$$G'(\{y, y_2\}) = \{a_2, b\}, \; G'(\{y_1, y_2\}) = \{g(x), b\}.$$

Assume for a moment that $X = \{x, y, y_1, y_2\}$. Let f be the function which takes the following values: 0 at y, 1 at y_1, $e^{i\pi/3}$ at y_2, and the geometric center $(1 + e^{i\pi/3})/3$ of the triangle conv$\{0, 1, e^{i\pi/3}\}$ at x. (Here conv stands for the convex hull.) Let χ denote the characteristic function of the singleton $\{x\}$. Since $\phi(\chi)$ is nonconstant, one of the values $\phi(\chi)(a_1), \phi(\chi)(a_2), \phi(\chi)(g(x))$

differs from $\phi(\chi)(b)$. Let it be $\phi(\chi)(g(x))$. Since $f \in \mathcal{S}(\{y_1, y_2\})$ and we have $\text{diam}(f(X)) = 1$, using the last equation in (1.3.6) it follows that $|\phi(f)(b) - \phi(f)(g(x))| = 1$. Pick a nonzero $\mu \in \mathbb{C}$ of modulus small enough to guarantee that $\mu + (1 + e^{i\pi/3})/3$ is still inside the triangle $\text{conv}\{0, 1, e^{i\pi/3}\}$ and for which

$$\mu(\phi(\chi)(b) - \phi(\chi)(g(x))) = \lambda(\phi(f)(b) - \phi(f)(g(x)))$$

holds with some positive scalar λ. Now, since $f + \mu\chi \in \mathcal{S}(\{y_1, y_2\})$ and $\text{diam}(f + \mu\chi) = 1$, by (1.3.6) we have

$$|\phi(f + \mu\chi)(b) - \phi(f + \mu\chi)(g(x))| = 1.$$

On the other hand, we can compute

$$|\phi(f + \mu\chi)(b) - \phi(f + \mu\chi)(g(x))| =$$
$$|(\phi(f)(b) - \phi(g(x))) + \mu(\phi(\chi)(b) - \phi(\chi)(g(x)))| = 1 + \lambda > 1$$

which is a contradiction. If X has at least five points, then by Step 5 we deduce that for any point $z \in X$, $z \neq x, y, y_1, y_2$, the set $G'(\{x, z\})$ has one common element with each of the sets $G'(\{x, y_1\})$, $G'(\{x, y_2\})$, $G'(\{x, y\})$. From examination of the first three equalities in (1.3.6), one can see that this again is a contradiction. Thus, we have proved that $g(x) \in G'(\{x, y\})$ is valid for every $x \neq y \in X$.

In view of Step 5, the uniqueness of $g(x)$ is obvious.

We next prove that g is injective. Suppose temporarily, that X has exactly three points x, y, z. If $g(x) = g(y) = a$, say, then we have

$$a \in G'(\{x, y\}) \cap G'(\{x, z\}) \cap G'(\{y, z\}).$$

Now, using the fact that each of the sets $G'(\{x, y\})$, $G'(\{x, z\})$, $G'(\{y, z\})$ has two different elements, we easily arrive at a contradiction. In the general case (when X has at least four points), we argue as follows. Let $x_1 \neq x_2$ and assume that $g(x_1) = g(x_2) = b$. Choose $\{y_1, y_2\} \in \tilde{X}$ such that $\{x_1, x_2\} \cap \{y_1, y_2\} = \emptyset$. We have $b \in G'(\{x_1, y_1\})$ and $b \in G'(\{x_2, y_2\})$. But, on the other hand, since the sets $\{x_1, y_1\}$ and $\{x_2, y_2\}$ are disjoint, from Step 6 it follows that $G'(\{x_1, y_1\}) \cap G'(\{x_2, y_2\}) = \emptyset$. Since again this is a contradiction, we obtain the injectivity of g.

We now show that g is surjective. Let $x_0 \in X$ be arbitrary and $y_0 \in X, y_0 \neq x_0$. By the surjectivity of G', there exists an $\{x, y\} \in \tilde{X}$ for which $\{g(x), g(y)\} = G'(\{x, y\}) = \{x_0, y_0\}$. Therefore, x_0 is in the range of g, thus verifying its surjectivity.

Step 8. *There exists a complex number τ of modulus 1 such that, for every $f \in \mathcal{T}(x, y, u)$, we have $\phi(f) \in \mathcal{T}(g(x), g(y), \tau u)$.*

Let $\{x, y\} \in \tilde{X}$ be arbitrary. By the surjectivity of ϕ there exists $f_0 \in \mathcal{T}(x, y, 1)$ for which $\phi(f_0) \in \mathcal{S}_s(\{g(x), g(y)\})$. Let

$$\tau(x, y) = \phi(f_0)(g(x)) - \phi(f_0)(g(y)).$$

Then we have

$$H(x, y, 1) \subset T(\phi(f_0)) = \{(g(x), g(y), \tau(x, y)), (g(y), g(x), -\tau(x, y))\}.$$

By the definition of H and its non-emptiness (see Step 1), we obtain

$$H(x, y, 1) = \{(g(x), g(y), \tau(x, y)), (g(y), g(x), -\tau(x, y))\}.$$

It is now easy to see that the implication

$$f \in \mathcal{T}(x, y, u) \implies \phi(f) \in \mathcal{T}(g(x), g(y), \tau(x, y)u) \qquad (1.3.7)$$

holds for every $u \in \mathbb{C}$. It remains to verify that τ does not depend on its variables x, y. To this end first observe that, by (1.3.7), τ is symmetric, i.e., $\tau(x, y) = \tau(y, x)$ ($\{x, y\} \in \tilde{X}$). Next, let $x, y_1, y_2 \in X$ be pairwise distinct points. Pick functions $f_1 \in \mathcal{T}(x, y_1, -1)$ and $f_2 \in \mathcal{T}(x, y_2, -1)$ with disjoint supports and with ranges in $[0,1]$. Define

$$f = f_1 + e^{i\pi/3}f_2 \in \mathcal{T}(x, y_1, -1) \cap \mathcal{T}(x, y_2, -e^{i\pi/3}) \cap \mathcal{T}(y_1, y_2, 1 - e^{i\pi/3}).$$

Then we have

$$\phi(f) \in \mathcal{T}(g(x), g(y_1)), -\tau(x, y_1)) \cap$$
$$\mathcal{T}(g(x), g(y_2), -e^{i\pi/3}\tau(x, y_2)) \cap \mathcal{T}(g(y_1), g(y_2), \tau(y_1, y_2)(1 - e^{i\pi/3})).$$

Hence, for the complex numbers $\phi(f)(g(x))$, $\phi(f)(g(y_1))$ and $\phi(f)(g(y_2))$ we have

$$|\phi(f)(g(x)) - \phi(f)(g(y_1))| = |\phi(f)(g(x)) - \phi(f)(g(y_2))| =$$
$$|\phi(f)(g(y_1)) - \phi(f)(g(y_2))| = 1.$$

It is easy to see that for arbitrary complex numbers a, b of modulus 1, if $|a - b| = 1$, then we have $a = e^{\pm i\pi/3}b$. Therefore, we deduce that

$$-\tau(x, y_1) = \phi(f)(g(x)) - \phi(f)(g(y_1)) =$$
$$e^{\pm i\pi/3}(\phi(f)(g(x)) - \phi(f)(g(y_2))) = e^{\pm i\pi/3}(-e^{i\pi/3}\tau(x, y_2)),$$

i.e., $\tau(x, y_1) = e^{\pm i\pi/3}e^{i\pi/3}\tau(x, y_2)$. Applying the same argument to the function $f_1 + e^{-i\pi/3}f_2$, we obtain $\tau(x, y_1) = e^{\pm i\pi/3}e^{-i\pi/3}\tau(x, y_2)$. Comparing these two equalities, we conclude that $\tau(x, y_1) = \tau(x, y_2)$. By symmetry it is now apparent that τ is a constant function whose value can be denoted by the same symbol τ. The assertion follows from (1.3.7).

Step 9. *For every $f \in C(X)$, the function $\phi(f) \circ g - \tau \cdot f$ is constant.*

Let $f \in \mathcal{T}(x, y, 1)$. Then, from Step 8 it follows that

$$\phi(f) \in \mathcal{T}(g(x), g(y), \tau).$$

Thus, we have

$$\phi(f)(g(x)) - \phi(f)(g(y)) = \tau = \tau(f(x) - f(y)),$$

which implies

$$\phi(f)(g(y)) - \tau f(y) = \phi(f)(g(x)) - \tau f(x). \tag{1.3.8}$$

Let $z \in X$ be such that $z \neq x, y$. Set $u = f(x) - f(z)$. Clearly, $|u| \leq 1$. If $u = 0$, then $f \in \mathcal{T}(z, y, 1)$, which further implies $\phi(f) \in \mathcal{T}(g(z), g(y), \tau)$. Analogously to the derivation of (1.3.8), we find that

$$\phi(f)(g(z)) - \tau f(z) = \phi(f)(g(y)) - \tau f(y) = \phi(f)(g(x)) - \tau f(x).$$

Suppose now that $u \neq 0$. Define

$$U = \left\{ p \in X \setminus \{y\} \; : \; f(p) \neq f(x) \text{ and } \left| \frac{f(x) - f(p)}{|f(x) - f(p)|} - \frac{u}{|u|} \right| < \frac{1}{2} \right\}.$$

Since f is continuous, U is an open neighbourhood of the point z. By Uryson's lemma, there exists a function $f_0 \in C(X)$ with range in $[0, 1]$ and support in U for which $f_0(z) = 1$ and $f_0(y) = 0$. Let

$$f_1 = \frac{|u| f_0}{\max\{|f(x) - f|, |u|\}} \quad \text{and} \quad f_2 = f_1(f(x) - f).$$

Clearly, the support of f_2 is included in that of f_0, the latter being a subset of U. By the definition of U and f_2, it is not hard to check that $\mathrm{diam}(f_2(X)) \leq |u|$. On the other hand, $f_2(z) - f_2(x) = u - 0$ and $f_2(z) - f_2(y) = u - 0$. Consequently, we have

$$f_2 \in \mathcal{T}(z, x, u) \cap \mathcal{T}(z, y, u). \tag{1.3.9}$$

Since the range of f_1 is a subset of $[0, 1]$, for an arbitrary $w \in X$ we have

$$(f + f_2)(w) = f(w) + f_1(w)(f(x) - f(w)) = (1 - f_1(w))f(w) + f_1(w)f(x).$$

This shows that $(f + f_2)(w)$ belongs to the convex hull of $f(X)$. The diameter of this convex hull is equal to $\mathrm{diam}(f(X))$. Therefore, we have

$$\mathrm{diam}(f + f_2)(X)) \leq \mathrm{diam}(f(X)) = 1.$$

On the other hand, we infer

$$(f + f_2)(z) - (f + f_2)(y) = f(z) - f(y) + f_2(z) - f_2(y) =$$
$$f(z) - f(y) + u = f(z) - f(y) + f(x) - f(z) = f(x) - f(y) = 1.$$

Consequently,

$$f + f_2 \in \mathcal{T}(z, y, 1). \tag{1.3.10}$$

By (1.3.9) and (1.3.10) we obtain that

$$\phi(f_2) \in \mathcal{T}(g(z), g(x), \tau u) \cap \mathcal{T}(g(z), g(y), \tau u), \ \phi(f + f_2) \in \mathcal{T}(g(z), g(y), \tau).$$

Hence, we compute

$$\phi(f)(g(x)) - \phi(f)(g(z)) =$$
$$(\phi(f)(g(x)) - \phi(f)(g(y))) - (\phi(f)(g(z)) - \phi(f)(g(y))) =$$
$$\tau - (\phi(f + f_2)(g(z)) - \phi(f + f_2)(g(y))) + (\phi(f_2)(g(z)) - \phi(f_2)(g(y))) =$$
$$\tau - \tau + \tau u = \tau(f(x) - f(z)).$$

Thus, we have proved that

$$\phi(f)(g(z)) - \tau f(z) = \phi(f)(g(x)) - \tau f(x)$$

is valid for every $z \in X$.

Now, the proof of the theorem is completed as follows. By the linearity of ϕ, there is a linear functional $t : C(X) \to \mathbb{C}$ such that

$$\phi(f) \circ g - \tau \cdot f = t(f)1 \qquad (f \in C(X)).$$

Since g is a bijection, with the notation $\varphi = g^{-1}$ we have

$$\phi(f) - \tau \cdot f \circ \varphi = t(f)1 \qquad (f \in C(X)). \tag{1.3.11}$$

It follows from (1.3.11) that $f \circ \varphi$ is continuous for every $f \in C(X)$. Using Uryson's lemma, we deduce that φ is continuous and hence, because it is a bijection between compact Hausdorff spaces, we obtain that it is a homeomorphism. Since the relation $t(1) \neq -\tau$ is obvious, the proof of the theorem is now complete. $\square$

1.3.4 Remarks

As it can be seen, in our proof we have used only very elementary arguments based on the classical Uryson's lemma. Applying more advanced techniques (involving, for example, abstract Banach spaces, dual spaces, extreme points, etc.), the proof can be essentially shortened and one can get rid of the assumption of first countability. Such proofs were presented in [39, 78]. Our result, which originally appeared in [89], motivated the study of diameter preservers on several other kinds of spaces of functions. For results in this direction we refer to [15, 40, 74, 78, 84, 217]. We also refer to the papers [42, 43, 75] for some interesting geometrical properties of the diameter seminorm. Here we mention only a particular case of a surprising result due to Cabello Sánchez [42]. It states that on the space $C_0^{\mathbb{R}}(\mathbb{R}^n)$ of all real-valued continuous functions on $\mathbb{R}^n$ ($n \geq 2$) vanishing at infinity, the sup-norm is not convex-transitive while the diameter semi-norm is so. This gives us the striking observation that the usual sup-norm is not the 'right' norm on $C_0^{\mathbb{R}}(\mathbb{R}^n)$ with regard to the isometries. We also recall a result of Font and Sanchis [74] in which diameter preserving maps were used to characterize a sort of topological equivalence among locally compact Hausdorff spaces. Namely, they proved that two such spaces X, Y have homeomorphic one-point compactifications if and only if there is a diameter preserving linear bijection from $C_0^{\mathbb{R}}(X)$ onto $C_0^{\mathbb{R}}(Y)$.

Our next remark comes from a completely different direction. Namely, since the spectrum of any element f in the Banach algebra $C(X)$ is just the range of f, one can obviously say that in this section we have determined the structure of all bijective linear transformations of the Banach algebra $C(X)$ which preserve the diameter of the spectrum. As we have already mentioned in Section 0.1 of the Introduction (see PROBLEMS I and II), there is much interest in linear preserver problems which concern the invertibility or the spectrum. Here, we refer only to the papers [107] and [28] where the authors described the form of all bijective linear transformations of the Banach algebra of all bounded linear operators on a given Banach space which preserve the spectrum and the spectral radius, respectively. Clearly, the spectral radius measures the spectrum in a certain sense. We are convinced that to measure it by its diameter is also important from many respects. Hence, we propose to study the following problem.

Problem. Let H be a Hilbert space and describe all the bijective linear transformations of $B(H)$ which preserve the diameter of the spectrum.

We feel that this problem is significantly harder than the problem concerning the spectral radius and is still open at the time of publication. However, if we replace here $B(H)$ by the Jordan-Banach algebra $B_s(H)$ of all self-adjoint bounded linear operators, then we have the solution of the corresponding problem. In fact, it will turn out in Section 2.6 that the solution of the preserver problem we consider there also provides the form of all linear bijections of $B_s(H)$ which preserves the diameter of the spectrum (see 2.6.4. Remarks).

1.4 *-Semigroup Endomorphisms of $B(H)$

1.4.1 Summary

In this section we describe the form of all *-semigroup endomorphisms ϕ of $B(H)$ (H being a complex separable infinite dimensional Hilbert space) which have the Lipschitz property on every commutative C^*-subalgebra. In particular, we obtain that if ϕ satisfies $\phi(0) = 0$, then ϕ is additive. The results of this section come from the paper [176].

1.4.2 Formulation of the Results

On the analogy of linear preservers, in [101] Hochwald initiated the study of multiplicative preserver problems (as the probably most important papers on the subject concerning matrix algebras, we refer to [52, 83]). In that paper Hochwald described the form of those multiplicative selfmaps of a matrix algebra which preserve the spectrum. Moreover, as a natural generalization, he raised the question of describing the structure of all spectrum preserving multiplicative maps on operator algebras even under the possible additional condition of surjectivity. However, by a purely algebraic result of Martindale [154, First Corollary], it follows that on many operator algebras (for example, on standard operator algebras over Banach spaces of dimension at least 2), every surjective multiplicative transformation which maps onto an arbitrary algebra is automatically additive. So, supposing surjectivity, we would get that the transformations under consideration are ring homomorphisms and hence the problem becomes less challenging. (Nevertheless, we refer to [9] where one can find a nice result on spectrum preserving multiplicative transformations which map from the group of invertibles of a semi-simple Banach algebra onto a group of the same kind.) The situation is much more exciting if surjectivity is not assumed.

In our paper [167] we considered certain multiplicative preserver problems on the operator algebra $B(H)$ which are the natural analogues of some of the most well-known and important linear preserver problems. We presented results on the structure of rank preserving, corank preserving and spectrum preserving multiplicative maps. (Our statements on rank and corank preservers have been refined by An and Hou in [4].) In particular, in [167, Theorem 4] we described the general form of all such multiplicative selfmaps ϕ of $B(H)$ (H being a separable infinite dimensional Hilbert space) which are continuous, *-preserving (i.e., $\phi(A^*) = \phi(A)^*$, $A \in B(H)$) and spectrum non-increasing (i.e., $\sigma(\phi(A)) \subset \sigma(A)$, $A \in B(H)$). In this section we extend this result rather significantly in a certain sense. Namely, we omit the condition concerning the spectrum non-increasing property. In fact, we describe the structure of all *-semigroup endomorphisms of $B(H)$ under a certain continuity assumption.

We emphasize that the main point in our result Theorem 1.4.1 below is that we do not assume surjectivity. We recall that in that case even the description of all linear *-semigroup endomorphisms (i.e. *-algebra endomorphisms) of $B(H)$ is a difficult question as it can be seen in [121, Section 10.4].

Beside the motivation coming from the territory of preserver problems there is another motivation for the present investigation which comes from the theory of operator algebras. In this respect we refer to the paper [94] of Hakeda and to the article [223] of Šemrl where *-semigroup or just semigroup isomorphisms of operator algebras were studied. In the first paper, the author dealt with the problem of the additivity of *-semigroup isomorphisms of operator algebras while in the second one, the general form of semigroup isomorphisms between standard operator algebras was determined. We also mention the paper [142] of Lu, where some results on the additivity of semi-group isomorphisms of nest algebras were presented. One can see in the proofs of those results that under the assumption of bijectivity, the problem is quite algebraic in character (for example, Šemrl refers at the very beginning of his proof to the above mentioned result of Martindale to obtain additivity). The situation is very much different in our case. Our argument relies on the deep result Theorem A.12 of Bunce and Wright that gives the solution of the so-called Mackey-Gleason problem.

The main result of the section reads as follows.

Theorem 1.4.1. *Let H be a complex separable infinite dimensional Hilbert space. Let $\phi : B(H) \to B(H)$ be a *-semigroup endomorphism which has the Lipschitz property on the commutative C^*-subalgebras of $B(H)$. Then ϕ can be written in the form*

$$
\phi(A) = \begin{bmatrix}
0 & 0 & \cdots & \cdots & \cdots & \cdots & \cdots & \cdots & \cdots \\
0 & I & 0 & \cdots & \cdots & \cdots & \cdots & \cdots & \cdots \\
0 & 0 & A & 0 & \cdots & \cdots & \cdots & \cdots & \cdots \\
0 & 0 & 0 & A & 0 & \cdots & \cdots & \cdots & \cdots \\
\vdots & \vdots & \vdots & \vdots & \ddots & \vdots & \vdots & \vdots & \vdots \\
0 & \cdots & \cdots & \cdots & 0 & A^{*tr} & 0 & \cdots & \cdots \\
0 & \cdots & \cdots & \cdots & \cdots & 0 & A^{*tr} & 0 & \cdots \\
\vdots & \vdots & \vdots & \vdots & \vdots & \vdots & \vdots & \ddots & \vdots
\end{bmatrix} \qquad (A \in B(H)).
$$

As for the additivity of *-semigroup endomorphisms of $B(H)$, we immediately obtain the following result.

Corollary 1.4.2. *Let H be a complex separable infinite dimensional Hilbert space. Let $\phi : B(H) \to B(H)$ be a *-semigroup endomorphism which has the Lipschitz property on the commutative C^*-subalgebras of $B(H)$. If $\phi(0) = 0$, then ϕ is additive.*

1.4.3 Proof

Proof of Theorem 1.4.1. Clearly, ϕ sends projections to projections. So, $\phi(I)$ and $\phi(0)$ are projections. Since

$$\phi(I)\phi(A) = \phi(A)\phi(I) = \phi(A)$$

and

$$\phi(0)\phi(A) = \phi(A)\phi(0) = \phi(0),$$

it is easy to see that ϕ can be written in the form

$$
\phi(A) = \begin{bmatrix}
0 & 0 & 0 \\
0 & I & 0 \\
0 & 0 & \phi'(A)
\end{bmatrix} \qquad (A \in B(H))
$$

where ϕ' is a *-semigroup endomorphism of $B(H)$ having the same continuity property as ϕ which sends 0 to 0 and maps I into I. Therefore, we can assume that our original map ϕ satisfies $\phi(0) = 0$ and $\phi(I) = I$. The main step of the proof which follows is to prove that ϕ is orthoadditive on the set of all projections. This means that $\phi(P + Q) = \phi(P) + \phi(Q)$ for all mutually orthogonal projections $P, Q \in B(H)$. To see this, let $P \in B(H)$ be an arbitrary projection and set $Q = I - P$. Consider the map

$$\lambda \longmapsto \phi(e^{\lambda P}) = \phi(Q + e^{\lambda} P)$$

from $\mathbb{R}$ into the group of all invertible operators in $B(H)$. This is a continuous one-parameter group and hence there is an operator $T \in B(H)$ such that

$$\phi(Q + e^{\lambda} P) = e^{\lambda T} \qquad (\lambda \in \mathbb{R})$$

(see, for example, [206, 6.4.6 Proposition]). Since ϕ is *-preserving, we obtain that $e^{\lambda T}$ is self-adjoint for all $\lambda \in \mathbb{R}$. This yields that T is also self-adjoint. The norm and the spectral radius of any self-adjoint operator coincide. From the conditions in our theorem it follows that

$$\sup_{t \in \sigma(T)} |e^{\lambda t} - e^{\mu t}| < L |e^{\lambda} - e^{\mu}| \qquad (\lambda, \mu \in \mathbb{R})$$

holds for some positive number L. Therefore, the function $x \mapsto x^t$ has the Lipschitz property on the positive half-line for every $t \in \sigma(T)$. This gives us that $\sigma(T) \subset \{0, 1\}$. Consequently, T is a projection. Then we have $e^{\lambda T} = (I - T) + e^{\lambda} T$ and thus

$$\phi(Q + e^{\lambda} P) = (I - T) + e^{\lambda} T \qquad (\lambda \in \mathbb{R})$$

or, equivalently,

$$\phi(Q + \epsilon P) = (I - T) + \epsilon T \tag{1.4.1}$$

for every positive ϵ. By the continuity property of ϕ we obtain $\phi(Q) = I - T$. Putting $P = I$ into (1.4.1) and using $\phi(0) = 0$ we have

$$\phi(\epsilon I) = \epsilon I \qquad (\epsilon > 0). \tag{1.4.2}$$

Therefore, ϕ is positive homogeneous, and referring to (1.4.1) again, if we divide by ϵ and use the continuity property of ϕ, then we arrive at $\phi(P) = T$. Thus, we obtain

$$\phi(P) + \phi(I - P) = I \tag{1.4.3}$$

for every projection P in $B(H)$. If P, Q are arbitrary projections with $PQ = QP = 0$, then we infer from the multiplicativity of ϕ and (1.4.3) that

$$\phi(P) + \phi(Q) = \phi(P)\phi(P + Q) + \phi(I - P)\phi(P + Q) =$$

$$(\phi(P) + \phi(I - P))\phi(P + Q) = \phi(P + Q).$$

Consequently, ϕ is orthoadditive on the set of all projections.

Since $\phi(I) = I$, it follows that ϕ sends unitaries to unitaries. Consider the map $t \mapsto \phi(e^{it}I)$. Clearly, this is a continuous one-parameter unitary group. By Stone's theorem there is a self-adjoint operator $S \in B(H)$ such that

$$\phi(e^{it}I) = e^{itS} \qquad (t \in \mathbb{R}).$$

Since $\phi(I) = I$, we have $e^{2\pi iS} = I$. By spectral mapping theorem this yields that the spectrum of S consists of integers. So, S can be written in the form $S = \sum_{k=-n}^{n} kP_k$, where the P_k's are pairwise orthogonal projections with $\sum_{k=-n}^{n} P_k = I$ and n is a suitable positive integer. We compute

$$\phi(e^{it}I) = e^{it\sum_{k=-n}^{n} kP_k} =$$

$$\prod_{k=-n}^{n} e^{itkP_k} = \prod_{k=-n}^{n} (I + (e^{itk} - 1)P_k) =$$

$$I + \sum_{k=-n}^{n} (e^{itk} - 1)P_k = \sum_{k=-n}^{n} e^{itk}P_k \qquad (t \in \mathbb{R}).$$

Consequently, we have

$$\phi(\lambda I) = \sum_{k=-n}^{k=n} \lambda^k P_k \qquad (\lambda \in \mathbb{C}, |\lambda| = 1). \tag{1.4.4}$$

From (1.4.2) and (1.4.4) we infer that

$$\phi(\lambda I) = \sum_{k=-n}^{n} |\lambda| \left(\frac{\lambda}{|\lambda|}\right)^k P_k \qquad (\lambda \in \mathbb{C}, \lambda \neq 0). \tag{1.4.5}$$

We know that $\phi(\lambda I)$ commutes with $\phi(A)$ for every $\lambda \in \mathbb{C}$. Thus, for any $A \in B(H)$ using

$$\phi(A) = I\phi(A)I = \sum_{k,l=-n}^{n} P_k\phi(A)P_l$$

we have

$$\sum_{k,l=-n}^{n} \lambda^k P_k\phi(A)P_l = \sum_{k,l=-n}^{n} \lambda^k P_k P_k\phi(A)P_l =$$

$$\phi(\lambda I)\phi(A) = \phi(A)\phi(\lambda I) =$$

$$\sum_{k,l=-n}^{n} P_k\phi(A)P_l \lambda^l P_l = \sum_{k,l=-n}^{n} \lambda^l P_k\phi(A)P_l$$

for every $\lambda \in \mathbb{C}$ of modulus 1. This implies that $P_k\phi(A)P_l = 0$ if $k \neq l$. Consequently, ϕ can be written in the form

$$\phi(A) = \sum_{k=-n}^{k=n} P_k \phi(A) P_k \qquad (A \in B(H))$$

or, in another way,

$$\phi(A) = \begin{bmatrix} \phi_{-n}(A) & 0 & \cdots & 0 \\ 0 & \phi_{-n+1}(A) & \cdots & 0 \\ \vdots & \vdots & \ddots & \vdots \\ 0 & \cdots & \cdots & \phi_n(A) \end{bmatrix} \qquad (A \in B(H)).$$

Here, every ϕ_k $(k = -n, \ldots, n)$ is a *-semigroup endomorphism of $B(H)$ which has the Lipschitz property on the commutative C^*-subalgebras.

Every orthoadditive projection valued measure on the set of all projections in $B(H)$ can be extended to a continuous linear map on $B(H)$. This is a particular case of Theorem A.12 due to Bunce and Wright. Since this extension is continuous, linear and sends projections to projections, by Theorem A.4 we infer that it is in fact a Jordan *-homomorphism. Therefore, we have (continuous) Jordan *-homomorphisms $\psi_{-n}, \ldots, \psi_n$ of $B(H)$ such that

$$\phi(P) = \begin{bmatrix} \psi_{-n}(P) & 0 & \cdots & 0 \\ 0 & \psi_{-n+1}(P) & \cdots & 0 \\ \vdots & \vdots & \ddots & \vdots \\ 0 & \cdots & \cdots & \psi_n(P) \end{bmatrix}$$

for every projection $P \in B(H)$.

Let $R_1, \ldots, R_m$ be pairwise orthogonal projections whose sum is I and pick nonzero scalars $\lambda_1, \ldots, \lambda_m \in \mathbb{C}$. Using the orthoadditivity of ϕ and (1.4.5), for any $k = -n, \ldots, n$ we compute

$$\phi_k(\lambda_1 R_1 + \ldots + \lambda_m R_m) = \phi_k(\lambda_1 R_1 + \ldots + \lambda_m R_m)\phi_k(R_1 + \ldots + R_m) =$$

$$\phi_k(\lambda_1 R_1 + \ldots + \lambda_m R_m)(\phi_k(R_1) + \ldots + \phi_k(R_m)) =$$

$$\phi_k(\lambda_1 R_1) + \ldots + \phi_k(\lambda_m R_m) =$$

$$\phi_k(\lambda_1 I)\phi_k(R_1) + \ldots + \phi_k(\lambda_m I)\phi_k(R_m) =$$

$$|\lambda_1|\left(\frac{\lambda_1}{|\lambda_1|}\right)^k \phi_k(R_1) + \ldots + |\lambda_m|\left(\frac{\lambda_m}{|\lambda_m|}\right)^k \phi_k(R_m) =$$

$$|\lambda_1|\left(\frac{\lambda_1}{|\lambda_1|}\right)^k \psi_k(R_1) + \ldots + |\lambda_m|\left(\frac{\lambda_m}{|\lambda_m|}\right)^k \psi_k(R_m) =$$

$$\psi_k\left(|\lambda_1|\left(\frac{\lambda_1}{|\lambda_1|}\right)^k R_1 + \ldots + |\lambda_m|\left(\frac{\lambda_m}{|\lambda_m|}\right)^k R_m\right).$$

Using the continuity property of ϕ_k, the continuity of ψ_k and the spectral theorem of normal operators, we deduce that

$$\phi_k(N) = \psi_k(|N|(N|N|^{-1})^k) \tag{1.4.6}$$

holds for every invertible normal operator $N \in B(H)$ (note that N and the range of its spectral measure generate a commutative C^*-subalgebra).

Every Jordan *-homomorphism of $B(H)$ is the direct sum of a *-homomorphism and a *-antihomomorphism (see Theorem A.6). Let ψ_k^h denote the *-homomorphic and let ψ_k^a denote the *-antihomomorphic part of ψ_k. Let $N, M \in B(H)$ be invertible normal operators whose product is also normal. By the multiplicativity of ϕ_k and (1.4.6) we have

$$\psi_k(|NM|(NM|NM|^{-1})^k) = \phi_k(NM) = \phi_k(N)\phi_k(M) =$$

$$\psi_k(|N|(N|N|^{-1})^k)\psi_k(|M|(M|M|^{-1})^k).$$

This implies that

$$\psi_k^h(|NM|(NM|NM|^{-1})^k) =$$
$$\psi_k^h(|N|(N|N|^{-1})^k)\psi_k^h(|M|(M|M|^{-1})^k) = \tag{1.4.7}$$
$$\psi_k^h(|N|(N|N|^{-1})^k|M|(M|M|^{-1})^k)$$

and

$$\psi_k^a(|NM|(NM|NM|^{-1})^k) =$$
$$\psi_k^a(|N|(N|N|^{-1})^k)\psi_k^a(|M|(M|M|^{-1})^k) = \tag{1.4.8}$$
$$\psi_k^a(|M|(M|M|^{-1})^k|N|(N|N|^{-1})^k).$$

Any *-homomorphism or *-antihomomorphism of $B(H)$ is either injective or identically 0 which follows from the form of the *-representations of $B(H)$ on separable Hilbert spaces (see Theorem A.11). Now, taking (1.4.7) and (1.4.8) into account, one can verify that the only values of k for which ψ_k can be nonzero are -1 and 1. Moreover, because of the same reasons, for $k = 1$ we have $\psi_1^a = 0$ and for $k = -1$ we have $\psi_{-1}^h = 0$. Observe that $|N|(N|N|^{-1})^{-1} = N^*$. Therefore, ϕ can be written in the form

$$\phi(N) = \begin{bmatrix} \psi'(N) & 0 \\ 0 & \psi''(N^*) \end{bmatrix} \tag{1.4.9}$$

for every invertible normal operator N, where ψ' is a *-endomorphism and ψ'' is a *-antiendomorphism of $B(H)$. By continuity and spectral theorem we clearly have (1.4.9) for every normal operator in $B(H)$. Define

$$\psi(A) = \begin{bmatrix} \psi'(A) & 0 \\ 0 & \psi''(A^*) \end{bmatrix}$$

for every $A \in B(H)$. Clearly, ψ is an additive *-semigroup endomorphism of $B(H)$ (it is not linear unless ψ'' is missing).

It is easy to see that every rank-one operator is the product of at most three (rank-one) normal operators. This gives us that $\phi(A) = \psi(A)$ for every rank-one operator $A \in B(H)$. Now, let $A \in B(H)$ be arbitrary. Pick rank-one projections $P, Q \in B(H)$. Since PAQ is of rank at most 1, we compute

$$\phi(P)\phi(A)\phi(Q) = \phi(PAQ) = \psi(PAQ) = \qquad (1.4.10)$$

$$\psi(P)\psi(A)\psi(Q) = \phi(P)\psi(A)\phi(Q).$$

It is a straightforward consequence of the form of the linear *-endomorphisms of $B(H)$ (see Theorem A.11) that for every maximal family $(P_n)_n$ of pairwise orthogonal rank-one projections we have $\sum_n \phi(P_n) = \phi(I) = I$. Therefore, we infer from (1.4.10) that $\phi(A) = \psi(A)$ for every $A \in B(H)$. Finally, in order to get the explicit form of ϕ we refer again to the general form of the linear *-endomorphisms of $B(H)$ appearing in Theorem A.11 and note that one could get the form of linear *-antiendomorphisms of $B(H)$ in a similar way. $\qquad \square$

1.4.4 Remarks

In our results we have assumed that H is infinite dimensional. As for the finite dimensional case, we refer to the paper [113] where all so-called non-degenerate multiplicative maps of a matrix algebra were determined.

In the paper [95], Hakeda studied the problem of the additivity of multiplicative Jordan *-isomorphisms (i.e., bijective maps ϕ which satisfy $\phi((AB + BA)/2)) = (\phi(A)\phi(B) + \phi(B)\phi(A))/2$ and $\phi(A^*) = \phi(A)^*$ without supposing linearity) between operator algebras under the condition of uniform continuity on commutative C^*-subalgebras. It was conjectured by S. Sakai and later proved in [96] that this assumption is in fact redundant. Therefore, it would be interesting to study our problem without assuming any kind of continuity. However, taking into account the results in [113], one can get evidence that, due to the lack of surjectivity, the problem in that case is surely much more complicated.

Finally, concerning recent results on the additivity of bijective maps between operator algebras or general algebras which are multiplicative with respect to the Jordan product or to the Jordan triple product we refer to a series of papers of Lu and Ling [138, 143, 144, 145].

2

Preservers on Quantum Structures

2.1 Transformations on the Set of All n-Dimensional Subspaces of a Hilbert Space Preserving Principal Angles

2.1.1 Summary

Wigner's classical theorem on symmetry transformations can be formulated in several ways. One of the possibilities reads as follows. Every bijective transformation on the set $\mathcal{L}$ of all 1-dimensional subspaces of a Hilbert space H which preserves the angle between the elements of $\mathcal{L}$ is induced by either a unitary or an antiunitary operator on H. The aim of this section is to extend Wigner's result from the 1-dimensional case to the case of n-dimensional subspaces of H with $n \in \mathbb{N}$ fixed. The result appeared in the paper [174].

2.1.2 Formulation of the Main Result

Let H be a real or complex Hilbert space. For any $n \in \mathbb{N}$, $P_n(H)$ denotes the set of all rank-n projections on H, and $P_\infty(H)$ stands for the set of all infinite rank projections. Clearly, $P_n(H)$ can be identified with the set of all n-dimensional subspaces of H. As mentioned above (also see the corresponding part of Section 0.3 in the Introduction) Wigner's theorem can be viewed as the complete description of all bijective transformations on the set $\mathcal{L}$ of all 1-dimensional subspaces of H which preserve the angle between the elements of $\mathcal{L}$. Having this in mind, it seems to be a natural problem to try to extend the result from the 1-dimensional case to the case of higher dimensional subspaces. The first question which arises immediately is that how to define the angle between two higher dimensional subspaces of H. For our present purposes, the most adequate concept of angles is that of the so-called principal angles (or canonical angles, in a different terminology). This concept is a generalization of the usual notion of angles between 1-dimensional subspaces and reads as follows. If P, Q are finite dimensional projections, then the principal angles between them (or, equivalently, between their ranges as subspaces) is defined as the arccos of the square root of the eigenvalues (counted according to multiplicity) of the positive (self-adjoint) finite rank operator QPQ (see, for example, [19, Exercise VII.1.10] or [128, Problem 559]). We remark that this concept of angles was motivated by the classical work [115] of Jordan and it has serious applications for example in mathematical statistics (see the canonical correlation theory of Hotelling [103], and also see the introduction of [158]). The system of all principal angles between P and Q is denoted by $\angle(P, Q)$. Thus, we have the desired concept of angles between finite rank projections. But in what follows we would also like to extend Wigner's theorem for the case of infinite rank projections. Therefore, we also need the concept of principal angles between infinite rank projections. Using deep concepts of operator theory (like scalar-valued spectral measure and multiplicity function) this could be carried out, but in order to formulate a Wigner-type result we need only the equality of angles. Hence, we can avoid these complications saying that for arbitrary projections P, Q, P', Q' on H we have $\angle(P, Q) = \angle(P', Q')$ if and only if the positive operators QPQ and $Q'P'Q'$ are unitarily equivalent. This obviously generalizes the equality of principal angles between pairs of finite rank projections.

After this preparation we are in a position to formulate the main result of the section.

Theorem 2.1.1. *Let $n \in \mathbb{N}$. Let H be a real or complex Hilbert space with $\dim H \geq n$. Suppose that $\phi : P_n(H) \to P_n(H)$ is a transformation with the property that*

$$\angle(\phi(P), \phi(Q)) = \angle(P, Q) \qquad (P, Q \in P_n(H)).$$

If $n = 1$ or $n \neq \dim H/2$, then there exists a linear or conjugate-linear isometry V on H such that

$$\phi(P) = VPV^* \qquad (P \in P_n(H)).$$

If H is infinite dimensional, the transformation $\phi : P_\infty(H) \to P_\infty(H)$ satisfies

$$\angle(\phi(P), \phi(Q)) = \angle(P, Q) \qquad (P, Q \in P_\infty(H)),$$

and ϕ is surjective, then there exists a unitary or antiunitary operator U on H such that

$$\phi(P) = UPU^* \qquad (P \in P_\infty(H)).$$

As one can suspect from the formulation of our main result, there are exceptional cases, namely, when we have $\dim H = 2n, n > 1$. See the remark after the proof of Theorem 2.1.8 below.

2.1.3 Proof

This subsection is devoted to the proof of Theorem 2.1.1. In fact, this will follow from the statements below.

The idea of the proof can be summarized in a single sentence as follows (also see Section 0.4 in the Introduction). We extend our transformation ϕ from $P_n(H)$ to a Jordan homomorphism of the algebra $F(H)$ of all finite rank operators on H which preserves the rank-one operators. Fortunately, those maps turn out to have a form and using this we can achieve the desired conclusion. On the other hand, quite unfortunately, we have to work hard to carry out all the details of the proof and this is done in what follows.

In the remaining part of this section let H be a real or complex Hilbert space and let $n \in \mathbb{N}$. Since our statement obviously holds when $\dim H = n$, hence we suppose that $\dim H > n$.

Let tr be the usual trace functional on the trace-class operators. Clearly, every element of the ideal $F(H)$ of all finite rank operators in $B(H)$ has a finite trace. In what follows, we denote by $F_s(H)$ the set of all self-adjoint elements of $F(H)$.

We begin with two key lemmas. In order to understand why we consider the property (2.1.1) in Lemma 2.1.2, we note that if $\angle(P,Q) = \angle(P',Q')$ for some finite rank projections P, Q, P', Q', then, by definition, the positive operators QPQ and $Q'P'Q'$ are unitarily equivalent. This implies that $\operatorname{tr} QPQ = \operatorname{tr} Q'P'Q'$. But, by the properties of the trace, we have $\operatorname{tr} QPQ = \operatorname{tr} PQQ = \operatorname{tr} PQ$ and, similarly, $\operatorname{tr} Q'P'Q' = \operatorname{tr} P'Q'$. So, if our transformation preserves the principal angles between projections, then it necessarily preserves the trace of the product of the projections under consideration. This justifies the condition (2.1.1) in the next lemma.

Lemma 2.1.2. *Let $\mathcal{P}$ be any set of finite rank projections on H. If $\phi : \mathcal{P} \to \mathcal{P}$ is a transformation with the property that*

$$\operatorname{tr} \phi(P)\phi(Q) = \operatorname{tr} PQ \qquad (P, Q \in \mathcal{P}), \tag{2.1.1}$$

then ϕ has a unique real-linear extension Φ onto the real-linear span $\operatorname{span}_\mathbb{R} \mathcal{P}$ of $\mathcal{P}$. The transformation Φ is injective, preserves the trace and satisfies

$$\operatorname{tr} \Phi(A)\Phi(B) = \operatorname{tr} AB \qquad (A, B \in \operatorname{span}_\mathbb{R} \mathcal{P}). \tag{2.1.2}$$

Proof. For any finite sets $\{\lambda_i\} \subset \mathbb{R}$ and $\{P_i\} \subset \mathcal{P}$ we define

$$\Phi(\sum_i \lambda_i P_i) = \sum_i \lambda_i \phi(P_i).$$

We have to show that Φ is well-defined. If $\sum_i \lambda_i P_i = \sum_k \mu_k Q_k$, where $\{\mu_k\} \subset \mathbb{R}$ and $\{Q_k\} \subset \mathcal{P}$ are finite subsets, then for any $R \in \mathcal{P}$ we compute

$$\operatorname{tr}(\sum_i \lambda_i \phi(P_i)\phi(R)) = \sum_i \lambda_i \operatorname{tr}(\phi(P_i)\phi(R)) = \sum_i \lambda_i \operatorname{tr}(P_i R) =$$

$$\mathrm{tr}(\sum_i \lambda_i P_i R) = \mathrm{tr}(\sum_k \mu_k Q_k R) = \sum_k \mu_k \, \mathrm{tr}(Q_k R) =$$

$$\sum_k \mu_k \, \mathrm{tr}(\phi(Q_k)\phi(R)) = \mathrm{tr}(\sum_k \mu_k \phi(Q_k)\phi(R)).$$

Therefore, we have

$$\mathrm{tr}((\sum_i \lambda_i \phi(P_i) - \sum_k \mu_k \phi(Q_k))\phi(R)) = 0$$

for every $R \in \mathcal{P}$. By the linearity of the trace functional it follows that we have similar equality if we replace $\phi(R)$ by any finite linear combination of $\phi(R)$'s. This gives us that

$$\mathrm{tr}((\sum_i \lambda_i \phi(P_i) - \sum_k \mu_k \phi(Q_k))(\sum_i \lambda_i \phi(P_i) - \sum_k \mu_k \phi(Q_k))) = 0.$$

The operator $(\sum_i \lambda_i \phi(P_i) - \sum_k \mu_k \phi(Q_k))^2$, being the square of a self-adjoint operator, is positive. Since its trace is zero, we obtain that

$$(\sum_i \lambda_i \phi(P_i) - \sum_k \mu_k \phi(Q_k))^2 = 0$$

which plainly implies that

$$\sum_i \lambda_i \phi(P_i) - \sum_k \mu_k \phi(Q_k) = 0.$$

This shows that Φ is well-defined. The real-linearity of Φ now follows from its definition. The uniqueness of Φ is also trivial to see. From (2.1.1) we immediately obtain (2.1.2). One can introduce an inner product on $F_s(H)$ by the formula

$$\langle A, B \rangle = \mathrm{tr}\, AB \qquad (A, B \in F_s(H))$$

(the norm induced by this inner product is called the Hilbert-Schmidt norm). The equality (2.1.2) shows that Φ is an isometry with respect to this norm. Thus, Φ is injective. It follows from (2.1.1) that

$$\mathrm{tr}\, \phi(P) = \mathrm{tr}\, \phi(P)^2 = \mathrm{tr}\, P^2 = \mathrm{tr}\, P \qquad (P \in \mathcal{P})$$

which clearly implies that

$$\mathrm{tr}\, \Phi(A) = \mathrm{tr}\, A \qquad (A \in \mathrm{span}_{\mathbb{R}} \mathcal{P}).$$

This completes the proof of the lemma. $\square$

Now we recall the concept of Jordan homomorphisms. If $\mathcal{A}$ and $\mathcal{B}$ are algebras, then a linear transformation $\Psi : \mathcal{A} \to \mathcal{B}$ is called a Jordan homomorphism if it satisfies

$$\Psi(A^2) = \Psi(A)^2 \qquad (A \in \mathcal{A}),$$

or, equivalently, if it satisfies

$$\Psi(AB + BA) = \Psi(A)\Psi(B) + \Psi(B)\Psi(A) \qquad (A, B \in \mathcal{A}).$$

Two projections P, Q on H are said to be orthogonal if $PQ = QP = 0$ (this means that the ranges of P and Q are orthogonal to each other). In this case we write $P \perp Q$. We denote $P \leq Q$ if $PQ = QP = P$ (this means that the range of P is included in the range of Q).

Lemma 2.1.3. *Let $\Phi : F_s(H) \to F_s(H)$ be a real-linear transformation which preserves the rank-one projections and the orthogonality between them. Then there is an either linear or conjugate-linear isometry V on H such that*

$$\Phi(A) = VAV^* \qquad (A \in F_s(H)).$$

Proof. Since every finite rank projection is the finite sum of pairwise orthogonal rank-one projections, it is obvious that Φ preserves the finite rank projections. This property of Φ implies that it is a Jordan homomorphism on $F_s(H)$, i.e., a linear transformation which preserves the square operation. In fact, this follows from the argument given in the proof of Theorem A.4.

We next prove that Φ can be extended to a Jordan homomorphism of $F(H)$. To see this, first suppose that H is complex and consider the transformation $\tilde{\Phi} : F(H) \to F(H)$ defined by

$$\tilde{\Phi}(A + iB) = \Phi(A) + i\Phi(B) \qquad (A, B \in F_s(H)).$$

It is easy to see that $\tilde{\Phi}$ is a linear transformation which satisfies $\tilde{\Phi}(T^2) = \tilde{\Phi}(T)^2$ ($T \in F(H)$). This shows that $\tilde{\Phi}$ is a Jordan homomorphism.

If H is real, then the situation is not so simple, but we can apply a deep algebraic result of Martindale as follows. Consider the unitalized algebra $F(H) \oplus \mathbb{R}I$ (of course, we have to add the identity only when H is infinite dimensional). Defining $\Phi(I) = I$, we can extend Φ to the set of all symmetric elements of the enlarged algebra in an obvious way. Now we are in a position to apply the results of Martindale [153] on the extendability of Jordan homomorphisms defined on the set of all symmetric elements of a ring with involution. To be precise, in [153] Jordan homomorphism means an additive map Ψ which, beside $\Psi(s^2) = \Psi(s)^2$, also satisfies $\Psi(sts) = \Psi(s)\Psi(t)\Psi(s)$. But if the ring in question is 2-torsion free (in particular, if it is a real or complex algebra), this second equality follows from the first one (see, for example, the proof of [206, 6.3.2 Lemma]). The statements [153, Theorem 1] in the case when $\dim H \geq 3$ and [153, Theorem 2] if $\dim H = 2$ imply that Φ can be uniquely extended to an associative homomorphism of $F(H) \oplus \mathbb{R}I$ into itself. To be honest, since the results of Martindale concern rings and hence linearity does not appear, we could guarantee only the additivity of the extension of Φ. However, the construction in [153] shows that in the case of algebras, linear Jordan homomorphisms have linear extensions.

To sum up, in every case we have a Jordan homomorphism of $F(H)$ extending Φ. In order to simplify the notation, we use the same symbol Φ for the extension as well.

As $F(H)$ is a locally matrix algebra (every finite subset of $F(H)$ can be included in a subalgebra of $F(H)$ which is isomorphic to a full matrix algebra), it follows from the classical result Theorem A.5 of Jacobson and Rickart that Φ can be written as $\Phi = \Phi_1 + \Phi_2$, where Φ_1 is a homomorphism and Φ_2 is an antihomomorphism. Let P be a rank-one projection on H. Since $\Phi(P)$ is also rank-one, we obtain that one of the idempotents $\Phi_1(P), \Phi_2(P)$ is zero. Since $F(H)$ is a simple ring, it is easy to see that this implies that either Φ_1 or Φ_2 is identically zero, that is, Φ is either a homomorphism or an antihomomorphism of $F(H)$. In what follows we can assume without loss of generality that Φ is a homomorphism. Since the kernel of Φ is an ideal in $F(H)$ and $F(H)$ is simple, we obtain that Φ is injective.

We show that Φ preserves the rank-one operators. Let $A \in F(H)$ be of rank 1. Then there is a rank-one projection P such that $PA = A$. We have $\Phi(A) = \Phi(PA) = \Phi(P)\Phi(A)$ which proves that $\Phi(A)$ is of rank at most 1. Since Φ is injective, we obtain that the rank of $\Phi(A)$ is exactly 1. From the conditions of our lemma it follows that ϕ sends rank-2 projections to rank-2 projections. Therefore, the range of Φ contains an operator with rank greater than 1. The form of Φ can now be described. It follows from Theorem A.1 that either there are linear operators T, S on H such that Φ is of the form

$$\Phi(x \otimes y) = (Tx) \otimes (Sy) \qquad (x, y \in H)$$

or there are conjugate-linear operators T', S' on H such that Φ is of the form

$$\Phi(x \otimes y) = (S'y) \otimes (T'x) \qquad (x, y \in H). \tag{2.1.3}$$

Suppose that we have the first possibility. By the multiplicativity of Φ we obtain that

$$\begin{aligned}
\langle u, y \rangle Tx \otimes Sv &= \langle u, y \rangle \Phi(x \otimes v) = \Phi(x \otimes y \cdot u \otimes v) = \\
&\Phi(x \otimes y)\Phi(u \otimes v) = \langle Tu, Sy \rangle Tx \otimes Sv.
\end{aligned} \tag{2.1.4}$$

This gives us that $\langle Tu, Sy \rangle = \langle u, y \rangle$ for every $u, y \in H$. On the other hand, since Φ sends rank-one projections to rank-one projections, we obtain that for every unit vector $x \in H$ we have $Tx = Sx$. These imply that $T = S$ is an isometry and with the notation $V = T = S$ we have

$$\Phi(A) = VAV^*$$

for every $A \in F_s(H)$.

We show that the possibility (2.1.3) cannot occur. In fact, similarly to (2.1.4) we have

$$\langle u, y \rangle S'v \otimes T'x = \langle S'v, T'x \rangle S'y \otimes T'u \qquad (x, y, u, v \in H).$$

Fixing unit vectors $x = y = u$ in H and considering the operators above at $T'x$, we find that

$$S'v = \langle S'v, T'x \rangle \langle T'x, T'u \rangle S'y \qquad (v \in H)$$

giving us that S' is of rank 1. Since Φ sends rank-2 projections to rank-2 projections, we arrive at a contradiction. This completes the proof of the lemma. □

We are now in a position to present a new proof of the non-surjective version of Wigner's theorem which is equivalent to the statement of our main theorem in the case when $n = 1$. For other proofs see [14, 230].

To begin, observe that if P, Q are finite rank projections such that $\operatorname{tr} PQ = 0$, then we have $\operatorname{tr}(PQ)^*PQ = \operatorname{tr} QPQ = \operatorname{tr} PQQ = \operatorname{tr} PQ = 0$ which implies that $(PQ)^*(PQ) = 0$. This gives us that $PQ = 0 = QP$. Therefore, P is orthogonal to Q if and only if $\operatorname{tr} PQ = 0$.

Theorem 2.1.4. *Let $\phi : P_1(H) \to P_1(H)$ be a transformation with the property that*

$$\operatorname{tr} \phi(P)\phi(Q) = \operatorname{tr} PQ \qquad (P, Q \in P_1(H)). \tag{2.1.5}$$

Then there is an either linear or conjugate-linear isometry V on H such that

$$\phi(P) = VPV^* \qquad (P \in P_1(H)).$$

Proof. By the spectral theorem it is obvious that the real linear span of $P_1(H)$ is $F_s(H)$. Then, by Lemma 2.1.2 we see that there is a unique real-linear extension Φ of ϕ onto $F_s(H)$ which preserves the rank-one projections and, by (2.1.5), Φ also preserves the orthogonality between the elements of $P_1(H)$. Lemma 2.1.3 applies to complete the proof. □

As for the cases when $n > 1$ we need the following lemma. Recall that we have previously supposed that $\dim H > n$.

Lemma 2.1.5. *Let $1 < n \in \mathbb{N}$. Then $\operatorname{span}_{\mathbb{R}} P_n(H)$ coincides with $F_s(H)$.*

Proof. Since the real-linear span of $P_1(H)$ is $F_s(H)$, it is sufficient to show that every rank-one projection is a real-linear combination of rank-n projections. To see this, choose orthonormal vectors $e_1, \ldots, e_{n+1}$ in H. Let $E = e_1 \otimes e_1 + \ldots + e_{n+1} \otimes e_{n+1}$ and define

$$P_k = E - e_k \otimes e_k \qquad (k = 1, \ldots, n+1).$$

Clearly, every P_k can be represented by a $(n+1) \times (n+1)$ diagonal matrix whose diagonal entries are all 1's with the exception of the kth one which is 0. The equation

$$\lambda_1 P_1 + \ldots + \lambda_{n+1} P_{n+1} = e_1 \otimes e_1$$

gives rise to a system of linear equations with unknown scalars $\lambda_1, \ldots, \lambda_{n+1}$. The matrix of this system of equations is an $(n+1) \times (n+1)$ matrix whose diagonal consists of 0's and its off-diagonal entries are all 1's. It is easy to see that this matrix is nonsingular, and hence $e_1 \otimes e_1$ (and, similarly, every other $e_k \otimes e_k$) is a real-linear combination of $P_1, \ldots, P_{n+1}$. This completes the proof. $\qquad\square$

We continue with a technical lemma.

Lemma 2.1.6. *Let P, Q be projections on H. If QPQ is a projection, then there are pairwise orthogonal projections R, R', R'' such that $P = R + R'$, $Q = R + R''$. In particular, we obtain that QPQ is a projection if and only if $PQ = QP$.*

Proof. Let $R = QPQ$. Since R is a projection whose range is contained in the range of Q, it follows that $R'' = Q - R$ is a projection which is orthogonal to R.

If x is a unit vector in the range of R, then we have $\|QPQx\| = 1$. Since PQx is a vector whose norm is at most 1 and its image under the projection Q has norm 1, we obtain that PQx is a unit vector in the range of Q. Similarly, we obtain that Qx is a unit vector in the range of P and, finally, that x is a unit vector in the range of Q. Therefore, x belongs to the range of P and Q. Since x was arbitrary, we can infer that the range of R is included in the range of P. Thus, we obtain that $R' = P - R$ is a projection which is orthogonal to R.

Next, using the obvious relations

$$PR = RP = R, \quad QR = RQ = R$$

we deduce

$$(Q - R)(P - R)(Q - R) =$$
$$QPQ - QPR - QRQ + QR - RPQ + RPR + RQ - R = \qquad (2.1.6)$$
$$R - R - R + R - R + R + R - R = 0.$$

Since $A^*A = 0$ implies $A = 0$ for any $A \in B(H)$, we obtain from (2.1.6) that $R'R'' = (P - R)(Q - R) = 0$.

The second part of the assertion is now easy to check. $\qquad\square$

We next prove the assertion of our main result Theorem 2.1.1 in the case when $1 < n \in \mathbb{N}$ and H is infinite dimensional.

Theorem 2.1.7. *Suppose $1 < n \in \mathbb{N}$ and H is infinite dimensional. If $\phi : P_n(H) \to P_n(H)$ is a transformation such that*

$$\angle(\phi(P), \phi(Q)) = \angle(P, Q) \qquad (P, Q \in P_n(H)),$$

then there exists a linear or conjugate-linear isometry V on H such that

$$\phi(P) = VPV^* \qquad (P \in P_n(H)).$$

Proof. By Lemma 2.1.2 and Lemma 2.1.5, ϕ can be uniquely extended to an injective real-linear transformation Φ on $F_s(H)$. The main point of the proof is to show that Φ preserves the rank-one projections. In order to verify this, just as in the proof of Lemma 2.1.5, we consider orthonormal vectors $e_1, \ldots, e_{n+1}$ in H, define $E = e_1 \otimes e_1 + \ldots + e_{n+1} \otimes e_{n+1}$ and set

$$P_k = E - e_k \otimes e_k \qquad (k = 1, \ldots, n+1).$$

We show that the ranges of all $P_k' = \phi(P_k)$'s can be jointly included in an $(n+1)$-dimensional subspace of H. To see this, we first recall that Φ has the property that

$$\operatorname{tr} \Phi(A)\Phi(B) = \operatorname{tr} AB \qquad (A, B \in F_s(H))$$

(see Lemma 2.1.2). Next we verify that Φ has the following property: if P, Q are orthogonal rank-one projections, then $\Phi(P)\Phi(Q) = 0$. Indeed, if P, Q are orthogonal, then we can include them into two orthogonal rank-$(n+1)$ projections. Now, referring to the construction given in Lemma 2.1.5 and having in mind that Φ preserves the orthogonality between rank-n projections, we obtain that $\Phi(P)\Phi(Q) = 0$. (Clearly, the same argument works if $\dim H \geq 2(n+1)$.) Since the rank-n projections P_k are commuting, by the preserving property of ϕ and Lemma 2.1.6, it follows that the projections $\Phi(P_k)$ are also commuting. It is well-known that any finite commuting family of operators in $F_s(H)$ can be diagonalized by the same unitary transformation (or, in the real case, by the same orthogonal transformation). Therefore, if we restrict Φ onto the real-linear subspace in $F_s(H)$ generated by $P_1, \ldots, P_{n+1}$, then it can be identified with a real-linear operator from $\mathbb{R}^{n+1}$ to $\mathbb{R}^m$ for some $m \in \mathbb{N}$. Clearly, this restriction of Φ can be represented by an $m \times (n+1)$ real matrix $T = (t_{ij})$. Let us examine how the properties of Φ are reflected in those of the matrix T. Firstly, Φ is trace preserving. This gives us that for every $\lambda \in \mathbb{R}^{n+1}$ the sums of the coordinates of the vectors $T\lambda$ and λ are the same. This easily implies that the sum of the entries of T lying in a fixed column is always 1. As we have already noted, $\Phi(e_i \otimes e_i)\Phi(e_j \otimes e_j) = 0$ holds for every $i \neq j$. For the matrix T this means that the coordinatewise product of any two columns of T is zero. Consequently in every row of T there is at most one nonzero entry. Since Φ sends rank-n projections to rank-n projections, we see that this possibly nonzero entry is necessarily 1. So, every row contains at most one 1 and all the other entries in that row are 0's. Since the sum of the elements in every column is 1, we have that in every column there is exactly one 1 and all the other entries are 0's. These now easily imply that if $\lambda \in \mathbb{R}^{n+1}$ is such that its coordinates are all 0's with the exception of one which is 1, then $T\lambda$ is of the same kind. What concerns Φ, this means that Φ sends every $e_k \otimes e_k$ ($k = 1, \ldots, n+1$) to a rank-one projection.

So, we obtain that Φ preserves the rank-one projections and the orthogonality between them. Using Lemma 2.1.3 we conclude the proof. □

We turn to the case when H is finite dimensional.

Theorem 2.1.8. *Suppose* $1 < n \in \mathbb{N}$, H *is finite dimensional and* $n \neq \dim H/2$. *If* $\phi : P_n(H) \to P_n(H)$ *satisfies*

$$\angle(\phi(P), \phi(Q)) = \angle(P, Q) \qquad (P, Q \in P_n(H)),$$

then there exists a unitary or antiunitary operator U *on* H *such that*

$$\phi(P) = UPU^* \qquad (P \in P_n(H)). \tag{2.1.7}$$

Proof. First suppose that $\dim H = 2d$, $1 < d \in \mathbb{N}$. If $n = 1, \ldots, d - 1$, then we can apply the method followed in the proof of Theorem 2.1.7 concerning the infinite dimensional case. If $n = d + 1, \ldots, 2d - 1$, then consider the transformation $\psi : P \mapsto I - \phi(I - P)$ on $P_{2d-n}(H)$. We learn from [128, Problem 559] that if $\angle(P, Q) = \angle(P', Q')$, then there exists a unitary operator U such that $UPU^* = P'$ and $UQU^* = Q'$. It follows from the preserving property of ϕ that for any $P, Q \in P_{2d-n}(H)$ we have

$$\phi(I - P) = U(I - P)U^*, \quad \phi(I - Q) = U(I - Q)U^*$$

for some unitary operator U on H. This gives us that

$$\angle(\psi(P), \psi(Q)) = \angle(UPU^*, UQU^*) = \angle(P, Q).$$

In that way we can reduce the problem to the previous case. So, there is an either unitary or antiunitary operator U on H such that

$$\psi(P) = UPU^* \qquad (P \in P_{2d-n}(H)).$$

It follows that $\phi(I - P) = I - \psi(P) = I - UPU^* = U(I - P)U^*$, and hence we have the result for the considered case.

Next suppose that $\dim H = 2d + 1$, $d \in \mathbb{N}$. If $n = 1, \ldots, d - 1$, then once again we can apply the method followed in the proof of Theorem 2.1.7. If $n = d + 2, \ldots, 2d + 1$, then using the 'dual method' that we have applied just above we can reduce the problem to the previous case. If $n = d$, consider a fixed rank-d projection P_0. Clearly, if P is any rank-d projection orthogonal to P_0, then the rank-d projection $\phi(P)$ is orthogonal to $\phi(P_0)$. Therefore, ϕ induces a transformation ϕ_0 which concerns $d + 1$-dimensional Hilbert spaces (namely, the orthogonal complement of the range of P_0 and that of the range of $\phi(P_0)$) and preserves the principal angles between rank-d projections. Our 'dual method' and the result concerning 1-dimensional subspaces lead us to the conslusion that the linear extension of ϕ_0 maps rank-one projections to rank-one projections and preserves the orthogonality between them. This implies that the same holds true for our original transformation ϕ. Just as before, using Lemma 2.1.2 and Lemma 2.1.3 we can conclude the proof. In the remaining case $n = d + 1$ we apply the 'dual method' once again. $\square$

We now show that the case when $1 < n \in \mathbb{N}$, $n = \dim H/2$ is really exceptional. To see this, consider the transformation $\phi : P \mapsto I - P$ on

$P_n(H)$. This ϕ maps $P_n(H)$ into itself and preserves the principal angles. As for the complex case, the preserving property follows from [19, Exercise VII.1.11] while in the real case it was proved already by Jordan in [115] (see [205], p. 310). Let us suppose that the transformation ϕ can be written in the form (2.1.7). Pick a rank-one projection Q on H. We know that it is a real linear combination of some $P_1, \ldots, P_{n+1} \in P_n(H)$. It would follow from (2.1.7) that considering the same linear combination of $\phi(P_1), \ldots, \phi(P_{n+1})$, it is a rank-one projection as well. But due to the definition of ϕ, we get that this linear combination is a constant minus Q. By the trace preserving property we obtain that this constant is $1/n$. Since $n > 1$, the operator $(1/n)I - Q$ is obviously not a projection. Therefore, we have arrived at a contradiction. This shows that the transformation above can not be written in the form (2.1.7).

We now turn to our statement concerning infinite rank projections. In the proof we shall use the following simple lemma.

Lemma 2.1.9. *Let H be an infinite dimensional Hilbert space. Suppose P, Q are projections on H with the property that for any projection R with finite corank we have $RP = PR$ if and only if $RQ = QR$. Then either $P = Q$ or $P = I - Q$.*

Proof. Let R be any projection on H commuting with P. By Lemma 2.1.6, it is easy to see that we can choose a monotone decreasing net (R_α) of projections with finite corank such that (R_α) converges weakly to R and R_α commutes with P for every α. Since R_α commutes with Q for every α, we obtain that R commutes with Q. Interchanging the role of P and Q, we obtain that any projection commutes with P if and only if it commutes with Q.

Let x be any unit vector from the range of P. Consider $R = x \otimes x$. Since R commutes with P, it must commute with Q as well. By Lemma 2.1.6 we obtain that x belongs either to the range of Q or to its orthogonal complement. It follows that either $d(x, \text{rng}\, Q) = 0$, or $d(x, \text{rng}\, Q) = 1$. Since the set of all unit vectors in the range of P is connected and the distance function is continuous, we get that either every unit vector in $\text{rng}\, P$ belongs to $\text{rng}\, Q$ or every unit vector in $\text{rng}\, P$ belongs to $(\text{rng}\, Q)^\perp$. Interchanging the role of P and Q and examining all the possible cases, we can easily conclude that either $P = Q$ or $P = I - Q$. □

Theorem 2.1.10. *Let H be an infinite dimensional Hilbert space. Suppose that $\phi : P_\infty(H) \to P_\infty(H)$ is a surjective transformation with the property that*

$$\angle(\phi(P), \phi(Q)) = \angle(P, Q) \qquad (P, Q \in P_\infty(H)).$$

Then there exists a unitary or antiunitary operator U on H such that

$$\phi(P) = UPU^* \qquad (P \in P_\infty(H)).$$

Proof. We first prove that ϕ is injective. If $P, P' \in P_\infty(H)$ and $\phi(P) = \phi(P')$, then by the preserving property of ϕ we have

$$\angle(P,Q) = \angle(P',Q) \qquad (Q \in P_\infty(H)). \qquad (2.1.8)$$

Putting $Q = I$, we see that P is unitarily equivalent to P'. We distinguish two cases. First, let P be of infinite corank. By (2.1.8), we deduce that for every $Q \in P_\infty(H)$ we have $Q \perp P$ if and only if $Q \perp P'$. This gives us that $P = P'$. As the second possibility, let P be of finite corank. Then P, P' can be written in the form $P = I - P_0$ and $P' = I - P_0'$, where, by the equivalence of P, P', the projections P_0 and P_0' have finite and equal rank. Let Q_0 be any finite rank projection on H. It follows from

$$\angle(I - P_0, I - Q_0) = \angle(I - P_0', I - Q_0)$$

that there is a unitary operator W on H such that

$$W(I - Q_0)(I - P_0)(I - Q_0)W^* = (I - Q_0)(I - P_0')(I - Q_0).$$

This implies that

$$W(-Q_0 - P_0 + P_0Q_0 + Q_0P_0 - Q_0P_0Q_0)W^* =$$
$$-Q_0 - P_0' + P_0'Q_0 + Q_0P_0' - Q_0P_0'Q_0.$$

Taking traces, by the equality of the rank of P_0 and P_0', we obtain that

$$\operatorname{tr} P_0Q_0 = \operatorname{tr} P_0'Q_0. \qquad (2.1.9)$$

Since this holds for every finite rank projection Q_0 on H, it follows that $P_0 = P_0'$ and hence we have $P = P'$. This proves the injectivity of ϕ.

Let $P \in P_\infty(H)$ be of infinite corank. Then there is a projection $Q \in P_\infty(H)$ such that $Q \perp P$. By the preserving property of ϕ, this implies that $\phi(Q) \perp \phi(P)$ which means that $\phi(P)$ is of infinite corank. One can similarly prove that if $\phi(P)$ is of infinite corank, then the same must hold for P. This yields that $P \in P_\infty(H)$ is of finite corank if and only if so is $\phi(P)$.

Denote by $P_f(H)$ the set of all finite rank projections on H. It follows that the transformation $\psi : P_f(H) \to P_f(H)$ defined by

$$\psi(P) = I - \phi(I - P) \qquad (P \in P_f(H))$$

is well-defined and bijective. Since $\phi(I - P)$ is unitarily equivalent to $I - P$ for every $P \in P_f(H)$ (this is because $\angle(\phi(I - P), \phi(I - P)) = \angle(I - P, I - P)$), it follows that ψ is rank preserving.

We next show that

$$\operatorname{tr} \psi(P)\psi(Q) = \operatorname{tr} PQ \qquad (P, Q \in P_f(H)). \qquad (2.1.10)$$

This can be done following the argument leading to (2.1.9). In fact, by the preserving property of ϕ there is a unitary operator W on H such that

$$W(I - \psi(Q))(I - \psi(P))(I - \psi(Q))W^* = (I - Q)(I - P)(I - Q).$$

This gives us that

$$W(-\psi(Q) - \psi(P) + \psi(P)\psi(Q) + \psi(Q)\psi(P) - \psi(Q)\psi(P)\psi(Q))W^* =$$
$$-Q - P + PQ + QP - QPQ.$$

Taking traces on both sides and referring to the rank preserving property of ψ, we obtain (2.1.10). According to Lemma 2.1.2, let $\Psi : F_s(H) \to F_s(H)$ denote the unique real-linear extension of ψ onto $\text{span}_{\mathbb{R}} P_f(H) = F_s(H)$. We know that Ψ is injective. Since $P_f(H)$ is in the range of Ψ, we obtain that Ψ is surjective as well. It is easy to see that Lemma 2.1.3 can be applied and we infer that there exists an either unitary or antiunitary operator U on H such that

$$\Psi(A) = UAU^* \qquad (A \in F_s(H)).$$

Therefore, we have

$$\phi(P) = UPU^*$$

for every projection $P \in P_\infty(H)$ with finite corank. It remains to prove that the same holds true for every $P \in P_\infty(H)$ with infinite corank as well. This could be quite easy to show if we know that ϕ preserves the order between the elements of $P_\infty(H)$. But this property is far away from being easy to verify. So we choose a different approach to attack the problem.

Let $P \in P_\infty(H)$ be a projection of infinite corank. By the preserving property of ϕ we see that for every $Q \in P_\infty(H)$ the operator $\phi(Q)\phi(P)\phi(Q)$ is a projection if and only if QPQ is a projection. By Lemma 2.1.6, this means that $\phi(Q)$ commutes with $\phi(P)$ if and only if Q commutes with P. Therefore, for any $Q \in P_\infty(H)$ of finite corank, we obtain that Q commutes with $U^*\phi(P)U$ (this is equivalent to that $\phi(Q) = UQU^*$ commutes with $\phi(P)$) if and only if Q commutes with P.

By Lemma 2.1.9 we have two possibilities, namely, either $U^*\phi(P)U = P$ or $U^*\phi(P)U = I - P$. Suppose that $U^*\phi(P)U = I - P$. Consider an orthonormal basis e_0, e_γ ($\gamma \in \Gamma$) in the range of P and, similarly, choose an orthonormal basis f_0, f_δ ($\delta \in \Delta$) in the range of $I - P$. Pick nonzero scalars λ, μ with the properties that $|\lambda|^2 + |\mu|^2 = 1$ and $|\lambda| \neq |\mu|$. Define

$$Q = (\lambda e_0 + \mu f_0) \otimes (\lambda e_0 + \mu f_0) + \sum_\gamma e_\gamma \otimes e_\gamma + \sum_\delta f_\delta \otimes f_\delta.$$

Clearly, Q is of finite corank (in fact, its corank is 1). Since $\phi(Q)\phi(P)\phi(Q) = UQU^*\phi(P)UQU^*$ is unitarily equivalent to QPQ, it follows that the spectrum of $QU^*\phi(P)UQ$ is equal to the spectrum of QPQ. This gives us that the spectrum of $Q(I - P)Q$ is equal to the spectrum of QPQ. By the construction of Q this means that

$$\{0, 1, |\mu|^2\} = \{0, 1, |\lambda|^2\}$$

which is an obvious contradiction. Consequently, we have $U^*\phi(P)U = P$, that is, $\phi(P) = UPU^*$. Thus, we have proved that this latter equality holds for every $P \in P_\infty(H)$ and the proof is complete. $\square$

2.1.4 Remarks

To conclude this section we make some remarks. Firstly, we note that as an Uhlhorn-type version of the main result above (see Uhlhorn's theorem in Section 0.3 of the Introduction), in his recent paper [228] Šemrl has described the structure of all bijective maps $\phi : P_n(H) \to P_n(H)$ (H is infinite dimensional separable and $n \in \mathbb{N}$) which preserve the orthogonality in both directions i.e., which have the property $PQ = 0 \Leftrightarrow \phi(P)\phi(Q) = 0$. He has obtained that every such transformation is of the usual form

$$\phi(P) = UPU^* \qquad (P \in P_n(H))$$

with some unitary or antiunitary operator U on H. (Also see the paper [87] of Győry.)

Secondly, observe that Theorem 2.1.1 has a noteworthy geometrical content. It says that every transformation on the set of all n-dimensional subspaces of an N-dimensional Euclidean space ($N \neq 2n$ or $n = 1$) which preserves the principal angles is induced by a unitary or antiunitary operator on the underlying space. The result can be reformulated saying that disregarding the exceptional case $N = 2n, n > 1$, the transformation group of the Grassmann space $P_n(H)$ of index n which leaves the principal angles invariant is isomorphic to the symmetry group of H that is the group of all unitary and antiunitary operators factorized by the normal subgroup of all scalar elements. (See the discussion on Klein's Erlanger Programm in Section 0.1 of the Introduction.)

We note that it would be a nice result if one could prove that in the exceptional case (i.e., when $1 < n, n = \dim H/2$) up to unitary-antiunitary equivalence there are exactly two transformations on $P_n(H)$ preserving the principal angles, namely, $P \mapsto P$ and $P \mapsto I - P$. This is left as an open problem.

The concept of principal angles what we have used above is informative enough to describe completely the mutual position of subspaces. We mean that, as mentioned in the proof of Theorem 2.1.8, for any $n \in \mathbb{N}$ and $P, Q, P', Q' \in P_n(H)$ we have $\angle(P, Q) = \angle(P', Q')$ if and only if there exists a linear or conjugate-linear isometry V on H such that $VPV^* = P'$ and $VQV^* = Q'$. (In fact, in the corresponding part of the proof of Theorem 2.1.8 only linear isometries appear but the reason is trivial: if V is conjugate-linear and satisfies $VPV^* = P', VQV^* = Q'$ for some $P, Q, P', Q' \in P_n(H)$, then V can be substituted by a linear isometry having the same properties.) Now our result Theorem 2.1.1 can be reformulated in the following interesting way. In the non-exceptional cases if $\phi : P_n(H) \to P_n(H)$ is a transformation such that for every pair $P, Q \in P_n(H)$ of projections we have a linear or conjugate-linear isometry V_{PQ} on H (that might depend on P, Q) such that $\phi(P) = V_{PQ}PV_{PQ}^*$ and $\phi(Q) = V_{PQ}QV_{PQ}^*$, then there is a global choice for such an isometry, i.e., there is a linear or conjugate-linear isometry V on H such that $\phi(P) = VPV^*$

holds for every $P \in P_n(H)$. We shall meet similar phenomena in Chapter 3 when dealing with 2-local automorphisms.

In conclusion we mention some of our further works on Wigner's theorem. Using our algebraic approach to the problem, in [164, 166] we extended Wigner's result for the case of Hilbert modules over matrix algebras. In those papers we were concerned with transformations defined on such a module which preserve the absolute value of the generalized inner product. Recently, these works have been extended by Bakić and Guljaš first for Hilbert modules over the algebra $K(H)$ of all compact operators [11] and then for full Hilbert C^*-modules over C^*-subalgebras of $K(H)$ [12]. (The latter structures are especially important because they are exactly the full Hilbert C^*-modules in which every closed submodule is complemented; see [147]). In [169] we obtained a generalization of Wigner's theorem for indefinite inner product spaces. (In the last decades it has turned out that the indefinite inner product spaces have important applications in the discussion of several physical problems.) The result was formulated in the language of ray transformations and we pointed out that it is a far-reaching generalization of a corresponding result of Bracci, Morchio and Strocchi in [22]. In [171] we proved a Wigner-type result for transformations on the set of all finite projections in a type II factor and in [172] we presented a similar result for transformations on the set of all rank-one idempotents on a real or complex Banach space or on a finite dimensional vector space over any field of characteristic different from 2. Finally, we note that in [181] we obtained a Wigner-type result concerning the set of all rank-one partial isometries which operators are the minimal tripotents and hence the natural analogues of rank-one projections in $B(H)$ as a JBW^*-algebra.

2.2 Orthogonality Preserving Transformations on Indefinite Inner Product Spaces: Generalization of Uhlhorn's Version of Wigner's Theorem

2.2.1 Summary

In this section we present an analogue of Uhlhorn's version of Wigner's theorem on symmetry transformations for the case of indefinite inner product spaces. This significantly generalizes a result of Van den Broek. The proof is based on a theorem which describes the form of all bijective transformations on the set of all rank-one idempotents on a Banach space preserving zero product in both directions. The results of this section appeared in the paper [183].

2.2.2 Formulation of the Results

As we have already mentioned in the Introduction (see Section 0.3 there), Wigner's theorem on symmetry transformations plays a fundamental role in the probabilistic aspects of quantum mechanics. Its physical content is that any quantum mechanical invariance transformation (symmetry transformation) can be represented by a unitary or antiunitary operator on a complex Hilbert space and that, conversely, any operator of that kind represents an invariance transformation. In mathematical language, we can formulate the result for example in the following way. If H is a complex Hilbert space and T is a bijective transformation on the set of all 1-dimensional linear subspaces of H which preserves the angle between every pair of such subspaces, then T is induced by either a unitary or an antiunitary operator U on H. This means that for every 1-dimensional subspace L of H we have $T(L) = U[L] = \{Ux : x \in L\}$. In his famous paper [237], Uhlhorn generalized this result by requiring only that T preserves the orthogonality between the 1-dimensional subspaces of H. This is a significant achievement since Uhlhorn's transformation preserves only the logical structure of the quantum mechanical system behind while Wigner's transformation preserves its complete probabilistic structure. However, in the case when the dimension of H is not less than 3, Uhlhorn was able to obtain the same conclusion as Wigner.

In the last decades it has become quite clear that indefinite inner product spaces are even more useful than definite ones in describing several physical problems (see, for example, the introduction in [22]). This has raised the need to study Wigner's theorem in the indefinite setting as well (see [22] and [32]). Our paper [169] was devoted to a generalization of Wigner's original theorem for indefinite inner product spaces. In the present section we treat Uhlhorn's version in that setting. Our approach here is different from the one followed in [169]. Namely, it is based on a beautiful result of Ovchinnikov [204] (see Theorem A.13) describing the automorphisms of the poset of all idempotents on a Hilbert space of dimension at least 3, which result can be regarded as a 'skew version' of the fundamental theorem of projective geometry. This result enables us to use operator algebraic tools to attack the problem. We emphasize that in the literature there does exist an Uhlhorn-type result concerning symmetry transformations on indefinite inner product spaces. In fact, this is due to Van den Broek [32] (an application of his result can be found in [31], also see [33]). In the paper [32] he considered indefinite inner product spaces induced by nonsingular self-adjoint operators on finite dimensional complex Hilbert spaces. Moreover, in the proof of the main result he basically followed the original idea of Uhlhorn. In the present section we apply a completely different approach and obtain a much more general result which concerns indefinite inner product spaces induced by any invertible bounded linear operator on a real or complex Hilbert space of any dimension not less than 3. (Quantum logics on spaces with such a general indefinite metric have been investigated by, for example, Matvejchuk in [156].) Our result will follow from the main theorem of the section which describes the form of all bijective transformations of the set of all rank-one idempotents on a Banach space which preserve zero product in both directions.

If X is a real or complex Banach space, then the set of all idempotents in $B(X)$ is denoted by $I(X)$ and $I_1(X)$ stands for the set of all rank-one elements of $I(X)$.

The main result of the section reads as follows.

Theorem 2.2.1. *Let X be a real or complex Banach space of dimension at least 3. Let $\phi : I_1(X) \to I_1(X)$ be a bijective transformation with the property that*

$$PQ = 0 \Longleftrightarrow \phi(P)\phi(Q) = 0$$

for all $P, Q \in I_1(X)$.

If X is real, then there exists an invertible bounded linear operator $A : X \to X$ such that ϕ is of the form

$$\phi(P) = APA^{-1} \qquad (P \in I_1(X)). \tag{2.2.1}$$

If X is complex and infinite dimensional, then there exists an invertible bounded linear or conjugate-linear operator $A : X \to X$ such that ϕ is of the form (2.2.1).

If X is complex and finite dimensional, then we can suppose that our transformation ϕ acts on the space of $n \times n$ complex matrices ($n = \dim X$). In this case there is a nonsingular matrix $A \in M_n(\mathbb{C})$ and a ring automorphism h of $\mathbb{C}$ such that ϕ is of the form

$$\phi(P) = Ah(P)A^{-1} \qquad (P \in I_1(\mathbb{C}^n)). \tag{2.2.2}$$

Here $h(P)$ denotes the matrix obtained from P by applying h to every entry of it.

Our main theorem can be summarized by saying that every bijective transformation on $I_1(X)$ which preserves zero product in both directions comes from a linear or conjugate-linear algebra automorphism of $B(X)$ if X is real or complex and infinite dimensional, and it comes from a semilinear algebra automorphism of $B(X)$ if X is complex and finite dimensional. We should note that this result has probably no physical content. This is because the poset of all idempotents on a Banach space (the partial order among idempotents is defined in Subsection 2.2.3) does not form a lattice in general and hence it is not a geometry or a logic in the sense of quantum mechanics (see [243]). In fact, the poset of idempotents is not to be confused with the lattice of subspaces of a linear space as the idempotents are determined not by one but two complementary subspaces. However, our main theorem will easily imply the result Corollary 2.2.3 generalizing Uhlhorn's version of Wigner's theorem for indefinite inner product spaces which statement we believe has physical meaning. On the other hand, it will be clear from the presented proof that one can readily get a very similar conclusion as in Theorem 2.2.1 concerning the form of zero product preserving transformations between the sets of rank-one idempotents on different Banach spaces (also see the remark after Corollary 2.2.3) which has an interesting mathematical consequence. Namely, it implies that the real Banach spaces as topological vector spaces are completely determined by the structure of their rank-one idempotents with the relation of zero product.

In the paper [169] we presented a Wigner-type result for pairs of ray transformations (see [169, Theorem 1]) which enabled us to generalize a result of Bracci, Morchio and Strocchi in [22] for indefinite inner product spaces generated by any invertible bounded linear (not necessarily self-adjoint) operator on a Hilbert space. The main result of the present section can be applied to obtain the following corollary which is a Banach space analogue and hence a remarkable generalization of the main result in [169] that was formulated for (complex) Hilbert spaces.

To formulate our corollary we need some concepts and notation. Following the terminology of Uhlhorn, for any nonzero vector $x \in X$, the set $\underline{x}$ of all nonzero scalar multiples of x is called the ray generated by x. The set of all rays in X is denoted by $\underline{X}$. The dual space of X (that is the set of all bounded linear functionals on X) is denoted by $X^\sharp$. For any $x \in X, f \in X^\sharp$ we use the common and convenient notation $\langle x, f \rangle$ for $f(x)$. We say that the rays $\underline{x} \in \underline{X}$ and $\underline{f} \in \underline{X^\sharp}$ are orthogonal to each other, in notation $\underline{x} \cdot \underline{f} = 0$, if we have $\langle y, g \rangle = 0$ for all $y \in \underline{x}$ and $g \in \underline{f}$. The Banach space adjoint of an operator $A \in B(X)$ is denoted by $A^\sharp$. We extend the concept of adjoints also for conjugate-linear operators. If A is a bounded conjugate-linear operator on the complex Banach space X, then its adjoint $A^\sharp : X^\sharp \to X^\sharp$ (which is also a bounded conjugate-linear operator) is defined by $A^\sharp f = \overline{f \circ A}$ ($f \in X^\sharp$). If X is a linear space over $\mathbb{K}$ ($\mathbb{K}$ denotes the real or complex field) and h is a ring automorphism of $\mathbb{K}$, then the function $A : X \to X$ is called h-semilinear if it is additive and $A(\lambda x) = h(\lambda)Ax$ holds for every $x \in X$ and $\lambda \in \mathbb{K}$. If X is a finite dimensional complex linear space and h is a ring automorphism of $\mathbb{C}$, then for any h-semilinear operator A, the adjoint $A^\sharp$ of A is defined by $A^\sharp f = h^{-1} \circ f \circ A$ ($f \in X^\sharp$). Clearly, $A^\sharp : X^\sharp \to X^\sharp$ is an h^{-1}-semilinear operator.

After this preparation we can formulate our first corollary as follows.

Corollary 2.2.2. *Let X be a real or complex Banach space of dimension not less than 3. Let $T : \underline{X} \to \underline{X}$ and $S : \underline{X^\sharp} \to \underline{X^\sharp}$ be bijective transformations with the property that*

$$T\underline{x} \cdot S\underline{f} = 0 \quad \text{if and only if} \quad \underline{x} \cdot \underline{f} = 0$$

for every $\underline{x} \in \underline{X}$ and $\underline{f} \in \underline{X^\sharp}$.

If X is real, then there exists an invertible bounded linear operator $A : X \to X$ such that T, S are of the forms

$$T\underline{x} = \underline{Ax} \quad \text{and} \quad S\underline{f} = \underline{A^{-1^\sharp}f} \quad (0 \neq x \in X, 0 \neq f \in X^\sharp). \tag{2.2.3}$$

If X is complex and infinite dimensional, then there exists an invertible bounded linear or conjugate-linear operator $A : X \to X$ such that T, S are of the forms (2.2.3).

If X is complex and finite dimensional, then there exist a ring automorphism h of $\mathbb{C}$ and an invertible h-semilinear operator $A : X \to X$ such that T, S are of the forms (2.2.3).

The operator A above is unique up to multiplication by a scalar.

Finally, as a consequence of the above result Corollary 2.2.2, we shall present our Uhlhorn-type version of Wigner's theorem for indefinite inner product spaces. As mentioned above, our result is a far-reaching generalization of the main result in [32], where a similar assertion in the particular case when H is finite dimensional and the generating invertible operator η is self-adjoint was presented.

Let η be an invertible bounded linear operator on a Hilbert space H. Denote by $(x, y)_\eta$ the quantity $\langle \eta x, y \rangle$ $(x, y \in H)$. We write $\underline{x} \cdot_\eta \underline{y} = 0$ if $\langle \eta x_0, y_0 \rangle = 0$ holds for every $x_0 \in \underline{x}$ and $y_0 \in \underline{y}$. We call the ray transformation $T : \underline{H} \to \underline{H}$ a generalized symmetry transformation on the indefinite inner product space H generated by η if it satisfies

$$T\underline{x} \cdot_\eta T\underline{y} = 0 \quad \Longleftrightarrow \quad \underline{x} \cdot_\eta \underline{y} = 0$$

for all $\underline{x}, \underline{y} \in \underline{H}$. We say that the transformation $T : \underline{H} \to \underline{H}$ is induced by the invertible linear or conjugate-linear operator $U : H \to H$ if $T\underline{x} = \underline{Ux}$ holds for every $0 \neq x \in H$.

Corollary 2.2.3. *Let H be a real or complex Hilbert space of dimension not less than 3 and let $\eta \in B(H)$ be invertible. Suppose that $T : \underline{H} \to \underline{H}$ is a bijective transformation with the property that*

$$T\underline{x} \cdot_\eta T\underline{y} = 0 \quad \text{if and only if} \quad \underline{x} \cdot_\eta \underline{y} = 0$$

holds for every $\underline{x}, \underline{y} \in \underline{H}$.

If H is real, then T is induced by an invertible bounded linear operator U on H. If H is complex, then T is induced by an invertible bounded linear or conjugate-linear operator U on H.

The operator U inducing T is unique up to multiplication by a scalar.

If H is real, then the invertible bounded linear operator $U : H \to H$ induces a generalized symmetry transformation on $\underline{H}$ if and only if

$$(Ux, Uy)_\eta = c(x, y)_\eta \qquad (x, y \in H)$$

holds for some constant $c \in \mathbb{R}$.

If H is complex, then the invertible bounded linear operator $U : H \to H$ induces a generalized symmetry transformation on $\underline{H}$ if and only if

$$(Ux, Uy)_\eta = c(x, y)_\eta \qquad (x, y \in H)$$

holds for some constant $c \in \mathbb{C}$. Similarly, the invertible bounded conjugate-linear operator $U : H \to H$ induces a generalized symmetry transformation on $\underline{H}$ if and only if

$$(Ux, Uy)_\eta = d(y, x)_{\eta^*} \qquad (x, y \in H)$$

holds for some constant $d \in \mathbb{C}$. Here, η^* denotes the Hilbert space adjoint of η.

Remark 2.2.4. It may be interesting to observe that in contrast with Theorem 2.2.1 and Corollary 2.2.2, in Corollary 2.2.3 above general semilinear operators do not appear.

In Uhlhorn's paper [237] it was mentioned that, for physical reasons, one should consider ray transformations between different spaces. It will be clear from the proofs below that one can generalize our result in that direction easily.

We should point out that, as will be clear from the proofs, in Corollary 2.2.2 and Corollary 2.2.3 there is in fact no need to assume the injectivity of the transformations T, S. We have posed this condition only for the sake of 'symmetricity'.

2.2.3 Proofs

In the proofs we need some additional notation and definitions.

Let X be a real or complex Banach space. The ideal of all finite rank operators in $B(X)$ is denoted by $F(X)$. Two idempotents P, Q in $B(X)$ are said to be (algebraically) orthogonal if $PQ = QP = 0$. There is a natural partial order on $I(X)$. Namely, for any $P, Q \in I(X)$ we write $P \leq Q$ if $PQ = QP = P$. Clearly, $P \leq Q$ holds if and only if the range rng P of P is a subset of the range of Q and the kernel ker P of P contains the kernel of Q. The symbol $I_f(X)$ stands for the collection of all finite rank idempotents in $B(X)$. The natural embedding of X into its second dual $X^{\sharp\sharp}$ is denoted by κ. If $x \in X$ and $f \in X^\sharp$, then $x \otimes f$ stands for the operator (of rank at most 1) defined by

$$(x \otimes f)(z) = \langle z, f \rangle x \qquad (z \in X).$$

Clearly, $x \otimes f$ is a rank-one idempotent if and only if $\langle x, f \rangle = 1$. It is easy to see that the elements of $F(X)$ are exactly the operators $A \in B(X)$ which can be written as finite sums of the form

$$A = \sum_i x_i \otimes f_i \qquad (2.2.4)$$

with $x_1, \ldots, x_n \in X$ and $f_1, \ldots, f_n \in X^\sharp$. Using this representation, the trace of A is defined by

$$\operatorname{tr} A = \sum_i \langle x_i, f_i \rangle.$$

It is known that $\operatorname{tr} A$ is well-defined, i.e., it does not depend on the particular representation (2.2.4) of A. Denote by $M_n(\mathbb{K})$ the algebra of all $n \times n$ matrices with entries in $\mathbb{K}$.

In the proof of the main result of the section we shall need the following lemma.

Lemma 2.2.5. *For any $P_1, P_2 \in I_f(X)$ there exists a $P \in I_f(X)$ such that $P_1, P_2 \leq P$.*

Proof. The assertion will follow from the following observation. Let $M, N \subset X$ be closed subspaces. Suppose that M is of finite codimension and N is of finite dimension. Then there exists an idempotent $P \in I_f(X)$ such that $\ker P \subset M$ and $\operatorname{rng} P \supset N$. Indeed, since every finite dimensional subspace of a Banach space is complemented, we can find a closed subspace K in X such that $K \oplus (M \cap N) = M$. Since the sum of a closed and a finite dimensional subspace is closed, it follows that $M + N$ is closed and has finite codimension. So, there is a finite dimensional subspace L in X such that $(M + N) \oplus L = X$. We clearly have

$$K \oplus (N \oplus L) = X.$$

Now, there exists an idempotent $P \in I_f(X)$ such that $\ker P = K$ and $\operatorname{rng} P = N \oplus L$. This verifies our observation.

If $P_1, P_2 \in I_f(X)$, then $\ker P_1 \cap \ker P_2$ is of finite codimension and $\operatorname{rng} P_1 + \operatorname{rng} P_2$ is of finite dimension. Now, the idempotent $P \in I_f(X)$ obtained according to the observation above clearly has the property that $P_1, P_2 \leq P$. This completes the proof. $\square$

Proof of Theorem 2.2.1. We first extend ϕ to the set $I_f(X)$ of all finite rank idempotents in $B(X)$. If $0 \neq P \in I_f(X)$, then there are mutually (algebraically) orthogonal rank-one idempotents $P_1, \ldots, P_n \in B(X)$ such that $P = \sum_i P_i$. Clearly, $\phi(P_1), \ldots, \phi(P_n)$ are also mutually orthogonal rank-one idempotents. Let us define

$$\tilde{\phi}(P) = \sum_i \phi(P_i).$$

We have to show that $\tilde{\phi}$ is well-defined. In order to do this, let $Q_1, \ldots, Q_n \in B(X)$ be mutually orthogonal rank-one idempotents with sum P. Pick any $R \in I_1(X)$. We have

$$\left(\sum_i \phi(P_i)\right)\phi(R) = 0 \iff \phi(P_i)\phi(R) = 0 \ (i = 1, \ldots, n) \iff$$

$$P_i R = 0 \ (i = 1, \ldots, n) \iff \left(\sum_i P_i\right)R = 0.$$

Similarly, we obtain

$$(\sum_i \phi(Q_i))\phi(R) = 0 \Longleftrightarrow (\sum_i Q_i)R = 0.$$

Since $\sum_i P_i = \sum_i Q_i$, these imply that

$$(\sum_i \phi(P_i))\phi(R) = 0 \Longleftrightarrow (\sum_i \phi(Q_i))\phi(R) = 0.$$

As $\phi(R)$ runs through the set $I_1(X)$, we deduce that the kernels of the idempotents $\sum_i \phi(P_i)$ and $\sum_i \phi(Q_i)$ are the same. A similar argument shows that the ranges of these two idempotents are also equal. Therefore, we have

$$\sum_i \phi(P_i) = \sum_i \phi(Q_i).$$

This shows that the transformation $\tilde{\phi}$ is well-defined. It is now easy to verify that $\tilde{\phi} : I_f(X) \to I_f(X)$ is a bijection which preserves the order, the orthogonality and the rank in both directions. In fact, only the injectivity is not trivial but it follows from an argument quite similar to the one proving that $\tilde{\phi}$ is well-defined.

Pick a finite rank idempotent $P_0 \in B(X)$ whose rank is at least 3. Consider the set $I_{P_0}(X)$ of all idempotents $P \in B(X)$ for which $P \le P_0$. Let $M = \ker P_0$ and $N = \operatorname{rng} P_0$. We have $M \oplus N = X$. Denote by $B(X, M, N)$ the set of all operators A in $B(X)$ for which $A(N) \subset N$ and $A(M) = \{0\}$. Clearly, we have $I_{P_0}(X) \subset B(X, M, N)$. Considering the transformation $A \longmapsto A_{|N}$ we get an algebra isomorphism from $B(X, M, N)$ onto $B(N)$. Moreover, $B(N)$ is obviously isomorphic to $M_n(\mathbb{K})$ for some $n \in \mathbb{N}$. Denote the so-obtained algebra isomorphism from $B(X, M, N)$ onto $M_n(\mathbb{K})$ by ψ. Similarly, we have an algebra isomorphism ψ' from $B(X, \ker \phi(P_0), \operatorname{rng} \phi(P_0))$ onto $M_n(\mathbb{K})$. Therefore, the transformation $P \mapsto \Psi(P) = \psi'(\tilde{\phi}(\psi^{-1}(P)))$ is a bijection of the set of all idempotents in $M_n(\mathbb{K})$ which preserves the order $\le$ in both directions. The form of all such transformations is described in Theorem A.13 due to Ovchinnikov. In particular, it follows from that form that there is a ring-automorphism h_{P_0} of $\mathbb{K}$ such that

$$\operatorname{tr} \Psi(P)\Psi(Q) = h_{P_0}(\operatorname{tr} PQ)$$

holds for all idempotents P, Q in $M_n(\mathbb{K})$. Since ψ, ψ' are algebra isomorphisms, it follows that they preserve the rank-one idempotents. This implies that ψ, ψ' preserve the traces of rank-one operators, from which we conclude that they are generally trace-preserving. It follows that

$$\operatorname{tr} \tilde{\phi}(P)\tilde{\phi}(Q) = h_{P_0}(\operatorname{tr} PQ) \qquad (P, Q \in I_{P_0}(X)). \qquad (2.2.5)$$

We claim that in fact h_{P_0} does not depend on P_0. Indeed, let $P_1 \in I_f(X)$ be such that $P_0 \le P_1$. Considering the corresponding ring automorphism h_{P_1} of $\mathbb{K}$, by (2.2.5) we get that

$$h_{P_0}(\operatorname{tr} PQ) = h_{P_1}(\operatorname{tr} PQ)$$

holds for every $P, Q \in I_{P_0}(X)$. Clearly, $\operatorname{tr} PQ$ runs through $\mathbb{K}$ as P, Q run through $I_{P_0}(X)$. This shows that $h_{P_0} = h_{P_1}$. Since for any two finite rank idempotents there is a finite rank idempotent majorizing both of them (this is the content of Lemma 2.2.5), we have the independence of h_{P_0} from P_0. Therefore, there exists a ring automorphism h of $\mathbb{K}$ such that

$$\operatorname{tr} \tilde{\phi}(P)\tilde{\phi}(Q) = h(\operatorname{tr} PQ) \qquad (P, Q \in I_f(X)). \qquad (2.2.6)$$

We now extend $\tilde{\phi}$ from $I_f(X)$ onto $F(X)$. For any $P_1, \ldots, P_n \in I_f(X)$ and $\lambda_1, \ldots, \lambda_n \in \mathbb{K}$ we define

$$\Phi(\sum_i \lambda_i P_i) = \sum_i h(\lambda_i)\tilde{\phi}(P_i).$$

(Compare this with the definition of the transformation Φ in the proof of Lemma 2.1.2 in Section 2.1.) We have to show that Φ is well-defined. Let $Q_1, \ldots, Q_m \in I_f(X)$ and $\mu_1, \ldots, \mu_m \in \mathbb{K}$ be such that

$$\sum_i \lambda_i P_i = \sum_j \mu_j Q_j.$$

It follows that

$$\sum_i \lambda_i P_i R = \sum_j \mu_j Q_j R$$

holds for every $R \in I_f(X)$. Taking traces we obtain

$$\sum_i \lambda_i \operatorname{tr} P_i R = \sum_j \mu_j \operatorname{tr} Q_j R.$$

By (2.2.6) it follows that

$$\sum_i \lambda_i h^{-1}(\operatorname{tr} \tilde{\phi}(P_i)\tilde{\phi}(R)) = \sum_j \mu_j h^{-1}(\operatorname{tr} \tilde{\phi}(Q_j)\tilde{\phi}(R)).$$

This implies that

$$h^{-1}(\sum_i h(\lambda_i) \operatorname{tr} \tilde{\phi}(P_i)\tilde{\phi}(R)) = h^{-1}(\sum_j h(\mu_j) \operatorname{tr} \tilde{\phi}(Q_j)\tilde{\phi}(R)),$$

that is,

$$h^{-1}(\operatorname{tr}(\sum_i h(\lambda_i)\tilde{\phi}(P_i))\tilde{\phi}(R)) = h^{-1}(\operatorname{tr}(\sum_j h(\mu_j)\tilde{\phi}(Q_j))\tilde{\phi}(R)).$$

This gives

$$\text{tr}(\sum_i h(\lambda_i)\tilde{\phi}(P_i))\tilde{\phi}(R) = \text{tr}(\sum_j h(\mu_j)\tilde{\phi}(Q_j))\tilde{\phi}(R).$$

Since $\tilde{\phi}(R)$ runs through the set $I_f(X)$, we obtain

$$\sum_i h(\lambda_i)\tilde{\phi}(P_i) = \sum_j h(\mu_j)\tilde{\phi}(Q_j).$$

Therefore, Φ is well-defined. Since the finite rank idempotents linearly generate $F(X)$, it follows that Φ is a surjective h-semilinear transformation on $F(X)$ which preserves the rank-one idempotents and their linear spans. The structure of surjective additive mappings of $F(X)$ which preserve the rank-one idempotents and their linear spans were described by Omladič and Šemrl in [203]. In fact, if X is real, then by [203, Main Result] either there exists an invertible bounded linear operator $A : X \to X$ such that

$$\phi(P) = APA^{-1} \qquad (P \in I_1(X)) \tag{2.2.7}$$

or there exists an invertible bounded linear operator $B : X^\sharp \to X$ such that

$$\phi(P) = BP^\sharp B^{-1} \qquad (P \in I_1(X)).$$

If we had this second possibility, then we would get that

$$\phi(P)\phi(Q) = 0 \iff BP^\sharp Q^\sharp B^{-1} = 0 \iff P^\sharp Q^\sharp = 0 \iff QP = 0$$

for every $P, Q \in I_1(X)$. On the other hand, we know that

$$\phi(P)\phi(Q) = 0 \iff PQ = 0.$$

So, we would have

$$PQ = 0 \iff QP = 0$$

for every $P, Q \in I_1(X)$ which is an obvious contradiction. Therefore, ϕ is of the form (2.2.7).

If X is complex, then one can argue in a very similar way referring to [203, Main Result] again (in the infinite dimensional case) or to [203, Theorem 4.5] (in the finite dimensional case). The proof is complete. □

Proof of Corollary 2.2.2. We define a bijective map $\phi : I_1(X) \to I_1(X)$ which preserves zero product in both directions.

Firstly, for every $0 \neq x \in X$ pick a nonzero vector from the ray $T\underline{x}$. In that way we get a transformation, which will be denoted by the same symbol T, from $X \setminus \{0\}$ into itself with the property that for every vector $0 \neq y \in X$, there exists a vector $0 \neq x \in X$ such that $y = \lambda Tx$ for some nonzero scalar $\lambda \in \mathbb{K}$. We do the same with the other transformation S. Clearly, we have

$$\langle Tx, Sf \rangle = 0 \quad \text{if and only if} \quad \langle x, f \rangle = 0 \tag{2.2.8}$$

for every nonzero $x \in X$ and nonzero $f \in X^{\sharp}$.

Let $x \in X$ and $f \in X^{\sharp}$ be such that $\langle x, f \rangle \neq 0$. Define

$$\phi\left(\frac{1}{\langle x, f \rangle} x \otimes f\right) = \frac{1}{\langle Tx, Sf \rangle} Tx \otimes Sf.$$

We show that ϕ is well-defined. Let $x_0 \in X$ and $f_0 \in X^{\sharp}$ be such that $\langle x_0, f_0 \rangle \neq 0$ and suppose that

$$\frac{1}{\langle x, f \rangle} x \otimes f = \frac{1}{\langle x_0, f_0 \rangle} x_0 \otimes f_0.$$

This implies that x, x_0 belong to the same ray in X and the same holds true for f, f_0 in $X^{\sharp}$. Consequently, Tx, Tx_0 and Sf, Sf_0 generate equal rays in X and $X^{\sharp}$, respectively. Therefore, the ranges and the kernels of the idempotents $\frac{1}{\langle Tx, Sf \rangle} Tx \otimes Sf$ and $\frac{1}{\langle Tx_0, Sf_0 \rangle} Tx_0 \otimes Sf_0$ are equal which implies the equality of these two idempotents. Hence, we obtain that ϕ is well-defined.

By the "almost surjectivity" property of the vector-vector transformations T, S we obtain the surjectivity of ϕ. The injectivity of ϕ can be proved by an argument like the one we used to prove that ϕ is well-defined. The transformation ϕ preserves zero product in both directions which is a consequence of (2.2.8).

Now, we can apply our main theorem. Suppose first that X is real. Then our transformation ϕ is of the form (2.2.1) with some invertible bounded linear operator A on X. If $x \in X$ and $f \in X^{\sharp}$ are such that $\langle x, f \rangle \neq 0$, then from the equality

$$\frac{1}{\langle Tx, Sf \rangle} Tx \otimes Sf = \phi\left(\frac{1}{\langle x, f \rangle} x \otimes f\right) =$$
$$A \cdot \frac{1}{\langle x, f \rangle} x \otimes f \cdot A^{-1} = \left(\frac{1}{\langle x, f \rangle} Ax\right) \otimes (A^{-1^{\sharp}} f)$$

(2.2.9)

we deduce that Tx is a scalar multiple of Ax and Sf is a scalar multiple of $A^{-1^{\sharp}} f$. This gives us that $T\underline{x} = \underline{Ax}$ and $S\underline{f} = \underline{A^{-1^{\sharp}} f}$.

If X is complex infinite dimensional, then one can argue in a very similar way.

Finally, let X be complex and finite dimensional. In that case there exist a ring automorphism h of $\mathbb{C}$ and an invertible h-semilinear operator $A : X \to X$ such that ϕ is of the form

$$\phi(P) = APA^{-1} \qquad (P \in I_1(X)).$$

This comes from a rewriting of the form (2.2.2) appearing in the formulation of our main theorem. Now, one can easily verify that we have the following equality very similar to (2.2.9):

$$\frac{1}{\langle Tx, Sf \rangle} Tx \otimes Sf = \left(\frac{1}{h(\langle x, f \rangle)} Ax\right) \otimes (A^{-1^{\sharp}} f).$$

This yields $T\underline{x} = \underline{Ax}$ and $S\underline{f} = A^{-1^\sharp}\underline{f}$ $(\underline{x} \in \underline{X}, \underline{f} \in \underline{X}^\sharp)$.

The assertion concerning essential uniqueness is a consequence of the result Theorem A.14 on locally linearly dependent operators. This completes the proof of Corollary 2.2.2. $\square$

Proof of Corollary 2.2.3. Just as in the proof of Corollary 2.2.2, we can define an "almost surjective" transformation (that is, one that has values in every ray) on the underlying Hilbert space H, denoted by the same symbol T, such that

$$\langle \eta Tx, Ty \rangle = 0 \quad \text{if and only if} \quad \langle \eta x, y \rangle = 0 \quad (x, y \in H \setminus \{0\}).$$

We can rewrite this equivalence first as

$$\langle \eta T\eta^{-1}x, Ty \rangle = 0 \quad \text{if and only if} \quad \langle x, y \rangle = 0 \quad (x, y \in H \setminus \{0\})$$

and next as

$$\langle Tx, \eta T\eta^{-1}y \rangle = 0 \quad \text{if and only if} \quad \langle x, y \rangle = 0 \quad (x, y \in H \setminus \{0\}).$$

Now, we apply Corollary 2.2.2. To be honest, we should point out that although that result is formulated for Banach spaces and hence dual spaces and Banach space adjoints of operators appear there, the very same argument can be applied to conclude that our present transformation T is generated by some invertible operator U on H. We learn from Corollary 2.2.2 that U is linear if H is real, it is either linear or conjugate-linear if H is complex infinite dimensional and, finally, U is semilinear if H is complex finite dimensional. From the proof of the remaining part of our corollary it will be clear that this general semilinear case in fact does not occur.

The essential uniqueness of U can be verified as in the proof of Corollary 2.2.2. As for the third part of the statement, we present the proof only in the complex finite dimensional case. In all other cases one can argue in a quite similar way. So, let h be a ring automorphism of $\mathbb{C}$. Suppose that the invertible h-semilinear operator $U : H \to H$ induces a generalized symmetry transformation. Then we have

$$\langle \eta Ux, Uy \rangle = 0 \quad \Longleftrightarrow \quad \langle \eta x, y \rangle = 0$$

for every $x, y \in H$. This implies that

$$h^{-1}(\langle \eta Ux, Uy \rangle) = 0 \quad \Longleftrightarrow \quad \langle \eta x, y \rangle = 0 \quad (x, y \in H).$$

If we fix $y \in H$, then the functions $x \mapsto h^{-1}(\langle \eta Ux, Uy \rangle)$ and $x \mapsto \langle \eta x, y \rangle$ are linear functionals with the same kernel. We deduce that these functionals differ only by a scalar multiple. Hence, there exists a $c(y) \in \mathbb{C}$ such that

$$h^{-1}(\langle \eta Ux, Uy \rangle) = c(y)\langle \eta x, y \rangle \tag{2.2.10}$$

for every $x, y \in H$. Similarly, for every $x \in H$ there exists a scalar $d(x) \in \mathbb{C}$ such that

$$h^{-1}(\langle Uy, \eta Ux \rangle) = d(x)\langle y, \eta x \rangle \qquad (x, y \in H).$$

Defining $g : \mathbb{C} \to \mathbb{C}$ by $g(\lambda) = \overline{h(\overline{\lambda})}$ $(\lambda \in \mathbb{C})$, we can write this last equality as

$$g^{-1}(\langle \eta Ux, Uy \rangle) = \overline{d(x)}\langle \eta x, y \rangle \qquad (x, y \in H). \tag{2.2.11}$$

It follows from (2.2.10) and (2.2.11) that

$$\langle \eta Ux, Uy \rangle = C(y)h(\langle \eta x, y \rangle) \quad \text{and} \quad \langle \eta Ux, Uy \rangle = D(x)g(\langle \eta x, y \rangle)$$

for every $x, y \in H$, where C, D are complex-valued functions on H. We then have

$$C(y)h(\langle \eta x, y \rangle) = D(x)g(\langle \eta x, y \rangle)$$

for every $x, y \in H$. It is easy to see that C, D are in fact constant functions. Indeed, pick nonzero orthogonal vectors $y_1, y_2 \in H$. Then there exists a vector $z \in H$ for which $\langle z, y_1 \rangle = \langle z, y_2 \rangle \neq 0$. Let $x \in H$ be such that $\eta x = z$. It follows from the equality above that $C(y_1) = C(y_2)$. In case $y_1, y_2 \in H \setminus \{0\}$ are arbitrary, we can choose a nonzero vector $y_3 \in H$ which is orthogonal to both y_1 and y_2 and hence we have $C(y_1) = C(y_3) = C(y_2)$. Since $C(0)$ does not count, we obtain that C is really constant. A similar argument applies to D. It follows that we have constants $C, D \in \mathbb{C}$ such that

$$\langle \eta Ux, Uy \rangle = Ch(\langle \eta x, y \rangle)$$

and

$$\langle \eta Ux, Uy \rangle = D\overline{h(\overline{\langle \eta x, y \rangle})}.$$

Since these hold for every $x, y \in H$ and we have $h(1) = 1$, it follows that $C = D$. This implies that h is self-adjoint in the sense that $h(\overline{\lambda}) = \overline{h(\lambda)}$ $(\lambda \in \mathbb{C})$. It is well-known that the only ring automorphisms of $\mathbb{C}$ with this property are the identity and the conjugation. In fact, this is an easy consequence of the fact that the only ring automorphism of $\mathbb{R}$ is the identity. It now follows that either U is linear and we have

$$(Ux, Uy)_\eta = C(x, y)_\eta \qquad (x, y \in H) \tag{2.2.12}$$

or U is conjugate-linear and we have

$$(Ux, Uy)_\eta = C(y, x)_{\eta^*} \qquad (x, y \in H). \tag{2.2.13}$$

It is obvious that if $U : H \to H$ is either an invertible linear operator on H such that (2.2.12) holds or it is an invertible conjugate-linear operator such that (2.2.13) holds, then U induces a generalized symmetry transformation.

The remaining part of the proof can be carried out in a similar but simpler way. $\qquad\square$

2.2.4 Remarks

As we have seen above, the proof of the main result of this section was built on that nice result of Ovchinnikov in which he described the general form of all bijective maps on the set of all idempotents on a real or complex Hilbert with $\dim H \geq 3$ which preserve the order in both directions. This fact motivated a recent research by Šemrl in two different ways.

First, in [227] he presented a short proof of our result using projective geometrical tools directly instead of applying Ovchinnikov's result (recall that there is some relation between Ovchinnikov's result and projective geometry as we have noted in the subsection 2.2.2). Moreover, in that manner he could improve the finite dimensional parts of Theorem 2.2.1 and Corollary 2.2.3 by omitting the condition of surjectivity (and, in the case of Theorem 2.2.1, even assuming the preservation of orthogonality only in one direction). Mixing our approach involving transformations on rank-one idempotents which preserve zero product in both directions with his approach applying projective geometry, in another paper [225] Šemrl has obtained an Uhlhorn-type result for indefinite inner product spaces over the skew-field of quaternions.

As for second direction of research, our use of Ovchinnikov's result motivated Šemrl to generalize that result in several ways. In [226] he determined the general form of bijective maps on the set of all idempotents on an n-dimensional ($n \geq 3$) linear space over any field not of characteristic two which preserve the order only in one direction. In [229] he characterized the bijective maps of the set of all idempotents on an infinite-dimensional real or complex Banach space which preserve the commutativity in both directions. In fact, as it was proved there, every bijective map on the set of all idempotents which preserves the order (or zero product) in both directions also preserves the commutativity in both directions.

Finally, we mention that in [229] Šemrl also described those bijective maps on the set of all idempotents with a fixed finite rank on an infinite-dimensional real or complex Banach space which preserve zero product in both directions. This result can be viewed as a Banach space analogue of his former result in [228] concerning maps on Hilbert space projections that we have already mentioned in Section 2.1.

2.3 Fidelity Preserving Maps on Quantum States

2.3.1 Summary

We prove that any bijective transformation on the set of all states or on the set of all positive trace-class operators on a given Hilbert space which preserves the fidelity is implemented by an either unitary or antiunitary operator. The results of this section, which appeared in the paper [178], can be considered as natural extensions of Wigner's famous theorem on symmetry transformations.

2.3.2 Formulation of the Results

Let H be a Hilbert space. The set of all positive trace-class operators on H is denoted by $C_1^+(H)$. Clearly, the normalized elements of this set (we mean the operators $A \in C_1^+(H)$ with tr $A = 1$) form exactly the set $S(H)$ of all (normal) states of H. In many cases when treating problems in quantum mechanics, it seems to be more convenient to consider all elements of $C_1^+(H)$ (which are sometimes called density operators) instead of only the normalized ones. This approach is emphasized in many works of Uhlmann (see our references) to whom the introduction of the concept of fidelity, the main object of the present section, is due.

According to Uhlmann [241, 242], for any $A, B \in C_1^+(H)$ we define the fidelity between A and B by

$$F(A, B) = \text{tr}(A^{1/2} B A^{1/2})^{1/2}. \tag{2.3.1}$$

This quantity is in fact the square-root of the transition probability introduced by Uhlmann beforehand in [238] for density operators which later Jozsa called fidelity and showed its use in quantum information theory [116]. (By now, fidelity is one of the most fundamental concepts in quantum computation and quantum information.) The reason that Uhlmann defined the fidelity by (2.3.1) is that after taking square-root the function F behaves significantly better.

It is easy to see that the fidelity is in intimate connection with the transition probability between pure states. Namely, if P, Q are rank-one projections onto the one-dimensional subspaces generated by the unit vectors x and y respectively, then we have

$$F(P, Q) = |\langle x, y \rangle| = (\text{tr } PQ)^{1/2}.$$

We know that Wigner's theorem describing the form of all bijective transformations on the set of all pure states which preserve the transition probability plays fundamental role in the probabilistic aspects of the theory of quantum systems. By analogy, Uhlmann raised the problem to describe all the bijective transformations on $S(H)$ or on $C_1^+(H)$ which preserve the fidelity and this is exactly what we are doing in the present section. In fact, we show that any such transformation is implemented by an either unitary or antiunitary operator on the underlying Hilbert space.

The precise formulation of the results is as follows.

Theorem 2.3.1. *Let* $\phi : C_1^+(H) \to C_1^+(H)$ *be a bijective transformation which preserves the fidelity, i.e., suppose that*

$$F(\phi(A), \phi(B)) = F(A, B) \qquad (A, B \in C_1^+(H)).$$

Then there is an either unitary or antiunitary operator $U : H \to H$ *such that*

$$\phi(A) = UAU^* \qquad (A \in C_1^+(H)). \tag{2.3.2}$$

Theorem 2.3.2. *Let $\phi : S(H) \to S(H)$ be a bijective transformation which preserves the fidelity. Then there is an either unitary or antiunitary operator $U : H \to H$ such that*

$$\phi(A) = UAU^* \qquad (A \in S(H)). \tag{2.3.3}$$

If the underlying Hilbert space is finite dimensional, then we can get rid of the assumption on bijectivity and obtain the following results.

Theorem 2.3.3. *Let H be a finite dimensional Hilbert space and let $\phi : C_1^+(H) \to C_1^+(H)$ be a transformation which preserves the fidelity. Then ϕ is of the form (2.3.2) with some unitary or antiunitary operator U on H.*

Theorem 2.3.4. *Let H be a finite dimensional Hilbert space and let $\phi : S(H) \to S(H)$ be a transformation which preserves the fidelity. Then there is an either unitary or antiunitary operator U on H such that ϕ is of the form (2.3.3).*

2.3.3 Proofs

Proof of Theorem 2.3.1. The main point of the proof is to reduce the problem to Wigner's classical result. In order to do so, we first prove that ϕ preserves the order $\leq$ (which comes from the usual order between self-adjoint bounded linear operators on H) on $C_1^+(H)$ in both directions. If $A, B \in C_1^+(H)$, $A \leq B$, then for any $C \in C_1^+(H)$ we have

$$C^{1/2} A C^{1/2} \leq C^{1/2} B C^{1/2}.$$

Since the square-root function is operator monotone, we have

$$(C^{1/2} A C^{1/2})^{1/2} \leq (C^{1/2} B C^{1/2})^{1/2}.$$

Taking trace we obtain
$$F(A, C) \leq F(B, C)$$

which implies
$$F(\phi(A), \phi(C)) \leq F(\phi(B), \phi(C))$$

for every $C \in C_1^+(H)$. Let $\phi(C)$ run through the set of all rank-one projections. If P is the rank-one projection onto the subspace generated by the unit vector $x \in H$, then we have

$$\langle \phi(A)x, x \rangle^{1/2} = F(\phi(A), P) \leq F(\phi(B), P) = \langle \phi(B)x, x \rangle^{1/2}.$$

Since this holds for every unit vector $x \in H$, we obtain $\phi(A) \leq \phi(B)$. Since ϕ^{-1} has the same properties as ϕ, it follows that ϕ preserves the order in both directions.

We next show that ϕ preserves the rank-one operators in both directions. In fact, one can easily see that an element $A \in C_1^+(H)$ is of rank one if and only if the set $\{T \in C_1^+(H) : T \leq A\}$ is infinite and total in the sense that any two elements in it are comparable with respect to the order $\leq$. By the order preserving property of ϕ it now follows that ϕ preserves the rank-one elements of $C_1^+(H)$ in both directions.

Clearly, a rank-one operator $A \in C_1^+(H)$ is a rank-one projection if and only if its trace is 1, that is, if $F(A, A) = 1$. It follows that ϕ preserves the rank-one projections.

As we have already observed, for any rank-one projections P, Q we have

$$F(P, Q) = (\operatorname{tr} PQ)^{1/2}.$$

Therefore, it follows that the restriction of ϕ onto the set of all rank-one projections is a bijective transformation which satisfies

$$\operatorname{tr} \phi(P)\phi(Q) = \operatorname{tr} PQ.$$

Hence, we can apply Wigner's theorem and get that there exists an either unitary or antiunitary operator $U : H \to H$ such that

$$\phi(P) = UPU^*$$

holds for every rank-one projection P. Replacing ϕ by the transformation

$$A \mapsto U^*\phi(A)U$$

if necessary, we can obviously assume that our original transformation ϕ satisfies $\phi(P) = P$ for every rank-one projection P. It remains to show that $\phi(A) = A$ holds for every $A \in C_1^+(H)$ as well. If $x \in H$ is a unit vector and P is the corresponding rank-one projection, then we compute

$$\langle \phi(A)x, x \rangle^{1/2} = F(\phi(A), P) =$$
$$F(\phi(A), \phi(P)) = F(A, P) = \langle Ax, x \rangle^{1/2}.$$

Since this holds for every unit vector $x \in H$, we conclude that $\phi(A) = A$ $(A \in C_1^+(H))$. This completes the proof. □

Proof of Theorem 2.3.2. Let ϕ be a bijective transformation on $S(H)$ which preserves the fidelity. Define $\tilde{\phi} : C_1^+(H) \to C_1^+(H)$ in the following way: let $\tilde{\phi}(0) = 0$ and for any $0 \neq A \in C_1^+(H)$ set

$$\tilde{\phi}(A) = (\operatorname{tr} A)\phi\left(\frac{A}{\operatorname{tr} A}\right).$$

It is easy to verify that $\tilde{\phi} : C_1^+(H) \to C_1^+(H)$ is a bijective transformation extending ϕ and it preserves the fidelity. Now, Theorem 2.3.1 applies. □

Proof of Theorem 2.3.3. If A, B are self-adjoint operators, then we say that A, B are mutually orthogonal if $AB = 0$. Clearly, A, B are mutually orthogonal if and only if they have mutually orthogonal ranges.

Let us assume that the dimension of H is $d \geq 2$. It is easy to see that one can characterize the positive rank-one operators in the following way: a positive operator A is of rank one if and only if $A \neq 0$ and there exists a system $A_1, \ldots, A_{d-1}$ of nonzero positive operators such that the elements of $A, A_1, \ldots, A_{d-1}$ are mutually orthogonal.

It is clear that ϕ preserves the nonzero operators. Indeed, this follows form the equality $F(A, A) = \operatorname{tr} A$.

Now, let A, B be positive operators with $AB = 0$. Then A, B are commuting and by the properties of the positive square-root of positive operators, we have the same for $A, B^{1/2}$. Therefore, we infer

$$B^{1/2}AB^{1/2} = AB = 0.$$

It follows that $\operatorname{tr}(\phi(B)^{1/2}\phi(A)\phi(B)^{1/2})^{1/2} = 0$. But this implies that

$$(\phi(B)^{1/2}\phi(A)\phi(B)^{1/2})^{1/2} = 0.$$

Hence, we have

$$\phi(B)^{1/2}\phi(A)\phi(B)^{1/2} = 0$$

from which we get

$$(\phi(A)^{1/2}\phi(B))^*(\phi(A)^{1/2}\phi(B)) = \phi(B)\phi(A)\phi(B) = 0.$$

Consequently, we have $\phi(A)^{1/2}\phi(B) = 0$ which implies $\phi(A)\phi(B) = 0$. This shows that ϕ preserves orthogonality. By the characterization of rank-one operators given in the beginning of the proof, we infer that ϕ sends rank-one operators to rank-one operators. Now, just as in the corresponding part of the proof of Theorem 2.3.1 one can check that ϕ sends rank-one projections to rank-one projections and that

$$\operatorname{tr}\phi(P)\phi(Q) = \operatorname{tr} PQ$$

holds for arbitary rank-one projections P, Q. By the non-surjective version of Wigner's theorem [14, 230] (also see Theorem 2.1.4), we have a linear or conjugate-linear isometry $U : H \to H$ such that

$$\phi(P) = UPU^*$$

holds for every rank-one projection P. Since H is finite dimensional, U is in fact a unitary or antiunitary operator. The proof can now be completed very similarly to the proof of Theorem 2.3.1. □

Proof of Theorem 2.3.4. Similarly to the proof of Theorem 2.3.2, the conclusion can be easily deduced from the corresponding result Theorem 2.3.3 concerning transformations defined on $C_1^+(H)$. □

2.3.4 Remarks

We conclude the section with some remarks. First we mention that one can easily generalize our results to obtain the same description of transformations on density operators which preserve not the 'full' fidelity but a certain part of it. We mean the quantity $F_m^+(A, B)$ denoting the sum of the m largest eigenvalues of the operator $(A^{1/2}BA^{1/2})^{1/2}$ $(A, B \in C_1^+(H))$ [242, Definition]. Here m is fixed and when we speak about eigenvalues we always take into account the multiplicities. Now, one can formulate the same assertions as in Theorems 2.3.1 and 2.3.3 with F_m^+ in the place of F. As for the proofs, one can follow quite the same argument. In fact, the only additional thing that should be observed concerns the order preserving property. Namely, one should verify that $A \leq B$ if and only if $F_m^+(A, C) \leq F_m^+(B, C)$ holds for every $C \in C_1^+(H)$. The sufficiency is clear if C runs through the set of all rank-one projections. The necessity follows from Weyl's monotonicity theorem stating that if $A \leq B$, then the the kth largest eigenvalue of A is less than or equal to the kth largest eigenvalue of B (cf. [77, Lemma 1.1, p. 26]).

We note that it would certainly be of interest to obtain similar results concerning the 'partial' fidelities introduced by Uhlmann in [241] (also see [2]).

Following the lines in the proofs above one can easily see that there is no need to assume the injectivity of the transformations ϕ. We set this condition only for the sake of 'symmetry' in the formulation. What concerns surjectivity, the question is more exciting. As Wigner's original theorem can be extended to the non-surjective case as well, it would certainly deserve some effort to try to describe all the transformations $\phi : C_1^+(H) \to C_1^+(H)$ which preserve the fidelity.

2.4 Isometries of Quantum States

2.4.1 Summary

In this section we are concerned with the isometries of metric spaces of quantum states. We consider two metrics on the set all states, namely the Bures metric and the one which comes from the trace-norm. We describe all the corresponding (non-linear) isometries and also present similar results concerning the space of all positive trace-class operators. The results appeared in the paper [194].

2.4.2 Formulation of the Results

In the literature one can find several metrics defined on the set of all states the consideration of which are motivated by certain physical problems. A short summary of such problems and the corresponding metrics can be found in the introduction of the paper [93]. It turns out from the discussion given there that the metrics under consideration can in fact be derived from two fundamental distance functions which are the so-called Bures metric and the metric induced by the trace-norm.

Recently, A. Uhlmann whose research work is closely connected with the study of the Bures metric and the transition probability (see, for example, [3, 238, 239, 240]) has posed the following questions. Is it possible to describe the structure of all transformations which preserve the Bures distance or, in other words, to describe the structure of all isometries of the space $S(H)$ of all states (or the larger space $C_1^+(H)$ of all density operators) equipped with the Bures metric? Moreover, how those isometries are related to the symmetry transformations of the pure states? In this section we answer these questions by showing that every isometry under consideration is implemented by an either unitary or antiunitary operator on the underlying Hilbert space. Furthermore, we obtain results of the same spirit concerning the other fundamental metric, i.e., the one which comes from the trace-norm. (We remark that interesting results and some physical applications can be found in the paper [36] of Busch on linear but not necessarily surjective isometries with respect to this latter metric. In fact, in what follows we shall use two of the results in [36].) As a conclusion, all the isometries with respect to all the metrics appearing in the Introduction of [93] are described.

As for the notation, just as before, $C_1^+(H)$ denotes the set all positive trace-class operators on the Hilbert space H and $S(H)$ stands for the set of all of its normalized elements. We call the elements of $C_1^+(H)$ density operators and the elements of $S(H)$ states.

For obvious reasons, we define our two basic metrics for the larger space $C_1^+(H)$. We begin with the Bures metric to which we need the concept of Uhlmann's fidelity. Just as in Section 2.3, the fidelity $F(A, B)$ between the operators $A, B \in C_1^+(H)$ is defined by

$$F(A, B) = \mathrm{tr}(A^{1/2} B A^{1/2})^{1/2}.$$

Using this, the Bures metric d_b on $C_1^+(H)$ is expressed by the formula

$$d_b(A, B) = (\mathrm{tr}\, A + \mathrm{tr}\, B - 2F(A, B))^{1/2} \qquad (A, B \in C_1^+(H)).$$

The other metric we are interested in comes from the trace-norm. If A is a trace-class operator on H, then its trace-norm (or, in other words, 1-norm) is

$$\|A\|_1 = \mathrm{tr}\,|A|,$$

where $|A|$ stands for the absolute value of A. Our second metric denoted by d_1 is defined by

$$d_1(A, B) = \|A - B\|_1 = \mathrm{tr}\,|A - B| \qquad (A, B \in C_1^+(H)).$$

As for the metrics on $S(H)$, they are just the restrictions of d_b and d_1 onto $S(H)$.

Turning to the results of the present section we note that they can be formulated in one single statement as follows. The isometries of both of the spaces $S(H)$, $C_1^+(H)$ with respect to both of the metrics d_b, d_1 are induced by unitary or antiunitary operators of the underlying Hilbert space. However, for convenience, we divide this statement into parts as seen below.

We emphasize that the transformations in our results are not assumed to be linear in any sense.

Theorem 2.4.1. *Let $\phi : C_1^+(H) \to C_1^+(H)$ be a bijective map which preserves the Bures metric, that is, suppose that*

$$d_b(\phi(A), \phi(B)) = d_b(A, B) \qquad (A, B \in C_1^+(H)).$$

Then there is an either unitary or antiunitary operator U on H such that ϕ is of the form

$$\phi(A) = UAU^* \qquad (A \in C_1^+(H)). \tag{2.4.1}$$

Theorem 2.4.2. *Let $\phi : S(H) \to S(H)$ be a bijective map which preserves the Bures metric. Then there is an either unitary or antiunitary operator U on H such that ϕ is of the form*

$$\phi(A) = UAU^* \qquad (A \in S(H)). \tag{2.4.2}$$

Theorem 2.4.3. *If ϕ is a bijective map of $C_1^+(H)$ which preserves the metric d_1, then there is an either unitary or antiunitary operator U on H such that ϕ is of the form (2.4.1).*

Theorem 2.4.4. *If $\phi : S(H) \to S(H)$ is a bijective map which preserves the metric d_1, then there is an either unitary or antiunitary operator U on H such that ϕ is of the form (2.4.2).*

2.4.3 Proofs

As it will be clear from the proofs below, the cases when ϕ is defined on $C_1^+(H)$ are the more complicated ones. In fact, concerning both metrics it is an essential part of our arguments to show that the isometries corresponding to both metrics map 0 to 0. In order to see this, we have to characterize 0 in terms of the metric alone. As for the Bures metric this is done in the following lemma.

Let $A \in C_1^+(H)$ and $\epsilon > 0$. Denote by $B_\epsilon^b(A)$ (resp. $B_\epsilon^1(A)$) the closed ball with center A and radius ϵ in $C_1^+(H)$ when it is equipped with the Bures metric d_b (resp. the metric d_1).

Lemma 2.4.5. *Let $A \in C_1^+(H)$. We have $A = 0$ if and only if $\operatorname{diam} B_\epsilon^b(A) \leq \sqrt{2}\epsilon$ holds for every $\epsilon > 0$.*

Proof. First we show that diam $B_\epsilon^b(0) \leq \sqrt{2}\epsilon$. Let $\epsilon > 0$. Pick arbitrary $X, Y \in B_\epsilon^b(0)$. We have

$$(\operatorname{tr} X)^{1/2} = d_b(X, 0) \leq \epsilon$$

and the same inequality holds for Y as well. We compute

$$d_b(X, Y)^2 = \operatorname{tr} X + \operatorname{tr} Y - 2F(X, Y) \leq \operatorname{tr} X + \operatorname{tr} Y \leq 2\epsilon^2$$

and hence obtain the desired inequality for the diameter of $B_\epsilon^b(0)$.

We note that it is quite easy to see that if $\dim H \geq 2$, then diam $B_\epsilon^b(0)$ is exactly $\sqrt{2}\epsilon$ (just take two rank-one projections P, Q which are orthogonal to each other and consider the operators $X = \epsilon^2 P$, $Y = \epsilon^2 Q$), while in the case when $\dim H = 1$ we have diam $B_\epsilon^b(0) = \epsilon$.

Now, let $A \in C_1^+(H)$ be nonzero and define $\epsilon = \sqrt{\operatorname{tr} A} > 0$. It is easy to verify that $0, 4A \in B_\epsilon^b(A)$ and $d_b(0, 4A) = 2\epsilon$, so we have diam $B_\epsilon^b(A) = 2\epsilon > \sqrt{2}\epsilon$. □

Using this metric characterization of 0, the proof of Theorem 2.4.1 is easy. The main point is to show that our isometries preserve the fidelity.

Proof of Theorem 2.4.1. As ϕ preserves the Bures metric, we obtain that

$$\operatorname{diam} B_\epsilon^b(\phi(A)) = \operatorname{diam} B_\epsilon^b(A).$$

Applying the characterization of 0 given in Lemma 2.4.5, we easily deduce that $\phi(0) = 0$. Since

$$\operatorname{tr} A = d_b(A, 0)^2 = d_b(\phi(A), \phi(0))^2 = d_b(\phi(A), 0)^2 = \operatorname{tr} \phi(A),$$

we see that ϕ preserves the trace. Considering the definition of the Bures metric, it is now obvious that ϕ preserves the fidelity. The form of such transformations is described in Section 2.3. By Theorem 2.3.1 we have that ϕ is of the form (2.4.1). □

Proof of Theorem 2.4.2. In this case the proof is easier. Indeed, since ϕ sends trace-1 operators to trace-1 operators, we see at once from the definition of d_b that ϕ preserves the fidelity. Thus we can apply our corresponding result Theorem 2.3.2 on the form of fidelity preserving maps on $S(H)$ to complete the proof. □

Now we turn to the description of the isometries with respect to the metric d_1. Just as in the case of the Bures metric, we shall need a characterization of 0 expressed by the metric d_1 alone. This is the content of the next lemma.

Lemma 2.4.6. *Let $A \in C_1^+(H)$. Then $A = 0$ if and only if for every $\epsilon > 0$ and $X, Y \in C_1^+(H)$ with the properties that*

$$d_1(X, A) = \epsilon, \ d_1(Y, A) = \epsilon, \ d_1(X, Y) = 2\epsilon$$

we have

$$B_\epsilon^1(X) \cap B_\epsilon^1(Y) \supsetneq \{A\}.$$

Proof. First let $A = 0$. Let $\epsilon > 0$ be arbitrary. Take $X, Y \in C_1^+(H)$ such that $\|X\|_1, \|Y\|_1 = \epsilon$, $\|X - Y\|_1 = 2\epsilon$. Set $Z = \frac{1}{2}(X + Y)$. It is obvious that $Z \in C_1^+(H)$ and

$$\|X - Z\|_1 = \frac{1}{2}\|X - Y\|_1 = \epsilon$$

and, similarly, we have $\|Y - Z\|_1 = \epsilon$. So,

$$Z \in B_\epsilon^1(X) \cap B_\epsilon^1(Y).$$

Moreover, $Z \neq 0$ since in the opposite case (that is, when $X + Y = 0$) by the positivity of X, Y we would get $X = Y = 0$ and this is a contradiction. This proves the first part of our statement.

To the second part let A be a nonzero element of $C_1^+(H)$. Clearly, there are a positive scalar ϵ and a rank-one projection P such that $A + \epsilon P, A - \epsilon P \in C_1^+(H)$. Define $X = A + \epsilon P, Y = A - \epsilon P$. We have $d_1(X, A) = d_1(Y, A) = \epsilon$ and $d_1(X, Y) = 2\epsilon$. Let $Z \in C_1^+(H)$ be such that $d_1(X, Z), d_1(Y, Z) \leq \epsilon$. Set $T = X - Z$ and $S = Z - Y$. We clearly have

$$\|T\|_1, \|S\|_1 \leq \epsilon \tag{2.4.3}$$

and

$$\frac{1}{2}(T + S) = \frac{1}{2}(X - Y) = \epsilon P. \tag{2.4.4}$$

The result [102, (3.1) Theorem] of Holub tells us that the extreme points of the unit ball of the normed linear space $C_1(H)$ are exactly the rank-one operators of norm 1. Therefore, using (2.4.3) and (2.4.4) we obtain that $T = S = \epsilon P$. This gives us that $\epsilon P = T = X - Z = A + \epsilon P - Z$ which implies $Z = A$. Therefore, we have proved that

$$B_\epsilon^1(X) \cap B_\epsilon^1(Y) = \{A\}.$$

The proof is complete. $\square$

Now, we are in a position to prove Theorem 2.4.3. In the proof we use a nice result of Mankiewicz, namely, [148, Theorem 5] (also see the remark after that theorem) which states that if we have a bijective isometry between convex sets in normed linear spaces with nonempty interiors, then this isometry can be uniquely extended to a bijective affine isometry between the whole spaces. Moreover, we also use a characterization of the orthogonality of the elements of $C_1^+(H)$ which can be found in [36]. We say that the operators $X, Y \in C_1^+(H)$ are orthogonal if $XY = 0$. By (2.2) in [36], for every $X, Y \in C_1^+(H)$ we have

$$XY = 0 \iff \|X - Y\|_1 = \|X + Y\|_1. \tag{2.4.5}$$

Proof of Theorem 2.4.3. By the metric characterization of 0 in Lemma 2.4.6, we obtain that $\phi(0) = 0$.

We assert that ϕ preserves the orthogonality in both directions. In order to verify this, let $X, Y \in C_1^+(H)$. By the positivity of X, Y and $X + Y$ we have

$$\|X + Y\|_1 = \operatorname{tr}(X + Y) = \operatorname{tr} X + \operatorname{tr} Y = \|X\|_1 + \|Y\|_1.$$

It follows from the characterization (2.4.5) of the orthogonality that

$$XY = 0 \iff \|X - Y\|_1 = \|X\|_1 + \|Y\|_1 \iff$$

$$d_1(X, Y) = d_1(X, 0) + d_1(Y, 0).$$

Since ϕ preserves the metric d_1 and sends 0 to 0, we obtain that ϕ preserves the orthogonality in both directions.

For any set $\mathcal{M} \subset C_1^+(H)$, we denote by $\mathcal{M}^\perp$ the set of all elements of $C_1^+(H)$ which are orthogonal to every element of $\mathcal{M}$. It is easy to see that an operator $A \in C_1^+(H)$ is of rank n if and only if the set $\{A\}^{\perp\perp}$ contains n pairwise orthogonal nonzero elements but it does not contain more. As ϕ preserves the orthogonality in both directions and sends 0 to 0, it is now clear that ϕ preserves the rank of operators.

Let H_n be an arbitrary n-dimensional subspace of H. Pick an operator $A \in C_1^+(H)$ whose range is H_n and let H_n' denote the range of $\phi(A)$. We know that $\dim H_n' = n$. We say that a self-adjoint operator T acts on the closed subspace H_0 of H if $T(H_0) \subset H_0$ and $T(H_0^\perp) = \{0\}$. It is then easy to see that those elements of $C_1^+(H)$ which act on H_n are exactly the elements of $\{A\}^{\perp\perp}$. By the orthogonality preserving property of ϕ we have

$$\phi(\{A\}^{\perp\perp}) = \{\phi(A)\}^{\perp\perp}.$$

Hence, we get that ϕ maps isometrically the set of all elements of $C_1^+(H)$ which act on H_n onto the set of all elements of $C_1^+(H)$ which act on H_n'. In this way we can reduce the problem to the finite dimensional case.

It is obvious that in the finite dimensional case the convex set of all density operators has nonempty interior in the normed linear spaces of all self-adjoint operators. (In fact, the interior of this set consists of all invertible positive operators.) Consequently, the result of Mankiewicz applies.

Denote by $C_1(H)$ the Banach space of all trace-class operators and let $C_1(H)_s$ stand for the real linear space of all self-adjoint elements in $C_1(H)$. Define the map $\psi : C_1(H)_s \to C_1(H)_s$ by

$$\psi(T) = \phi(T_+) - \phi(T_-) \qquad (T \in C_1(H)_s).$$

Here T_+, T_- denote the positive and negative parts of $T \in C_1(H)_s$, respectively, that is, we have

$$T_+ = \frac{1}{2}(|T| + T), \quad T_- = \frac{1}{2}(|T| - T).$$

Using Mankiewicz's result and what we have proved above, we see that ψ, when restricted to the set of all self-adjoint operators which act on H_n, equals the Mankiewicz extension of ϕ and hence it is a linear isometry onto the set of all self-adjoint operators which act on H'_n. We recall that H_n was an arbitrary finite dimensional subspace of H. Therefore, we deduce that ψ is a linear isometry from the space of all self-adjoint finite rank operators on H onto itself. But this set is dense in $C_1(H)_s$ and ψ is continuous on $C_1(H)_s$. In fact, this follows from the continuity of ϕ and from the continuity of the absolute value in $C_1(H)$ (see [233, Example 1, p. 42]). It is now obvious that ψ is a surjective linear isometry of $C_1(H)_s$. Even more is true. Namely, as ϕ is an isometry and sends 0 to 0, it is clear that ψ sends positive operators to positive operators and preserves the trace. In the terminology of the paper [36], we can say that ψ is a surjective stochastic isometry. The structure of those maps is known. According to the result [36, Proposition 3.1], ψ is implemented by a unitary or antiunitary operator and this completes the proof. $\qquad\square$

Proof of Theorem 2.4.4. Let $\phi : S(H) \to S(H)$ be a bijective map which preserves the metric d_1.

Let $X, Y \in S(H)$. Since $\|X\|_1 = \|Y\|_1 = 1$ and

$$\|X + Y\|_1 = \operatorname{tr}(X + Y) = \operatorname{tr} X + \operatorname{tr} Y = 2,$$

using (2.4.5) we infer that

$$XY = 0 \iff \|X - Y\|_1 = 2 \iff d_1(X, Y) = 2.$$

Therefore, we obtain that ϕ preserves the orthogonality in both directions.

Now, we can borrow some steps from the proof of Theorem 2.4.3. Indeed, using the argument presented there we can prove that ϕ preserves the rank. Next we can show that for an arbitrary n-dimensional subspace H_n of H there exists an n-dimensional subspace H'_n of H with the property that $A \in S(H)$ acts on H_n if and only if $\phi(A)$ acts on H'_n. Hence, just as there we can reduce the problem to the finite dimensional case.

Let us see what we can do if H is finite dimensional. Denote by $T_0(H)$ the linear space of all trace-zero self-adjoint operators on H. Clearly, $T_0(H)$ is a normed linear space under the norm $\|.\|_1$. Let $n = \dim H$. We assert that the convex subset $D(H) = S(H) - \frac{I}{n}$ of $T_0(H)$ has nonempty interior. In fact, this is because the elements of that set can be characterized as those trace-zero self-adjoint operators on H whose eigenvalues lie in the interval $[-\frac{1}{n}, 1 - \frac{1}{n}]$. Now, one can verify that the interior of $D(H)$ consists of those trace-zero self-adjoint operators whose eigenvalues lie in $]-\frac{1}{n}, 1 - \frac{1}{n}[$. Consider the map

$$A \longmapsto \phi\left(A + \frac{I}{n}\right) - \frac{I}{n}.$$

It is clear that this is a bijective isometry of the convex set $D(H)$. Hence, Mankiewicz's result applies and we get that this map is affine. Obviously, we obtain that ϕ is also affine. This was about the finite dimensional case.

In the general case, similarly to the corresponding part of the proof of Theorem 2.4.3 we can deduce that ϕ is an affine bijection of the subset of all finite rank elements in $S(H)$. But this set is dense in $S(H)$ and ϕ is an isometry. Hence we infer that ϕ is a bijective affine map on $S(H)$, that is, an affine automorphism of $S(H)$. These transformations are well-known to be of the form (2.4.2) (see Section 0.3 of the Introduction or [45, 232]) and we are done.   □

2.4.4 Remarks

To conclude the part of our work concerning preservers on states, we refer in addition to the paper [195] where we have described all the bijective transformations of $S(H)$ which preserve (a general notion of) measure of compatibility between states. Compatibility has some loose connection to fidelity. The result in [195] says that every transformation in question is of the usual nice form, i.e., it is implemented by a unitary or antiunitary operator.

2.5 Order Automorphisms of the Set of Bounded Observables

2.5.1 Summary

Let H be a complex Hilbert space and denote by $B_s(H)$ the set of all self-adjoint bounded linear operators on H. In this section we describe the form of all bijective maps (no linearity or continuity is assumed) on $B_s(H)$ which preserve the usual order $\leq$ in both directions. The results appeared in the paper [177].

2.5.2 Formulation of the Results

As mentioned in the Introduction (see Section 0.3 there), in the Hilbert space framework of quantum mechanics the bounded observables are represented by self-adjoint bounded linear operators. If H denotes the underlying Hilbert space, then these operators form the set $B_s(H)$ on which usually several operations and relations are considered. The automorphisms of $B_s(H)$ with respect to those operations and/or relations are, just as with any algebraic structure in mathematics, of remarkable importance.

Undoubtedly, the most important class of automorphisms of $B_s(H)$ is obtained when we consider $B_s(H)$ as a Jordan algebra with the usual addition, multiplication by scalars and Jordan product $(AB + BA)/2$. The corresponding automorphisms of $B_s(H)$ (called Jordan automorphisms or, in the terminology of [232], Segal automorphisms) are well-known to be implemented by unitary or antiunitary operators of H [45, 232].

The aim of this section is to determine another class of automorphisms of $B_s(H)$. Namely, we equip the set $B_s(H)$ with the usual order among self-adjoint operators. That is, for any $A, B \in B_s(H)$, we write $A \leq B$ if $\langle Ax, x \rangle \leq \langle Bx, x \rangle$ holds for every $x \in H$. Alternatively, in the language of quantum mechanics, the bounded observable A is said to be less than or equal to the bounded observable B if the mean value (or, in other words, expectation) of A in any state is less than or equal to the mean value of B in the same state (also see the subsection 2.6.2). The relation $\leq$ is no doubt an important one among observables.

It is easy to see that the Jordan automorphisms of $B_s(H)$ preserve the order in both directions. This can be expressed in short by saying that they are order automorphisms. In what follows we shall see how close the general order automorphisms of $B_s(H)$ are to the Jordan automorphisms. In fact, we determine all the bijective maps of $B_s(H)$ which preserve the order $\leq$ in both directions (this is done in the main result Theorem 2.5.2 of the section) and also present some corollaries (Corollary 2.5.3, Corollary 2.5.4) that have physical meaning.

We begin with the following proposition on which the proof of our main result rests. Let H be a complex Hilbert space and let $B(H)^+$ denote the cone of all positive operators on H (that is, the set of all $A \in B_s(H)$ for which $\langle Ax, x \rangle \geq 0$ holds for every $x \in H$). Our first result describes the form of all bijective maps on $B(H)^+$ which preserve the order $\leq$ in both directions.

Theorem 2.5.1. *Assume that* $\dim H > 1$. *Let* $\phi : B(H)^+ \rightarrow B(H)^+$ *be a bijective map with the property that*

$$A \leq B \iff \phi(A) \leq \phi(B)$$

holds whenever $A, B \in B(H)^+$. *Then there exists an invertible bounded either linear or conjugate-linear operator* $T : H \rightarrow H$ *such that* ϕ *is of the form*

$$\phi(A) = TAT^* \qquad (A \in B(H)^+).$$

After having proved this result, it will be easy to deduce the main result of the section that follows.

Theorem 2.5.2. *Suppose that* $\dim H > 1$. *Let* $\phi : B_s(H) \rightarrow B_s(H)$ *be a bijective map with the property that*

$$A \leq B \iff \phi(A) \leq \phi(B)$$

holds whenever $A, B \in B_s(H)$. *Then there exists an operator* $X \in B_s(H)$ *and an invertible bounded either linear or conjugate-linear operator* $T : H \rightarrow H$ *such that* ϕ *is of the form*

$$\phi(A) = TAT^* + X \qquad (A \in B_s(H)).$$

This result has some corollaries that seem worth mentioning. In the first one we determine the form of all bijective transformations on $B_s(H)$ which preserve the order and the commutativity in both directions (in quantum mechanics, instead of commutativity they usually use the word 'compatibility' for this important concept).

Corollary 2.5.3. *Assume that* $\dim H > 1$. *Let* $\phi : B_s(H) \to B_s(H)$ *be a bijective map which preserves the order and the commutativity in both directions. Then there is an either unitary or antiunitary operator* $U : H \to H$, *a positive scalar* λ, *and a real number* μ *such that* ϕ *is of the form*

$$\phi(A) = \lambda U A U^* + \mu I \qquad (A \in B_s(H)).$$

The next corollary describes all the bijective maps on $B_s(H)$ which preserve the order and the complementarity in both directions (two observables are called complementary if the range of any nontrivial projection from the range of the spectral measure of the first observable has zero intersection with the range of any nontrivial projection from the range of the spectral measure of the second observable). Although this latter concept is in some sense opposite to compatibility, as it turns out below we still have the same form for ϕ as above.

Corollary 2.5.4. *Suppose that* $\dim H > 1$. *Let* $\phi : B_s(H) \to B_s(H)$ *be a bijective map which preserves the order and the complementarity in both directions. Then there is an either unitary or antiunitary operator* $U : H \to H$, *a positive scalar* λ, *and a real number* μ *such that* ϕ *is of the form*

$$\phi(A) = \lambda U A U^* + \mu I \qquad (A \in B_s(H)).$$

Finally, our last corollary characterizes those bijective maps on $B_s(H)$ which preserve the order and the orthogonality in both directions (two operators $A, B \in B_s(H)$ are called orthogonal if $AB = 0$ which is equivalent to the mutual orthogonality of the ranges of A and B).

Corollary 2.5.5. *Assume that* $\dim H > 1$. *Let* $\phi : B_s(H) \to B_s(H)$ *be a bijective map which preserves the order and the orthogonality in both directions. Then there is an either unitary or antiunitary operator* $U : H \to H$, *and a positive scalar* λ *such that* ϕ *is of the form*

$$\phi(A) = \lambda U A U^* \qquad (A \in B_s(H)).$$

In closing the formulation of the results of this section we note that the condition $\dim H > 1$ in our results is trivially indispensable.

2.5.3 Proofs

This subsection is devoted to the proofs of the results formulated above. We begin with the following auxiliary results. If A is a bounded linear operator, then $\operatorname{rng} A$ denotes its range. The rank of A is, by definition, the algebraic dimension of $\operatorname{rng} A$ and it is denoted by $\operatorname{rank} A$. For any $A \in B(H)^+$, $\sqrt{A}$ stands for the unique positive linear operator whose square is A.

Lemma 2.5.6. *Let $A, B \in B(H)^+$ be such that $\operatorname{rank} A = 1, \operatorname{rank} B < \infty$. We have $\lambda A \leq B$ for some positive scalar λ if and only if $\operatorname{rng} A \subset \operatorname{rng} B$.*

Proof. We recall the following result of Busch and Gudder [37, Theorem 3]: if $B \in B(H)^+$, $x \in H$, and P is the rank-one projection onto the subspace generated by x, then we have $\lambda P \leq B$ for some positive scalar λ if and only if x is in the range of $\sqrt{B}$. As in our case B is a finite rank operator, it follows from the spectral theorem that $\operatorname{rng} B = \operatorname{rng} \sqrt{B}$. Since the positive rank-one operators are exactly the positive scalar multiples of rank-one projections, we obtain the assertion.   □

Lemma 2.5.7. *Let $A \in B(H)^+$ and $n \in \mathbb{N}$. We have $\operatorname{rank} A > n+1$ if and only if there are operators $E, F \in B(H)^+$ such that $E, F \leq A$, $\operatorname{rank} E = n, \operatorname{rank} F > 1$ and there is no $G \in B(H)^+$ of rank 1 with $G \leq E, F$.*

Proof. Suppose that $\operatorname{rank} A > n+1$. We assert that there exists a finite rank operator $A' \in B(H)^+$ such that $A' \leq A$ and $\operatorname{rank} A' > n+1$. In case A is of finite rank, this is trivial. If A is compact and not of finite rank, then by the spectral theorem of compact self-adjoint operators we can verify our claim very easily. Finally, if A is non-compact, then using the spectral theorem of self-adjoint operators and the properties of the spectral integral, we can find an infinite rank projection P on H and a positive scalar λ such that $\lambda P \leq A$ from which the existence of an appropriate operator A' follows.

Clearly, A' can be written as the sum of positive scalar multiples of pairwise orthogonal rank-one projections. Let E be the sum of the first n terms in this sum and let F be the sum of the remaining part. It is easy to see that E, F have the required property. In fact, the non-existence of G follows from Lemma 2.5.6.

To prove the converse, suppose that there are operators $E, F \in B(H)^+$ with the properties formulated in the lemma. It follows from the relation $E, F \leq A$ that E, F are of finite rank and $\operatorname{rng} E, \operatorname{rng} F \subset \operatorname{rng} A$. As there is no positive rank-one operator G with $G \leq E, F$, by Lemma 2.5.6 we have $\operatorname{rng} E \cap \operatorname{rng} F = \{0\}$. So, $\operatorname{rng} A$ contains two subspaces with trivial intersection the sum of whose dimensions is greater than $n+1$. This shows that $\operatorname{rank} A > n+1$, completing the proof of the lemma.   □

Now, we are in a position to prove Theorem 2.5.1.

Proof of Theorem 2.5.1. We first remark that our proof is based on a nice result of Rothaus [218] concerning the automatic linearity of bijective maps between closed convex cones in normed spaces preserving order in both directions. In the paper [218] conclusions of that kind were reached under some quite restrictive assumptions. In our present situation, that is, when the normed space in question is an operator algebra, one can check that those assumptions are fulfilled exactly when the underlying Hilbert space H is finite dimensional. Accordingly, the main point of our proof is to reduce the problem to the finite dimensional case. This is in fact what we are going to do below.

Clearly, $\phi(0) = 0$. We prove that ϕ preserves the rank of operators. In fact, we show that the assertion

$$\text{rank}\, A = k \Longleftrightarrow \text{rank}\, \phi(A) = k$$

($k = 1, \ldots, n$) holds for every $n \in \mathbb{N}$. To begin, as for the case $n = 1$, we remark that a nonzero operator $A \in B(H)^+$ is of rank 1 if and only if the operator interval $[0, A]$ is total under the partial ordering $\leq$, that is, every two elements of it are comparable. Suppose that our assertion is true for some $n \in \mathbb{N}$. We show that in that case it holds also for $n + 1$. Let $A \in B(H)^+$ be of rank $n + 1$. By our assumption of induction, it follows that the rank of $\phi(A)$ is at least $n + 1$. Suppose that $\text{rank}\, \phi(A) > n + 1$. Using Lemma 2.5.7 and the order preserving property of ϕ we obtain that $\text{rank}\, A > n + 1$ which is a contradiction. Therefore, we have $\text{rank}\, \phi(A) = n+1$. Referring to the fact that ϕ^{-1} possesses the same properties as ϕ, we obtain the desired assertion.

We now prove that if $A_1, \ldots, A_n \in B(H)^+$ are of rank 1, then their ranges are linearly independent if and only if the same holds for the ranges of $\phi(A_1), \ldots, \phi(A_n)$. (A system of 1-dimensional subspaces in H of n members is called linearly independent if they cannot be included in an $(n-1)$-dimensional subspace.) This statement is clear for $n = 1$. Suppose that it holds for n and prove that it then necessarily holds also for $n + 1$. Let $A_1, \ldots, A_n, A_{n+1}$ be rank-one operators with linearly independent ranges and assume that this is not the case with the ranges of $\phi(A_1), \ldots, \phi(A_n), \phi(A_{n+1})$. Then these ranges can be included in an at most n-dimensional subspace implying that there is a rank-n operator $B \in B(H)^+$ such that $\phi(A_1), \ldots, \phi(A_{n+1}) \leq B$. By the rank-preserving property of ϕ we have a rank-n operator $A \in B(H)^+$ such that $A_1, \ldots, A_n, A_{n+1} \leq A$. By Lemma 2.5.6 this implies that $\text{rng}\, A_1, \ldots, \text{rng}\, A_{n+1} \subset \text{rng}\, A$ and it follows that the ranges of $A_1, \ldots, A_{n+1}$ can be included in an n-dimensional subspace of H which is a contradiction. This verifies our claim.

Fix rank-one operators $A_1, \ldots, A_n \in B(H)^+$ with linearly independent ranges which generate the n-dimensional subspace H_n of H. Denote by H'_n the n-dimensional subspace of H generated by the ranges of the operators $\phi(A_1), \ldots, \phi(A_n)$. We assert that an operator $T \in B(H)^+$ acts on H_n if and only if $\phi(T)$ acts on H'_n. (Recall that an operator T acts on the closed subspace M of H if M is an invariant subspace of T and T is zero on the orthogonal complement of M.) This will follow from the following observation: the positive finite rank operator T acts on H_n if and only if for every rank-one operator A for which the ranges of $A_1, \ldots, A_n, A$ are linearly independent we have $A \not\leq T$. To see this, suppose that T acts on H_n. If $A \leq T$, then we have $\text{rng}\, A \subset \text{rng}\, T \subset H_n$ implying that the ranges of $A_1, \ldots, A_n, A$ cannot be linearly independent. This gives us the necessity. As for the sufficiency, suppose that T does not act on H_n. Then there exists a unit vector x in the range of T which does not belong to H_n. On the other hand, as $x \in \mathrm{rng}\, T$, by Lemma 2.5.6 it follows that a positive scalar multiple of the rank-one projection onto the subspace generated by x is less than or equal to T. This gives us a rank-one operator A for which the ranges of $A_1, \ldots, A_n, A$ are linearly independent and we have $A \leq T$. This proves our claim.

So, we know that for any n-dimensional subspace H_n of H, there exists an n-dimensional subspace H'_n of H such that for every $T \in B(H)^+$, T acts on H_n if and only if $\phi(T)$ acts on H'_n. This gives rise to a bijective transformation ψ on the cone $M_n(\mathbb{C})^+$ of all positive $n \times n$ complex matrices which preserves the order in both direction. (Here positivity is used in the operator theoretical sense, that is, our concept of positivity is just the same as positive semidefiniteness in matrix theory.)

Since ϕ preserves the rank, it follows that ψ preserves the rank-n matrices in both directions. The set of all such matrices is just the interior of $M_n(\mathbb{C})^+$ in the real normed space of all $n \times n$ Hermitian matrices. Now, the result [218, Proposition 2] of Rothaus on the linearity of order preserving maps can be applied and it gives us that ψ is linear on the set of all rank-n elements in $M_n(\mathbb{C})^+$. We show that ψ is linear on the whole set $M_n(\mathbb{C})^+$. Pick $A, B \in M_n(\mathbb{C})^+$. Then there are sequences $(A_k), (B_k)$ of rank-n elements in $M_n(\mathbb{C})^+$ which are monotone decreasing with respect to the order $\leq$ and $A_k \to A$, $B_k \to B$. It is clear that the equalities $A = \inf_k A_k$, $B = \inf_k B_k$ and $A + B = \inf_k(A_k + B_k)$ hold in the partially ordered set $M_n(\mathbb{C})^+$. By the order preserving property of ψ we obtain that $\psi(A) = \inf_k \psi(A_k)$, $\psi(B) = \inf_k \psi(B_k)$ and $\psi(A + B) = \inf_k \psi(A_k + B_k)$. The sequences $\psi(A_k)$, $\psi(B_k)$, $\psi(A_k + B_k)$ are monotone decreasing and bounded below. Therefore, Vigier's theorem (e.g. [198, 4.1.1. Theorem]) applies and gives us that the sequences above converge (strongly) to their infima. Now, by the partial additivity property of ψ which has been obtained above as a consequence of Rothaus's result, we have

$$\psi(A + B) = \lim_k \psi(A_k + B_k) = \lim_k \psi(A_k) + \lim_k \psi(B_k) = \psi(A) + \psi(B).$$

So, ψ is additive on $M_n(\mathbb{C})^+$ and one can prove in the same way that it is positive homogeneous as well. Since every pair of finite rank elements in $B(H)^+$ can be embedded into a matrix space $M_n(\mathbb{C})^+$, we deduce that ϕ is additive and positive homogeneous on the set of all finite rank elements in $B(H)^+$.

Since every finite sum $\sum_i \lambda_i P_i$, where the λ_i's are positive numbers and the P_i's are projections of not necessarily finite rank, is the strong limit of a monotone increasing net of finite rank elements in $B(H)^+$, one can prove in a very similar way as above that ϕ is additive and positive homogeneous on the set of all such finite sums. Finally, using the fact that every operator in $B(H)^+$ is the norm limit of a monotone increasing sequence of operators of the form $\sum_i \lambda_i P_i$ (this follows form the spectral theorem), repeating the above argument once again, we obtain that ϕ is additive and positive homogeneous.

Extend ϕ from $B(H)^+$ to $B_s(H)$ in the obvious way, that is, define $\tilde{\phi}(T) = \phi(A) - \phi(B)$ for every $T \in B_s(H)$ and $A, B \in B(H)^+$ for which $T = A - B$. It is easy to check that $\tilde{\phi} : B_s(H) \to B_s(H)$ is a linear transformation which preserves the order in both directions. To see the less trivial part of this last assertion, suppose that $T \in B_s(H)$, $T = A - B$, $A, B \in B(H)^+$ are such that $0 \leq \tilde{\phi}(T) = \phi(A) - \phi(B)$. This implies that $\phi(B) \leq \phi(A)$ which yields $B \leq A$, that is, we have $0 \leq T$. The linear transformation $\tilde{\phi}$ is surjective since $B(H)^+$ is included in its range. Moreover, it is injective as well which follows from the fact that $\tilde{\phi}$ preserves the order in both directions. Now, if one further extends $\tilde{\phi}$ to a linear transformation on the algebra $B(H)$ of all bounded linear operators on H, one gets a linear bijection of the C^*-algebra $B(H)$ which preserves the order in both directions. Due to the well-known result Theorem A.3 of Kadison, every such transformation sending the identity to itself is a Jordan *-automorphism. Therefore, the linear transformation

$$A \longmapsto \sqrt{\phi(I)}^{-1} \phi(A) \sqrt{\phi(I)}^{-1}$$

is a Jordan *-automorphism of $B(H)$. By Herstein's theorem (see Theorem A.7), every Jordan *-automorphism of $B(H)$ is either a *-automorphism or a *-antiautomorphism. Applying Theorem A.8 it then follows that for any Jordan *-automorphism J of $B(H)$ we have

$$J(A) = UAU^* \qquad (A \in B_s(H))$$

where U is either a unitary or an antiunitary operator on H. It is now easy to infer that ϕ is of the desired form. This completes the proof of the theorem. $\square$

The main result of the section is now easy to prove.

Proof of Theorem 2.5.2. Let $X = \phi(0)$ and consider the transformation

$$\psi : A \mapsto \phi(A) - X.$$

Clearly, ψ is a bijection of $B_s(H)$ preserving the order in both directions. So, without loss of generality we can assume that $\phi(0) = 0$. Now, restricting ϕ onto $B(H)^+$ we have a bijection of $B(H)^+$ which preserves the order in both directions. So, we can apply Theorem 2.5.1 and obtain that there exists an invertible bounded either linear or conjugate-linear operator $T : H \to H$ for which we have

$$\phi(A) = TAT^* \qquad (A \in B(H)^+). \tag{2.5.1}$$

It remains to show that this equality holds also for every $A \in B_s(H)$. Let $B \in B_s(H)$ be arbitrary but fixed. Then there exists a constant $K \in \mathbb{R}$ such that $K \leq B$ (for example, one can choose $K = -\|B\|$). Consider the transformation

$$A \mapsto \phi(A + K) - \phi(K)$$

on $B(H)^+$. Just as above, this transformation is a bijective map on $B(H)^+$ which preserves the order in both directions. Therefore, there exists an invertible bounded either linear or conjugate-linear operator $S : H \to H$ such that

$$\phi(A + K) - \phi(K) = SAS^* \qquad (A \in B(H)^+). \qquad (2.5.2)$$

If $A \geq -K, 0$, then by (2.5.1) we have

$$T(A + K)T^* - \phi(K) = SAS^*. \qquad (2.5.3)$$

Considering this equality for another A' with $A' \geq -K, 0$, we see that

$$T(A - A')T^* = S(A - A')S^*.$$

As the difference $A - A'$ can be an arbitrary self-adjoint operator, we obtain that $TCT^* = SCS^*$ holds for every $C \in B_s(H)$. It now follows from (2.5.3) that

$$T(A + K)T^* - \phi(K) = SAS^* = TAT^*$$

where $A \in B_s(H)$, $A \geq -K, 0$. This yields $\phi(K) = TKT^* = SKS^*$. We deduce from (2.5.2) that

$$\phi(A + K) = SAS^* + \phi(K) = TAT^* + TKT^* = T(A + K)T^*$$

holds for every $A \in B(H)^+$. Choosing $A = B - K \geq 0$, we have

$$\phi(B) = TBT^*.$$

This completes the proof. □

We now turn to the proofs of the corollaries.

Proof of Corollary 2.5.3. It follows from Theorem 2.5.2 that there is an invertible bounded either linear or conjugate-linear operator T on H such that

$$\phi(A) = TAT^* + \phi(0) \qquad (A \in B_s(H))$$

Since 0 is commuting with every $A \in B_s(H)$, the same is true for $\phi(0)$. This gives us that $\phi(0)$ is a scalar operator, that is, there is a $\mu \in \mathbb{R}$ such that $\phi(0) = \mu I$. Similarly, we have a constant $\lambda \in \mathbb{R}$ such that $TT^* = \phi(I) - \phi(0) = \lambda I$. It is trivial that λ is necessarily positive and then we obtain that the operator $T/\sqrt{\lambda}$ is either unitary or antiunitary. □

Proof of Corollary 2.5.4. It is easy to see that $A \in B_s(H)$ is complementary with every $B \in B_s(H)$ if and only if A is scalar. Hence, ϕ preserves the scalar operators and one can apply the argument in the proof of Corollary 2.5.3 to get the desired form of ϕ. □

For the proof of Corollary 2.5.5 we recall the following notation. If $x, y \in H$, then $x \otimes y$ denotes the operator defined by $(x \otimes y)z = \langle z, y \rangle x$ $(z \in H)$.

Proof of Corollary 2.5.5. Since 0 is the only operator in $B_s(H)$ which is orthogonal to every operator, we infer that $\phi(0) = 0$. By Theorem 2.5.2 we have an invertible bounded either linear or conjugate-linear operator T on H such that $\phi(A) = TAT^*$ holds for every $A \in B_s(H)$. Without serious loss of generality we can suppose that T is linear. It now follows that for every $A, B \in B_s(H)$ with $AB = 0$ we have $\phi(A)\phi(B) = 0$ which implies that $AT^*TB = 0$. Choosing nonzero orthogonal vectors $x, y \in H$, for $A = x \otimes x$ and $B = y \otimes y$ we get $x \otimes Tx \cdot Ty \otimes y = 0$ which yields $\langle T^*Tx, y \rangle = \langle Tx, Ty \rangle = 0$. So, we have $\langle T^*Tx, y \rangle = 0$ whenever $\langle x, y \rangle = 0$. This clearly implies that for every $x \in H$ there is a scalar λ_x such that $T^*Tx = \lambda_x x$. In other words, the operators T^*T and I are locally linearly dependent. By Theorem A.14 they are linearly dependent, i.e., there exists a scalar $\lambda \in \mathbb{R}$ such that $T^*T = \lambda I$. Now, the proof can be completed as in the proof of Corollary 2.5.3.   □

To conclude this section we refer the reader to Section 2.7 where some results concerning order preserving maps on the set of Hilbert space effects can be found.

2.6 Linear Maps on the Set of Bounded Observables Preserving Maximal Deviation

2.6.1 Summary

In this section we determine the structure of all bijective linear maps on the space of bounded observables which preserve a fixed moment or the variance. Non-linear versions of the corresponding results are also presented. The content of this section is taken from the paper [187].

2.6.2 Formulation of the Results

In this section we solve a linear preserver problem concerning linear transformations on the space $B_s(H)$ of all bounded observables (self-adjoint bounded linear operators) on the Hilbert space H. When studying LPPs, one can observe that in many cases the solutions lead to the form of the automorphisms of the underlying algebra or, at least, to a form which is quite close to it. This is a remarkable observation since in that way we can gain important new information on the automorphisms which show how they are determined by their various preserving properties. In the present section we shall see to what extent the preservation of certain probabilistic characteristics of observables determines the Jordan automorphisms of $B_s(H)$.

In classical probability theory the mean value (or, more generally, the moments) and the variance are among the most important characteristics of a random variable. Therefore, it is not surprising that the same is true for the quantum mechanical variables, i.e., for the observables. The main aim of this section is to show that the preservation of any of those quantities more or less completely characterizes the Jordan automorphisms among all linear transformations of $B_s(H)$.

In what follows, let H be a complex Hilbert space. Let $A \in B_s(H)$ and pick a unit vector $\varphi \in H$. The mean value $m(A, \varphi)$ of the observable A in the pure state represented by φ is defined as

$$m(A, \varphi) = \langle A\varphi, \varphi \rangle.$$

So, unlike in classical probability, in quantum theory there is a set of mean values of a single variable. We intend to determine all the bijective linear transformations ϕ of $B_s(H)$ which preserve this set in the sense that

$$\{m(\phi(A), \varphi) : \varphi \in H, \|\varphi\| = 1\} = \{\langle \phi(A)\varphi, \varphi \rangle : \varphi \in H, \|\varphi\| = 1\}$$

$$= \{\langle A\varphi, \varphi \rangle : \varphi \in H, \|\varphi\| = 1\} = \{m(A, \varphi) : \varphi \in H, \|\varphi\| = 1\}$$

holds for every $A \in B_s(H)$. Clearly, the set of all mean values of an observable $A \in B_s(H)$ is equal to the numerical range of the operator A. So, the above problem can be reformulated as the linear preserver problem concerning the numerical range on $B_s(H)$. Obviously, it is a more general problem to preserve the numerical radius $w(.)$ instead of the numerical range. It is well-known that for a self-adjoint operator A this former quantity $w(A)$ is equal to the operator norm $\|A\|$. Hence, we easily arrive at the problem of describing the surjective linear isometries of $B_s(H)$. The solution of this problem is well-known in the literature. For example, one can consult the paper [61]. The corresponding result reads as follows.

Theorem 2.6.1. *Let* $\phi : B_s(H) \to B_s(H)$ *be a bijective linear map which preserves the operator norm, that is, suppose that*

$$\|\phi(A)\| = \|A\| \qquad (A \in B_s(H)).$$

Then there is an either unitary or antiunitary operator U on H such that ϕ is either of the form

$$\phi(A) = UAU^* \qquad (A \in B_s(H)) \qquad (2.6.1)$$

or of the form

$$\phi(A) = -UAU^* \qquad (A \in B_s(H)). \qquad (2.6.2)$$

Although this is not a new result, below we present the sketch of a short proof that applies preserver techniques.

Observe that the above statement is a self-adjoint analogue of the well-know result of Kadison [117] on the surjective linear isometries of C^*-algebras (cf. Theorem A.9) and also that of a result of Brešar and Šemrl [29] describing the form of all bijective linear maps of the algebra of all bounded linear operators on a Banach space which preserve the spectral radius (recall that the norm of a self-adjoint operator is equal to its spectral radius). However, there is no doubt, those results are much deeper than the one we have formulated above.

By the help of Theorem 2.6.1 we can describe the bijective linear maps of $B_s(H)$ which preserve the set of mean values. In fact, as the second possibility (2.6.2) can be excluded, we obtain that the maps in question are exactly the Jordan automorphisms of the Jordan algebra $B_s(H)$ (see Section 0.3 of the Introduction). Moreover, observe that using the same result Theorem 2.6.1 we can solve also the problem of preserving a fixed moment of bounded observables. For any $n \in \mathbb{N}$, the nth moment of an observable $A \in B_s(H)$ is the set

$$\{m(A^n, \varphi) \, : \, \varphi \in H, \|\varphi\| = 1\} = \{\langle A^n \varphi, \varphi \rangle \, : \, \varphi \in H, \|\varphi\| = 1\}.$$

Now, the solution of the mentioned problem immediately follows as one can refer to the equality

$$\sup\{|\langle A^n \varphi, \varphi \rangle| \, : \, \varphi \in H, \|\varphi\| = 1\} = w(A^n) = \|A^n\| = \|A\|^n$$

which holds for every self-adjoint operator A on H.

Beside moments, the other very important probabilistic character of an observable is captured in its variance. Just as in the case of mean values, we have variance with respect to every (pure) state. Let $A \in B_s(H)$ and $\varphi \in H, \|\varphi\| = 1$. The variance $var(A, \varphi)$ of A in the state φ is defined by

$$\begin{aligned} var(A, \varphi) &= m((A - m(A, \varphi)I)^2, \varphi) \\ &= \langle (A - \langle A\varphi, \varphi \rangle I)^2 \varphi, \varphi \rangle \\ &= \langle A^2 \varphi, \varphi \rangle - \langle A\varphi, \varphi \rangle^2. \end{aligned}$$

We intend to determine all bijective linear maps on $B_s(H)$ which preserve the set of variances of observables. It is obvious that every linear map ϕ on $B_s(H)$ which preserves this set, i.e., which satisfies

$$\{var(\phi(A), \varphi) \, : \, \varphi \in H, \|\varphi\| = 1\} = \{var(A, \varphi) \, : \, \varphi \in H, \|\varphi\| = 1\}$$

for every $A \in B_s(H)$, also preserves the quantity

$$\|A\|_v = \sup_{\|\varphi\|=1} var(A, \varphi)^{1/2}, \tag{2.6.3}$$

i.e., satisfies

$$\|\phi(A)\|_v = \|A\|_v$$

for every $A \in B_s(H)$. The quantity $\|A\|_v$ is called the maximal deviation of the observable $A \in B_s(H)$. In its definition (2.6.3) we have used the square root of the variances since, as it will be clear from Lemma 2.6.5, the so-obtained quantity is a semi-norm on $B_s(H)$ which is quite convenient to handle.

Observe that every Jordan automorphism of $B_s(H)$ as well as its negative preserves the maximal deviation and that perturbations by scalar operators also do not change this quantity. Our result that follows (which can be considered as the main result of this section) states that from these two types of transformations we can construct all the linear preservers under consideration.

Theorem 2.6.2. *Let $\phi : B_s(H) \to B_s(H)$ be a bijective linear map which preserves the maximal deviation, that is, suppose that*

$$\|\phi(A)\|_v = \|A\|_v \qquad (A \in B_s(H)).$$

Then there exist an either unitary or antiunitary operator U on H and a linear functional $f : B_s(H) \to \mathbb{R}$ such that ϕ is either of the form

$$\phi(A) = UAU^* + f(A)I \qquad (A \in B_s(H)) \tag{2.6.4}$$

or of the form

$$\phi(A) = -UAU^* + f(A)I \qquad (A \in B_s(H)). \tag{2.6.5}$$

Unlike with the transformations preserving the set of mean values, for the bijective linear maps on $B_s(H)$ which preserve the set of variances, the second possibility (2.6.5) above can obviously occur. Hence, we obtain that every such preserver is 'a Jordan automorphism of $B_s(H)$ or its negative perturbed by a linear functional times I'.

Since, from the physical point of view, to assume the linearity of the considered transformations on the space of observables sometimes seems to be a strong assumption that can be quite difficult to check in the particular cases, in the remaining results we formulate non-linear versions of Theorems 2.6.1 and 2.6.2 as follows. First observe that

$$d_m(A, B) = \sup_{\|\varphi\|=1} |m(A - B, \varphi)| = \|A - B\| \quad (A, B \in B_s(H))$$

defines a metric on $B_s(H)$, while

$$d_v(A, B) = \sup_{\|\varphi\|=1} var(A - B, \varphi)^{1/2} = \|A - B\|_v \quad (A, B \in B_s(H))$$

defines a semimetric on $B_s(H)$. Both d_m and d_v represent certain stochastic distances between bounded observables. Using the first two results and the celebrated Mazur-Ulam theorem on surjective non-linear isometries of normed spaces [157], we can prove the following statements which show how close the 'stochastic isometries' with respect to either d_m or d_v are to the Jordan automorphisms of $B_s(H)$.

Theorem 2.6.3. *Let $\phi : B_s(H) \to B_s(H)$ be a bijective transformation (linearity is not assumed) with the property that*

$$d_m(\phi(A), \phi(B)) = d_m(A, B) \qquad (A \in B_s(H)).$$

Then there is an either unitary or antiunitary operator U on H and a fixed operator $X \in B_s(H)$ such that ϕ is either of the form

$$\phi(A) = UAU^* + X \qquad (A \in B_s(H))$$

or of the form

$$\phi(A) = -UAU^* + X \qquad (A \in B_s(H)).$$

The last result of the section describes the form of all 'stochastic isometries' with respect to the semimetric d_v.

Theorem 2.6.4. *Let $\phi : B_s(H) \to B_s(H)$ be a bijective transformation (linearity is not assumed) with the property that*

$$d_v(\phi(A), \phi(B)) = d_v(A, B) \qquad (A \in B_s(H)).$$

Then there exist an either unitary or antiunitary operator U on H, a fixed operator $X \in B_s(H)$, and a functional $f : B_s(H) \to \mathbb{R}$ (not linear in general) such that ϕ is either of the form

$$\phi(A) = UAU^* + X + f(A)I \qquad (A \in B_s(H))$$

or of the form

$$\phi(A) = -UAU^* + X + f(A)I \qquad (A \in B_s(H)).$$

2.6.3 Proofs

We first remark that *in the remaining part of this section whenever we speak about the preservation of an object or relation we always mean that it is preserved in both directions.*

We now present a short proof of Theorem 2.6.1.

Sketch of the proof of Theorem 2.6.1. Let $\phi : B_s(H) \to B_s(H)$ be a surjective linear isometry. Clearly, ϕ preserves the extreme points of the unit ball of $B_s(H)$ which are well-known (and easily seen) to be exactly the self-adjoint unitaries, i.e., the operators of the form $2P - I$ where P is a projection. Now, one can readily prove that among those extreme points, I and $-I$ are distinguished by the following property. The extreme point U is either I or $-I$ if and only if we have $\|U - V\| \in \{0, 2\}$ for every extreme point V. Therefore, we get $\phi(\{I, -I\}) = \{I, -I\}$. Clearly, we can suppose without loss of generality that $\phi(I) = I$. In that case we obtain that ϕ preserves the projections. This gives us that ϕ is a Jordan automorphism of $B_s(H)$, that is, it satisfies the equality $\phi(AB + BA) = \phi(A)\phi(B) + \phi(B)\phi(A)$ for every $A, B \in B_s(H)$. In fact, this follows from the argument given in the proof of Theorem A.4. Therefore, we have that ϕ is of the form

$$\phi(A) = UAU^* \qquad (A \in B_s(H))$$

with some unitary or antiunitary operator U on H (see Section 0.3 of the Introduction). $\qquad\qquad\qquad\qquad\qquad\qquad\qquad\qquad\qquad\qquad\qquad\qquad$ □

The proof of Theorem 2.6.2 is much more difficult than the one given above and is based on the following series of lemmas. Our first observation below will prove to be fundamental from the view-point of the proof of Theorem 2.6.2. It states that the maximal deviation of an operator T is equal to the so-called factor norm of T in the factor Banach space $B_s(H)/\mathbb{R}I$. (In particular, this result implies that the function $T \mapsto \|T\|_v$ is a semi-norm on $B_s(H)$.) Denote by $\overline{T}$ the equivalence class of T in $B_s(H)/\mathbb{R}I$. The factor norm $\|\overline{T}\|$ of T is defined by

$$\|\overline{T}\| = \inf_{\lambda \in \mathbb{R}} \|T + \lambda I\|.$$

As the spectral radius and the operator norm of a self-adjoint operator are the same, it easily follows that $\|\overline{T}\|$ is equal to the half of the diameter of the spectrum $\sigma(T)$ of T.

Lemma 2.6.5. *For all $T \in B_s(H)$ we have $\|T\|_v = \|\overline{T}\| = diam(\sigma(T))/2$.*

Proof. As we have already verified that $\|\overline{T}\| = \text{diam}(\sigma(T))/2$, we have to prove only the first equality. For a scalar operator T, both quantities $\|T\|_v$ and $\|\overline{T}\|$ are 0. Otherwise, we can assume that $0 \leq T \leq I$ and that $\{0, 1\} \subset \sigma(T) \subset [0, 1]$. This is because the factor norm and the maximal deviation are invariant under adding scalar operators and they are absolute homogeneous. In this case we have $\|\overline{T}\| = \frac{1}{2}$.

First we prove the easier inequality $\|T\|_v \leq \|\overline{T}\|$. For any $\lambda \in \mathbb{R}$ we have

$$\|T\|_v^2 = \|T + \lambda I\|_v^2 = \sup_{\|\varphi\|=1} \left(\langle (T + \lambda I)^2 \varphi, \varphi \rangle - \langle (T + \lambda I)\varphi, \varphi \rangle^2 \right) \leq$$
$$\sup_{\|\varphi\|=1} \langle (T + \lambda I)^2 \varphi, \varphi \rangle = \|(T + \lambda I)^2\| = \|T + \lambda I\|^2.$$

This yields $\|T\|_v \leq \|T + \lambda I\|$ for all $\lambda \in \mathbb{R}$ which implies that $\|T\|_v \leq \|\overline{T}\|$.

Now, we turn to the less obvious inequality $\frac{1}{2} = \|\overline{T}\| \leq \|T\|_v$. Let E_T be the spectral measure corresponding to T. Since 0 and 1 are in the spectrum of T, it follows that for any $0 < \delta \leq \frac{1}{2}$, the measures of $]-\delta, \delta[\cap\sigma(T)$ and $]1-\delta, 1+\delta[\cap\sigma(T)$ under E_T are mutually orthogonal nonzero projections. At this stage δ is not fixed, we shall specify it later. Denote these projections by P_0 and P_1, respectively.

Let x be a unit vector in the range of P_0 and y be a unit vector in the range of P_1. Define $\varphi = (x + y)/\sqrt{2}$. Then $\varphi \in H$ is a unit vector and we assert that the following inequality holds

$$\sqrt{\langle T^2 \varphi, \varphi \rangle - \langle T\varphi, \varphi \rangle^2} \geq \sqrt{\frac{(1 - 2\delta)^2}{2} - \frac{(1 + 2\delta)^2}{4}}. \qquad (2.6.6)$$

To see this, first observe that $Tx = TP_0x$ and $Ty = TP_1y$. Since

$$TP_0 = \int_{]-\delta,\delta[\cap\sigma(T)} t\,d\,E_T(t),$$

we deduce $\|TP_0\| \le \delta$. This yields that

$$\|Tx\| \le \delta.$$

A similar argument shows that $\|Ty - y\| = \|TP_1y - P_1y\| \le \delta$. Since $\|y\| = 1$, this gives us that

$$1 - \delta \le \|Ty\| \le 1 + \delta.$$

Now, to prove (2.6.6) we estimate $\langle T^2\varphi, \varphi\rangle = \|T\varphi\|^2$ from below and $\langle T\varphi, \varphi\rangle^2$ from above. Since $T\varphi = (Tx + Ty)/\sqrt{2}$, we have

$$\|T\varphi\| \ge \frac{-\|Tx\| + \|Ty\|}{\sqrt{2}} \ge \frac{-\delta + 1 - \delta}{\sqrt{2}}$$

and thus we get

$$\langle T^2\varphi, \varphi\rangle = \|T\varphi\|^2 \ge \frac{(1 - 2\delta)^2}{2}. \tag{2.6.7}$$

Using the equality $TP_0 = P_0T$ and the fact that P_0 and P_1 are mutually orthogonal projections, we have

$$\langle Tx, y\rangle = \langle TP_0x, P_1y\rangle = \langle P_0Tx, P_1y\rangle = \langle Tx, P_0P_1y\rangle = 0.$$

This also implies that $\langle Ty, x\rangle = 0$. Therefore, we infer

$$\langle T\varphi, \varphi\rangle = \frac{1}{2}\left(\langle Tx, x\rangle + \langle Ty, y\rangle\right).$$

Since $|\langle Tx, x\rangle| \le \|Tx\| \le \delta$ and $|\langle Ty, y\rangle| \le \|Ty\| \le 1 + \delta$, we obtain

$$\langle T\varphi, \varphi\rangle^2 \le \frac{(1 + 2\delta)^2}{4}.$$

This inequality together with (2.6.7) gives (2.6.6).

Now, for an arbitrary $\epsilon > 0$, choosing δ such that

$$\sqrt{\frac{(1 - 2\delta)^2}{2} - \frac{(1 + 2\delta)^2}{4}} \ge \frac{1}{2} - \epsilon,$$

it follows that we can pick a unit vector $\varphi \in H$ for which

$$\|T\|_v \ge \sqrt{\langle T^2\varphi, \varphi\rangle - \langle T\varphi, \varphi\rangle^2} \ge \frac{1}{2} - \epsilon.$$

This gives us that $\|T\|_v \ge \frac{1}{2} = \|\overline{T}\|$ which completes the proof of the lemma.
$\square$

Remark 2.6.6. As we have seen, the quantity $\|T\|_v = \|\overline{T}\|$ is exactly the half of the diameter of the spectrum of T. Therefore, if $T \geq 0$ and $0 \in \sigma(T)$, then $\|T\|_v = \|\overline{T}\| \leq \frac{1}{2}$ if and only if $0 \leq T \leq I$.

This observation will be used in the proof of our next lemma which determines the extreme points of the (closed) $\frac{1}{2}$-ball of the Banach space $B_s(H)/\mathbb{R}I$.

Lemma 2.6.7. *The extreme points of the ball $\{\overline{A} \in B_s(H)/\mathbb{R}I : \|\overline{A}\| \leq \frac{1}{2}\}$ are the classes of nontrivial projections, that is, the elements $\overline{P} \in B_s(H)/\mathbb{R}I$, where P is a nontrivial projection $(P \neq 0, I)$ on H.*

Proof. The point in the proof is to reduce the problem concerning classes of operators to a problem concerning single operators.

First we check that the classes of nontrivial projections are extreme points of the ball in question. Suppose that P is a nontrivial projection and

$$\overline{P} = \mu\overline{T} + (1 - \mu)\overline{S},$$

where $0 < \mu < 1$, $\|\overline{T}\| \leq \frac{1}{2}$, $\|\overline{S}\| \leq \frac{1}{2}$, $T, S \in B_s(H)$. Adding scalar operators if necessary, we can suppose that $T, S \geq 0$, $0 \in \sigma(T)$, $0 \in \sigma(S)$. Clearly,

$$P = \mu T + (1 - \mu)S + \lambda I$$

holds for some $\lambda \in \mathbb{R}$.

We claim that $\lambda = 0$. If $\varphi \in H$ is a unit vector in the kernel of P, we infer that

$$0 = \langle P\varphi, \varphi \rangle = \mu\langle T\varphi, \varphi \rangle + (1 - \mu)\langle S\varphi, \varphi \rangle + \lambda.$$

Since $\langle T\varphi, \varphi \rangle \geq 0$ and $\langle S\varphi, \varphi \rangle \geq 0$, the above equality yields $\lambda \leq 0$.

It follows from $\sigma(P) = \{0, 1\}$ that $\|\overline{P}\| = \frac{1}{2}$. We compute

$$\frac{1}{2} = \|\overline{P}\| = \|\mu\overline{T} + (1 - \mu)\overline{S}\| \leq \mu\|\overline{T}\| + (1 - \mu)\|\overline{S}\| \leq (\mu + 1 - \mu)\frac{1}{2} = \frac{1}{2},$$

from which we deduce that $\|\overline{T}\| = \|\overline{S}\| = \frac{1}{2}$. Using Remark 2.6.6 we get $0 \leq T, S \leq I$. So, if φ is a unit vector in the range of P, then we have

$$1 = \langle P\varphi, \varphi \rangle = \mu\langle T\varphi, \varphi \rangle + (1 - \mu)\langle S\varphi, \varphi \rangle + \lambda \leq \mu + (1 - \mu) + \lambda,$$

which gives us that $\lambda \geq 0$. Therefore, it follows that $\lambda = 0$ as we have claimed.

Consequently, we have $P = \mu T + (1 - \mu)S$. This means that P is a nontrivial convex combination of two elements of the operator interval $[0, I]$. However, it is well-known that the extreme points of this operator interval are exactly the projections. Hence, we get $P = T = S$. This proves that the classes of nontrivial projections are really extreme points.

It remains to prove that these classes are the only extreme points. In order to see this, let B be a self-adjoint operator with $\|\overline{B}\| = \frac{1}{2}$ and suppose that the equivalence class of B does not coincide with the equivalence class of any nontrivial projection. We show that $\overline{B}$ is not an extreme point of the ball in question. Clearly, just as above, we can assume that $B \geq 0$ and $0 \in \sigma(B)$. Then we have $0 \leq B \leq I$. As $\|\overline{B}\| = \frac{1}{2}$, it also follows that $1 \in \sigma(B)$. We are going to show that there exist two operators B_1, B_2 in the operator interval $[0, I]$ such that $B = (B_1 + B_2)/2$ and $B \neq B_1, B_2$. In the present situation this will imply that $\overline{B} \neq \overline{B_1}, \overline{B_2}$. Then, as $\overline{B} = (\overline{B_1} + \overline{B_2})/2$, $\|\overline{B_1}\|, \|\overline{B_2}\| \leq \frac{1}{2}$ (see Lemma 2.6.5), we can infer that $\overline{B}$ is not an extreme point. So, in order to construct such operators B_1, B_2, choose $\lambda_0 \in \sigma(B) \cap]0, 1[$. (The existence of such a λ_0 follows from the facts that the equivalence class of B does not coincide with the equivalence class of any nontrivial projection and that $\|\overline{B}\| \neq 0$.) Now, one can easily find continuous real valued functions $f_1, f_2 : [0, 1] \to [0, 1]$ such that $(f_1 + f_2)/2$ is the identity on $[0, 1]$ and $f_1(\lambda_0) \neq \lambda_0 \neq f_2(\lambda_0)$. Defining $B_1 = f_1(B), B_2 = f_2(B)$, it follows from the properties of the continuous function calculus that we obtain operators with the desired properties. This completes the proof of the lemma. $\qquad\square$

In what follows, we intend to characterize the unitary equivalence of nontrivial projections P, Q by means of some correspondence between the classes $\overline{P}$ and $\overline{Q}$ that can be expressed in terms of the metric induced by the factor norm. The first step in this direction is made in the following lemma.

Lemma 2.6.8. *Let P and Q be projections on H. Suppose that P is nontrivial and $\|\overline{P} - \overline{Q}\| < \frac{1}{2}$. Then P is unitarily equivalent to Q.*

Proof. First observe that $Q \neq 0, I$. In fact, in the opposite case we would have $\|\overline{P}\| < 1/2$. But this means that the diameter of $\sigma(P)$ is less than 1, which gives us that P is a trivial projection, a contradiction.

Because of the definition of the factor norm there exists a $\mu \in \mathbb{R}$ such that $\|P - (Q + \mu I)\| < \frac{1}{2}$. Let R be a projection of rank at most 2 whose range contains a unit vector from the range of P and a unit vector from the range of Q, respectively. The operators RPR and $R(Q + \mu I)R$ are of finite rank, 1 is the largest eigenvalue of RPR and $1 + \mu$ is the largest eigenvalue of $R(Q + \mu I)R$. To prove this latter statement, observe that

$$R(Q + \mu I)R \leq R(I + \mu I)R = (1 + \mu)R \leq (1 + \mu)I.$$

This shows that the spectrum of $R(Q + \mu I)R$ is a subset of the interval $] - \infty, 1 + \mu]$. On the other hand, $1 + \mu$ is an eigenvalue of the operator $R(Q + \mu I)R$ because the range of R contains a unit vector from the range of Q.

Since the largest eigenvalue of RPR is 1 and the largest eigenvalue of $R(Q + \mu I)R$ is $1 + \mu$, by Weyl's perturbation theorem (see, for example, [19, Corollary III.2.6]) we have $|1 - (1 + \mu)| \leq \|RPR - R(Q + \mu I)R\|$. We deduce

$$|\mu| = |1 - (1 + \mu)| \leq \|RPR - R(Q + \mu I)R\|$$

$$\leq \|R\| \|P - (Q + \mu I)\| \|R\| < \frac{1}{2},$$

and so we have

$$\|P - Q\| \le \|P - (Q + \mu I)\| + |\mu| < \frac{1}{2} + \frac{1}{2} = 1.$$

But it is a well-known result that if the distance between two projections in the operator norm is less than 1, then they are unitarily equivalent. This completes the proof of the lemma. □

A useful solution of the problem mentioned before Lemma 2.6.8 is given in the next result.

Lemma 2.6.9. *Let P and Q be projections on H and suppose that P is nontrivial. Then P is unitarily equivalent to Q if and only if there exists a continuous function $\gamma : [0, 1] \to \overline{P(H)}$ such that $\gamma(0) = \overline{P}$ and $\gamma(1) = \overline{Q}$.*

(Here $\overline{P(H)}$ denotes the set of classes in $B_s(H)/\mathbb{R}I$ which correspond to projections.)

Proof. The necessity is easy to see. Indeed, this follows from the well-known fact that if P, Q are equivalent projections then they can be connected by a continuous curve (continuity is meant in the operator norm topology) in the set of projections and from the fact that the operator norm majorizes the factor norm.

Now, conversely, suppose that there exists a continuous mapping $\gamma : [0, 1] \to \overline{P(H)}$ such that $\gamma(0) = \overline{P}$ and $\gamma(1) = \overline{Q}$. As γ is defined on a compact set, it is uniformly continuous. Hence, we can choose a positive δ such that

$$\|\gamma(t) - \gamma(s)\| < \frac{1}{2} \qquad \text{if} \quad |s - t| < \delta, \ s, t \in [0, 1].$$

It follows that there exist projections $P_1, \ldots, P_n$ with the property that

$$\|\overline{P} - \overline{P_1}\| < \frac{1}{2} \quad , \ldots, \quad \|\overline{P_n} - \overline{Q}\| < \frac{1}{2}.$$

By Lemma 2.6.8, we obtain that P and P_1 are unitarily equivalent (and, consequently, P_1 is nontrivial). Using this argument again and again we can conclude that P is unitarily equivalent to Q. □

The meaning of our last lemma which follows is a metric characterization of the equality of nontrivial projections in $B_s(H)$ with respect to the semi-norm $\|.\|_v$. Denote by $F_s(H)$ the set of all finite rank elements in $B_s(H)$.

Lemma 2.6.10. *Let P and Q be nontrivial projections on H such that*

$$\|P + A\|_v = \|Q + A\|_v$$

holds for all $A \in F_s(H)$. Then we have $P = Q$.

Proof. Let R be a rank-one subprojection of the projection P. Then the diameter of the spectrum of $P + R$ is 2, so by Lemma 2.6.5 we have

$$1 = \|P + R\|_v = \|Q + R\|_v,$$

that is, the diameter of $\sigma(Q + R)$ is also equal to 2. Since $0 \leq Q + R \leq 2I$, thus $\sigma(Q + R)$ is a subset of the closed interval $[0, 2]$. Therefore, we have $0, 2 \in \sigma(Q + R)$.

It is well-known that the spectrum of any normal operator coincides with its approximate point spectrum. Consequently, we can find unit vectors x_n in H $(n \in \mathbb{N})$ such that

$$\|Qx_n + Rx_n - 2x_n\| \to 0 \quad \text{as} \quad n \to \infty.$$

This yields that

$$\|Qx_n + Rx_n\| \to 2. \tag{2.6.8}$$

Denote $u_n = Qx_n$ and $v_n = Rx_n$. We have $\|u_n\| \leq 1, \|v_n\| \leq 1$. Since v_n is in the range of R which is 1-dimensional, there must exist a convergent subsequence of (v_n). Without any loss of generality we can assume that this subsequence is (v_n) itself. So, there exists a vector v in the range of R such that $\|v_n - v\| \to 0$. Since

$$\big| \|u_n + v\| - \|u_n + v_n\| \big| \leq \|v - v_n\| \to 0$$

and $\|u_n + v_n\| \to 2$, we have $\|u_n + v\| \to 2$. On the other hand, by the parallelogram identity we obtain

$$\|u_n - v\|^2 = 2\|u_n\|^2 + 2\|v\|^2 - \|u_n + v\|^2.$$

Therefore, we have

$$\limsup_{n \to \infty} \|u_n - v\|^2 \leq 2 + 2 - 4 = 0,$$

which implies that $\|u_n - v\| \to 0$. So, both $(u_n), (v_n)$ converge to v. Taking (2.6.8) into account, it is clear that $v \neq 0$.

Since the sequence (u_n) is in the range of Q which is a closed subspace, it follows that its limit v also belongs to this range. But v generates the range of R and hence R is a subprojection of Q. So, we have proved the following: every rank-one subprojection of P is a subprojection of Q. Therefore, P is a subprojection of Q. Changing the role of P and Q, we get that Q is also a subprojection of P and hence we obtain $P = Q$. □

Now, we are in a position to prove the main result of the section.

Proof of Theorem 2.6.2. The brief summary of the proof is as follows. Our transformation ϕ which preserves the maximal deviation induces a surjective linear isometry Φ on the factor space $B_s(H)/\mathbb{R}I$. This Φ necessarily preserves the extreme points of the $\frac{1}{2}$-ball which points are well characterized in Lemma 2.6.7. This implies a certain preserving property of the original transformation ϕ. Namely, we obtain that ϕ preserves the operators of the form 'nontrivial projection + scalar$\cdot I$'. This will imply that ϕ preserves the commutativity on $F_s(H) + \mathbb{R}I$. Extending ϕ from this set to its complex linear span $F(H) + \mathbb{C}I$ ($F(H)$ stands for the set of all finite rank bounded linear operators on H), we obtain a complex-linear transformation which preserves normal operators. Applying the technique of the proof of a nice result of Brešar and Šemrl on normal-preserving linear transformations of $B(H)$ given in [26], we can conclude the proof in the case when dim $H \geq 3$. If dim $H = 2$, then rather surprisingly we can reduce our problem quite easily to Wigner's classical unitary-antiunitary theorem. So, this is the plan what we now carry out.

Define a map $\Phi : B_s(H)/\mathbb{R}I \rightarrow B_s(H)/\mathbb{R}I$ in the following way

$$\Phi(\overline{A}) = \overline{\phi(A)} \qquad (A \in B_s(H)).$$

The transformation ϕ is a linear bijection of $B_s(H)$ which preserves the maximal deviation. By Lemma 2.6.5, we easily obtain that ϕ preserves the scalar operators and then that Φ is a well-defined linear bijection on $B_s(H)/\mathbb{R}I$ which preserves the factor norm. It follows that Φ preserves all closed balls around $\overline{0}$ as well as their extreme points. Therefore, by Lemma 2.6.7, we deduce that ϕ preserves the set of all operators of the form $P + \lambda I$, where P is a nontrivial projection and $\lambda \in \mathbb{R}$.

We shall show that ϕ preserves the commutativity on $F_s(H) + \mathbb{R}I$. Let P' and Q' be mutually orthogonal projections. We known that there exist projections P, Q, R and real numbers $\lambda_1, \lambda_2, \lambda_3$ such that

$$\phi(P') = P + \lambda_1 I$$
$$\phi(Q') = Q + \lambda_2 I$$
$$\phi(P' + Q') = R + \lambda_3 I.$$

By the linearity of ϕ this implies that $P + Q = R + tI$ for some real number t (in fact, $t = \lambda_3 - \lambda_1 - \lambda_2$). We assert that P and Q are either commuting or the projections P, Q, R are unitarily equivalent to each other.

In order to prove this, we distinguish the following cases.

CASE I. Suppose that R is scalar. Then $P + Q$ is also scalar which implies that P, Q commute.

CASE II. Suppose that R is not scalar, that is, R is a nontrivial projection. Consider the orthogonal decomposition of H induced by the range and the kernel of R. Every operator has a matrix representation with respect to this decomposition. As for $P + Q$, we can write

$$P + Q = R + tI = \begin{bmatrix} (1+t)I & 0 \\ 0 & tI \end{bmatrix}. \tag{2.6.9}$$

The inequality $0 \leq P + Q \leq 2I$ implies that $0 \leq t \leq 1$. According to the possible values of t we have the following sub-cases.

CASE II/1. Suppose that $t = 0$. Then $P + Q = R$ is a projection and hence $(P + Q)^2 = P + Q$. From this equality we easily deduce $PQ = QP = 0$ which implies that P, Q commute.

CASE II/2. Suppose that $t = 1$. Then $P + Q = R + I$, which implies that $R + (I - Q) = P$ is a projection. Just as above, we obtain that $R, I - Q$ are commuting projections. This implies that R, Q commute and, finally, it follows from the equality $R + (I - Q) = P$ that P, Q also commute.

CASE II/3. Suppose that $0 < t < 1$. In this case we use the result that any two projections in generic position (i.e., with no common eigenvectors) are unitarily equivalent (see [64, 98]). As the spectrum of $P + Q = R + tI$ is contained in $\{t, 1 + t\}$, the numbers 0,1,2 are not in the spectrum of $P + Q$. This implies that P, Q are in a generic position and hence they are unitarily equivalent. Similarly, as the spectrum of $R - P$ is contained in $\{-t, 1 - t\}$ which does not contain -1,0,1, we infer that P, R are in a generic position and hence they are unitarily equivalent. It follows that the projections P, Q, R are pairwise unitarily equivalent. What does this mean for our original projections P', Q'? Obviously, in the present case P, Q, R are nontrivial. Using Lemma 2.6.9 and the isometric property of Φ with respect to the factor norm, we obtain that the projections $P', Q', P' + Q'$ are pairwise unitarily equivalent. But if P', Q' are nonzero mutually orthogonal finite rank projections, then this can not happen.

Therefore, we have proved that for any finite rank projections P', Q' with $P'Q' = Q'P' = 0$ it follows that $\phi(P')\phi(Q') = \phi(Q')\phi(P')$. If we pick operators $A, B \in F_s(H)$ which commute, then they can be diagonalized simultaneously. Using the just proved property of ϕ one can easily deduce that $\phi(A), \phi(B)$ also commute.

We show that

$$\phi(F_s(H) + \mathbb{R}I) = F_s(H) + \mathbb{R}I.$$

If $\dim H < \infty$, this is obvious. So, let H be infinite dimensional. Pick a nonzero finite rank projection P'. Then $\phi(P') = P + \lambda I$ holds for some nontrivial projection P and real number λ. If P is of finite rank or of finite corank, then we obtain $\phi(P') \in F_s(H) + \mathbb{R}I$. So, let us see what happens if P is of infinite rank and infinite corank.

First suppose that $\dim \operatorname{rng} P \leq \dim \operatorname{rng} P^{\perp}$. Then we can find nontrivial projections P_1 and P_2 such that $P = P_1 + P_2$ and P, P_1, P_2 are mutually unitarily equivalent. Now, referring to Lemma 2.6.9, there are nontrivial projections P_1', P_2' such that

$$P' + \mu I = P_1' + P_2'$$

holds for some $\mu \in \mathbb{R}$ and the projections P', P_1', P_2' are mutually unitarily equivalent. So, the projections P_1', P_2' are of finite rank and we see that on the right hand side of the equality above there is a finite rank operator. This gives us that μ must be zero and then we have $P' = P_1' + P_2'$. Like in the argument

given in CASE II/1., we obtain that P'_1, P'_2 are mutually orthogonal projections. We now conclude that, because of unitary equivalence and orthogonality,
the equality $P' = P'_1 + P'_2$ is untenable which is a contradiction.

Next suppose that $\dim \operatorname{rng} P \geq \dim \operatorname{rng} P^\perp$. Then we can apply the ar-
gument above for $P^\perp$ to find nontrivial projections P_1 and P_2 such that $P^\perp = P_1 + P_2$ and $P^\perp, P_1, P_2$ are mutually unitarily equivalent. This implies that
there are nontrivial projections P'_1, P'_2 such that

$$P'^\perp + \nu I = P'_1 + P'_2 \tag{2.6.10}$$

holds for some $\nu \in \mathbb{R}$ and the projections $P'^\perp, P'_1, P'_2$ are mutually unitarily
equivalent. (Observe that, as $\Phi(\overline{P'}) = \overline{P}$, we have $\Phi(\overline{P'^\perp}) = \overline{P^\perp}$.) It follows
that the projections P'_1, P'_2 are of finite corank and hence their ranges have
nontrivial intersection. Therefore, we obtain that 2 belongs to the spectrum
of the operator $P'_1 + P'_2$, and by (2.6.10) this implies that $\nu = 1$. Now, the
equation (2.6.10) can be rewritten in the form

$$P' = (I - P'_1) + (I - P'_2) = P'^\perp_1 + P'^\perp_2,$$

where the nontrivial projections $P', P'^\perp_1, P'^\perp_2$ are pairwise unitarily equivalent.
Just as in the previous paragraph we arrive at a contradiction.

Therefore, we have $\phi(P') \in F_s(H) + \mathbb{R}I$ for every finite rank projection P'.
Applying the spectral theorem for self-adjoint finite rank operators, it follows
that $\phi(F_s(H) + \mathbb{R}I) \subset F_s(H) + \mathbb{R}I$. As ϕ^{-1} has the same properties as ϕ,
considering the above relation for ϕ^{-1} in the place of ϕ, we conclude that

$$\phi(F_s(H) + \mathbb{R}I) = F_s(H) + \mathbb{R}I.$$

To sum up what we have already proved, it has turned out that ϕ when
restricted onto $F_s(H) + \mathbb{R}I$ is a bijective linear map which preserves commutativity. Consider the complex unital algebra $F(H) + \mathbb{C}I$. As the real and
imaginary parts of an operator in $F(H) + \mathbb{C}I$ belong to $F_s(H) + \mathbb{R}I$, one can
readily verify that the map $\tilde{\phi} : F(H) + \mathbb{C}I \rightarrow F(H) + \mathbb{C}I$ defined by

$$\tilde{\phi}(A + iB) = \phi(A) + i\phi(B) \qquad (A, B \in F_s(H) + \mathbb{R}I)$$

is a bijective complex-linear transformation. It is an elementary fact that a
bounded linear operator is normal if and only if its real and imaginary parts
are commuting. As ϕ preserves commutativity between self-adjoint finite rank
operators, it follows that $\tilde{\phi}$ preserves normality. If $\dim H \geq 3$, then this latter
preserving property is strong enough to imply that $\tilde{\phi}$ is of a certain particular
form. In fact, there is a nice result of Brešar and Šemrl [26, Theorem 2] which,
in the case when $\dim H \geq 3$, characterizes the bijective linear mappings on
$B(H)$ that preserve the normal operators. Although the algebra on which our
transformation $\tilde{\phi}$ is defined differs from $B(H)$ in general, it is not hard to see
that the technique used in [26] can be applied to our present situation as well.
This gives us the following two possibilities for the form of $\tilde{\phi}$:

(i) there exist a unitary operator U on H, a linear functional $f : F(H) + \mathbb{C}I \to \mathbb{C}$ and a scalar $c \in \mathbb{C}$ such that

$$\tilde{\phi}(T) = cUTU^* + f(T)I \qquad (T \in F(H) + \mathbb{C}I)$$

(ii) there exist an antiunitary operator U on H, a linear functional $f : F(H) + \mathbb{C}I \to \mathbb{C}$ and a scalar $c \in \mathbb{C}$ such that

$$\tilde{\phi}(T) = cUT^*U^* + f(T)I \qquad (T \in F(H) + \mathbb{C}I).$$

Concerning ϕ, this means that there is an either unitary or antiunitary operator U on H, a real-linear function $f : F_s(H) + \mathbb{R}I \to \mathbb{C}$, and a constant $c \in \mathbb{C}$ such that

$$\phi(A) = cUAU^* + f(A)I \qquad (A \in F_s(H) + \mathbb{R}I).$$

As $\phi(A)$ is self-adjoint, we have

$$\bar{c}UAU^* + \overline{f(A)}I = cUAU^* + f(A)I \qquad (2.6.11)$$

for every $A \in F_s(H) + \mathbb{R}I$. If A is not a scalar operator, then it follows from this equality that $\bar{c} = c$. Next, we obtain from (2.6.11) that f is real valued. As ϕ preserves maximal deviation, we obtain that $|c| = 1$. Therefore, $c = \pm 1$ and we have the desired form for our transformation ϕ on $F_s(H) + \mathbb{R}I$. It remains to show that the same formula holds also on the whole space $B_s(H)$.

In order to see this, observe that composing ϕ by the transformation $A \mapsto cU^*AU$, we can assume without loss of generality that

$$\phi(A) = A + l(A)I$$

holds for every $A \in F_s(H) + \mathbb{R}I$, where $l : F_s(H) + \mathbb{R}I \to \mathbb{R}$ is a linear functional. Let P be a nontrivial projection on H. We know that $\phi(P) = Q + \mu I$ for some nontrivial projection Q and real number μ. Pick an arbitrary $A \in F_s(H)$. Since $\phi(A)$ is a scalar perturbation of A, we have

$$\|Q + A\|_v = \|\phi(P) + A\|_v = \|\phi(P) + \phi(A)\|_v = \|\phi(P + A)\|_v = \|P + A\|_v.$$

Since this holds true for every self-adjoint finite rank operator A, it follows from Lemma 2.6.10 that $Q = P$. This gives us that $\phi(P) - P \in \mathbb{R}I$ which holds also in the case when P is trivial. So, we have $\Phi(\overline{P}) = \overline{P}$ for every projection P. Since the linear transformations $A \mapsto \Phi(\overline{A})$ and $A \mapsto \overline{A}$ are continuous (on $B_s(H)$ we consider the operator norm while $B_s(H)/\mathbb{R}I$ is equipped with the factor norm), they are equal on the projections, it follows from the spectral theorem of self-adjoint operators and from the properties of the spectral integral that we have $\Phi(\overline{A}) = \overline{A}$ for every $A \in B_s(H)$. This gives us that

$$\phi(A) - A \in \mathbb{R}I \qquad (A \in B_s(H))$$

which obviously implies that there is a linear functional $h : B_s(H) \to \mathbb{R}$ such that

$$\phi(A) = A + h(A)I \qquad (A \in B_s(H)).$$

This completes the proof in the case when $\dim H \geq 3$.

As the statement of the theorem is trivial for $\dim H = 1$, it remains to consider the case when $\dim H = 2$. In this case the nontrivial projections are exactly the rank-one projections. Pick a rank-one projection P. We know that there is a rank-one projection P' such that $\phi(P)$ is equal to the sum of P' and a scalar operator. It is easy to see that this P' is unique. (In fact, one can prove independently from the dimension of H that in the class of every nontrivial projection there is only one projection.) Therefore, we can denote $P' = \psi(P)$ and obtain a bijective transformation ψ on the set of all rank-one projections. We assert that ψ has the property that

$$\operatorname{tr} PQ = \operatorname{tr} \psi(P)\psi(Q) \qquad (2.6.12)$$

holds for arbitrary rank-one projections P, Q on H. As ϕ preserves the maximal deviation, this will clearly follow from the equality

$$\|P - Q\|_v = \sqrt{1 - \operatorname{tr} PQ} \qquad (2.6.13)$$

that we are going to prove now. In fact, observe that the maximal deviation and the trace functional are invariant under the transformations $A \mapsto VAV^*$, where V is any unitary operator. Therefore, we can assume that

$$P = \begin{bmatrix} 1 & 0 \\ 0 & 0 \end{bmatrix}$$

while Q is an arbitrary self-adjoint idempotent 2 by 2 matrix. It is easy to check that Q is of the form

$$Q = \begin{bmatrix} a & \sqrt{a(1-a)}e^{i\theta} \\ \sqrt{a(1-a)}e^{-i\theta} & 1 - a \end{bmatrix}$$

where a, θ are real numbers and $0 \leq a \leq 1$. We have that the eigenvalues of $P - Q$ are $\pm\sqrt{1-a}$ and hence obtain that $\|P - Q\|_v = \sqrt{1-a}$. On the other hand, it is trivial to check that $\operatorname{tr} PQ = a$. This results in the desired equality (2.6.13).

So, we have a bijective transformation ψ on the set of all rank-one projections which satisfies (2.6.12). By Wigner's theorem we obtain that there exists an either unitary or antiunitary operator U on H such that

$$\psi(P) = UPU^*$$

holds for every rank-one projection P. As $\phi(P)$ differs from $\psi(P)$ only by a scalar operator, we obtain that $\phi(P) - UPU^* \in \mathbb{R}I$. By linearity this gives us that $\phi(A) - UAU^*$ is a scalar operator for every $A \in B_s(H)$. Now, one can easily complete the proof in the case when $\dim H = 2$. □

As it is seen, preserving commutativity has played important role in our proof above. In fact, as we have mentioned in the Introduction (see Section 0.1), preserver problems of this kind are among the most fundamental ones in the theory of LPPs.

Now we turn to the proofs of the non-linear results of the section.

Proof of Theorem 2.6.3. This follows immediately from Theorem 2.6.1 using the following important result of Mazur and Ulam [157]. If $\mathcal{V}$ is a real normed vector space and $T : \mathcal{V} \to \mathcal{V}$ is a bijective map which preserves the distance on $\mathcal{V}$ (i.e., T satisfies $\|T(x) - T(y)\| = \|x - y\|$ $(x, y \in \mathcal{V})$), then T can be written in the form $T(x) = L(x) + x_0$ $(x \in \mathcal{V})$, where $L : \mathcal{V} \to \mathcal{V}$ is a bijective linear isometry and $x_0 \in \mathcal{V}$ is a fixed vector. $\qquad\square$

As for the proof of Theorem 2.6.4, we have to work more than in the previous proof as $\|.\|_v$ is only a semi-norm.

Proof of Theorem 2.6.4. Considering the map $A \mapsto \phi(A) - \phi(0)$, it is obvious that we can assume that ϕ sends 0 to 0. In what follows we use this assumption.

Consider the linear functional $\lambda I \mapsto \lambda$ on $\mathbb{R}I$. Extend it to a linear functional l of the whole vector space $B_s(H)$. (We do not need any kind of continuity of l, so no need to use Hahn-Banach theorem.) Define the transformation $\phi_1 : B_s(H) \to B_s(H)$ in the following way

$$\phi_1(A) = \phi(A) - l(\phi(A))I + l(A)I \qquad (A \in B_s(H)).$$

We assert that $\phi_1 : B_s(H) \to B_s(H)$ is a bijective linear map, it preserves the distance (with respect to the semimetric d_v) and for every $A \in B_s(H)$, $\phi(A)$ and $\phi_1(A)$ differs only in a scalar operator. If this is really the case, then we can apply Theorem 2.6.2 for ϕ_1 and we are done. So, it remains to prove that ϕ_1 has the mentioned properties. As the last two ones are obvious from the definition, we have to prove only that ϕ_1 is linear and bijective. We begin with the linearity. As ϕ preserves the distance with respect to d_v and we have supposed that $\phi(0) = 0$, it follows that ϕ preserves the scalar operators (in fact, scalar operators can be characterized by the equality $\|A\|_v = 0$; see Lemma 2.6.5). Next, it is easy to show that the formula

$$\Phi(\overline{A}) = \overline{\phi(A)} \qquad (A \in B_s(H))$$

defines a bijective isometry (distance preserving map) on $B_s(H)/\mathbb{R}I$ with respect to the factor norm. We only prove the isometric property. Indeed,

$$\|\Phi(\overline{A}) - \Phi(\overline{B})\| = \|\phi(A) - \phi(B)\|_v = \|A - B\|_v = \|\overline{A} - \overline{B}\|$$

holds for every $A, B \in B_s(H)$. Since $\Phi(\overline{0}) = \overline{\phi(0)} = \overline{0}$, by Mazur-Ulam theorem we obtain that Φ is linear. Thus, for any $A, B \in B_s(H)$ we have

$$\Phi(\overline{A} + \overline{B}) = \Phi(\overline{A}) + \Phi(\overline{B}),$$

that is,
$$\overline{\phi(A + B)} = \overline{\phi(A) + \phi(B)}.$$

This gives us that $\phi(A + B) - (\phi(A) + \phi(B))$ is a scalar operator, say

$$\phi(A + B) - (\phi(A) + \phi(B)) = \lambda I.$$

We compute

$$\phi(A + B) - (\phi(A) + \phi(B)) = \lambda I =$$
$$l(\lambda I)I = l(\phi(A + B) - (\phi(A) + \phi(B)))I.$$

This implies that

$$\phi(A + B) - l(\phi(A + B))I = \phi(A) - l(\phi(A))I + \phi(B) - l(\phi(B))I.$$

Adding $l(A + B)I = l(A)I + l(B)I$ to this equality, we obtain the additivity of ϕ_1. The homogeneity can be proved in a similar way.

We next show that ϕ_1 is injective. Suppose that

$$0 = \phi_1(A) = \phi(A) - l(\phi(A))I + l(A)I$$

holds for some $A \in B_s(H)$. Then $\phi(A)$ is a scalar operator, say $\phi(A) = \lambda I$, and this implies that A is also scalar, say $A = \mu I$. It follows from the above equation that

$$0 = \lambda I - l(\lambda I)I + l(\mu I)I = (\lambda - \lambda + \mu)I$$

which yields $\mu = 0$, i.e., we have $A = 0$. This proves the injectivity of ϕ_1.

Finally, we prove that ϕ_1 is surjective. To show this, first observe that, by the definition of ϕ_1 and the surjectivity of ϕ, the range of ϕ_1 and $\mathbb{R}I$ generate the whole space $B_s(H)$. So, if ϕ_1 is not surjective, then we have $\mathrm{rng}\,\phi_1 \cap \mathbb{R}I = \{0\}$. This means that the only scalar operator in the range of ϕ_1 is 0. Now, as $\phi(I)$ is a scalar operator, it follows that $\phi_1(I)$ is also scalar. As $\phi_1(I) \in \mathrm{rng}\,\phi_1$, we obtain that $\phi_1(I) = 0$, which, by the injectivity of ϕ_1 implies that $I = 0$, a contradiction. Therefore, ϕ_1 must be surjective. So, we have proved all the asserted properties of ϕ_1 and hence the proof of the theorem is complete. □

2.6.4 Remarks

To conclude the section we give another interpretation of the main result Theorem 2.6.2 that has been already mentioned in the subsection 1.3.4. Remarks. Namely, in view of Lemma 2.6.5, our theorem describes the form of all bijective linear transformations of $B_s(H)$ which preserve the diameter of the spectrum. We repeat our opinion that, due to the much interest in linear preserver problems which concern the spectrum, the similar problem for the whole operator algebra $B(H)$ seems to be a prosperous and rather deep problem which deserves attention.

Furthermore, we note that in the very beginning of the proof of Theorem 2.6.2 we observed that the transformation ϕ under consideration preserves the operators of the form 'projection + scalar'. Hopefully, it has already turned out from this work that the preservation of projections (in one direction) is a rather important problem. Therefore, we feel that it would be useful and interesting to determine all the linear transformations which preserve the operators 'projection + scalar' in one direction. It seems that this is a more difficult problem than the preservation of projections. For a result on the similar problem preserving the Hermitian matrices 'rank-one + scalar' we refer to the paper [247].

2.7 Some Preservers on Hilbert Space Effects

2.7.1 Summary

In this section we study a class of transformations on the set $E(H)$ of all Hilbert space effects. This consists of the bijective maps which preserve the order and zero product in both directions. The main result of the section gives a complete description of the structure of those transformations. As applications we obtain additional new results and some former ones as easy corollaries. In particular, we obtain the general form of the ortho-order automorphisms as well as that of the sequential automorphisms of $E(H)$. Furthermore, we show that the automorphisms of these two kinds belong to our class of transformations even when their domain is the set of all effects in a general von Neumann algebra. The results of the present section appeared in the paper [185].

2.7.2 Formulation of the Results

As we already mentioned in the Introduction (see Section 0.3), quantum effects play important role in certain parts of quantum mechanics, for example in the quantum theory of measurement (see [38]). In the Hilbert space formalism, quantum effects are represented by positive bounded linear operators on a Hilbert space H which are bounded by the identity I. The set of all such operators on H which are sometimes called Hilbert space effects is denoted by $E(H)$.

There are several operations and relations defined on $E(H)$ which are important from different aspects of the theory. Here we are interested in the ortho-order structure and in the sequential structure on $E(H)$. The first one is obtained as follows. The usual ordering $\leq$ among self-adjoint bounded linear operators gives rise to a partial order on $E(H)$ and the operation $\perp\colon A \mapsto A^\perp = I - A$ defines a kind of orthocomplementation on $E(H)$. This relation and operation together give the ortho-order structure on $E(H)$ [146]. As for the second structure, it comes from the sequential product which is defined as follows. If $A, B \in E(H)$, then their sequential product is $A \circ B = A^{1/2} B A^{1/2}$ [81].

Supposing $\dim H \geq 3$, the automorphisms of $E(H)$ with respect to the ortho-order structure (called ortho-order automorphisms) as well as the ones with respect to the sequential product (called sequential automorphisms) are known to be implemented by unitary or antiunitary operators. This means that any such automorphism ϕ is of the form

$$\phi(A) = UAU^* \qquad (A \in E(H)) \qquad (2.7.1)$$

where U is a unitary or antiunitary operator. The result concerning ortho-order automorphisms was obtained by Ludwig in [146, Section V.5] (the proof was later clarified in [46]) and the corresponding result on sequential automorphisms was given by Gudder and Greechie in [79]. In the paper [190] we showed that Ludwig's theorem holds also in the 2-dimensional case.

In our paper [175] we presented some characterizations of the ortho-order automorphisms of $E(H)$ by means of their preserving properties. This investigation was motivated by the extensive study of preserver problems in matrix theory and in operator theory. In the present section we continue the investigation started in [175]. We consider the bijective maps $\phi : E(H) \to E(H)$ which preserve the order and zero product in both directions, i.e., which satisfy

$$A \leq B \Longleftrightarrow \phi(A) \leq \phi(B) \qquad (A, B \in E(H)) \qquad (a)$$

and

$$AB = 0 \Longleftrightarrow \phi(A)\phi(B) = 0 \qquad (A, B \in E(H)). \qquad (b)$$

In what follows we present a complete description of the structure of those maps and give several applications of the corresponding result. Among others, we obtain the form of all bijective transformations of $E(H)$ which

(i) preserve the order in both directions and
(ii) map one single nontrivial scalar operator λI ($\lambda \neq 0, 1$) to an operator of the same kind.

At the first glance, it might be rather surprising that these maps are of a nice form but it turns out that they necessarily satisfy (a) and (b). Next, we easily recover one of the main results in [175] on maps preserving the order and coexistency in both directions (the definition of coexistency is given below). What is probably more important, we also show that the main result of the present section readily implies former results on the structure of ortho-order automorphisms and sequential automorphisms of $E(H)$ that were given in [146] and in [79], respectively. In fact, we prove that the automorphisms in question satisfy (a) and (b). It should be emphasized that all statements presented in this section are valid also when the underlying Hilbert space is 2-dimensional. In particular, this holds for the result on the form of sequential automorphisms of $E(H)$ which extends the above mentioned theorem of Gudder and Greechie in [79].

Finally, in the subsection 2.7.4. Remarks we demonstrate that even on the set $E(\mathcal{A})$ of all effects belonging to a general von Neumann algebra $\mathcal{A}$, the ortho-order automorphisms and the sequential automorphisms (the definitions should be self-explanatory) belong to our new class of preservers, i.e., they preserve the order and zero product in both directions. This observation is worth mentioning since in that generality it is quite hard to see any connection between those two kinds of automorphisms.

Now we turn to the precise formulation of the results. *Throughout this section we assume that H is a (complex) Hilbert space with* $\dim H \geq 2$. Our main result reads as follows.

Theorem 2.7.1. *Let $\phi : E(H) \to E(H)$ be a bijective map which preserves the order and zero product in both directions. Then there is an either unitary or antiunitary operator U on H and a real number $p < 1$ such that with the function $f_p(x) = \frac{x}{xp+(1-p)}$ $(x \in [0,1])$ we have*

$$\phi(A) = U f_p(A) U^* \qquad (A \in E(H)).$$

Here $f_p(A)$ denotes the image of the function f_p under the continuous function calculus belonging to the self-adjoint operator A.

To get the form of all bijective maps on $E(H)$ having the properties (i), (ii), we need the following proposition which might be interesting on its own right.

Proposition 2.7.2. *Let $\phi : E(H) \to E(H)$ be a bijective map which preserves the order in both directions and suppose that there is a single pair of scalars $\lambda, \mu \in]0, 1[$ such that $\phi(\lambda I) = \mu I$. Then ϕ preserves zero product in both directions.*

The above results have the following immediate consequences. To the second statement in the corollary below observe that the scalar operators in $E(H)$ can be characterized as those effects which commute with every other effect.

Corollary 2.7.3. *Let $\phi : E(H) \to E(H)$ be a bijective map which preserves the order in both directions and suppose that there is a single pair of scalars $\lambda, \mu \in]0, 1[$ such that $\phi(\lambda I) = \mu I$. Then there exist an either unitary or antiunitary operator U on H and a real number $p < 1$ such that with the function $f_p(x) = \frac{x}{xp+(1-p)}$ $(x \in [0,1])$ we have*

$$\phi(A) = U f_p(A) U^* \qquad (A \in E(H)).$$

In particular, we obtain the same form for any bijection of $E(H)$ which preserves the order and commutativity in both directions.

It is easy to see that if a function f_p appearing above satisfies $f_p(\lambda) = \lambda$ for some $\lambda \in]0, 1[$, then we have $p = 0$ and hence f_p is the identity function. This gives us the following corollary stating that if an order preserving bijection ϕ of $E(H)$ fixes one single nontrivial scalar operator, then ϕ is implemented by an either unitary or antiunitary operator.

Corollary 2.7.4. *Let $\phi : E(H) \to E(H)$ be a bijective map which preserves the order in both directions and suppose that there is a scalar $\lambda \in]0, 1[$ such that $\phi(\lambda I) = \lambda I$. Then there exists an either unitary or antiunitary operator U on H such that*

$$\phi(A) = UAU^* \qquad (A \in E(H)).$$

The following corollary in the case when $\dim H \geq 3$ appeared in [175] as Theorem 1. Here we shall present a short new proof based on the main result of this section which functions also in the 2-dimensional case. Recall that two effects $A, B \in E(H)$ are called coexistent if they are in the range of a POV (positive operator valued) measure or, equivalently, if there are effects $E, F, G \in E(H)$ such that

$$A = E + G, \quad B = F + G, \quad E + F + G \in E(H).$$

Coexistency is probably the most important relation among effects (see, for example, [130]).

Corollary 2.7.5. *Let $\phi : E(H) \to E(H)$ be a bijective map which preserves the order and coexistency in both directions. Then there is an either unitary or antiunitary operator U on H such that*

$$\phi(A) = UAU^* \qquad (A \in E(H)).$$

As for the last two corollaries below, we shall see that the structure of the ortho-order automorphisms and that of the sequential automorphisms of $E(H)$ can be deduced from our main result Theorem 2.7.1 even in the 2-dimensional case. The proofs are based on the observation that these automorphisms necessarily preserve the order and zero product in both directions. (For a similar observation concerning effects in general von Neumann algebras, see the subsection 2.7.4. Remarks below.) Recall that a bijective map $\phi : E(H) \to E(H)$ is called an ortho-order automorphism if for any $A, B \in E(H)$ we have

$$A \leq B \Longleftrightarrow \phi(A) \leq \phi(B),$$

(i.e., ϕ preserves the order in both directions) and

$$\phi(A^\perp) = \phi(A)^\perp$$

(i.e., ϕ preserves the orthocomplementation). Moreover, a bijective map $\phi : E(H) \to E(H)$ is called a sequential automorphism if

$$\phi(A \circ B) = \phi(A) \circ \phi(B)$$

holds for every $A, B \in E(H)$.

Corollary 2.7.6. *Let* $\phi : E(H) \rightarrow E(H)$ *be an ortho-order automorphism. Then* ϕ *preserves the order and zero product in both directions. In fact, there is an either unitary or antiunitary operator* U *on* H *such that*

$$\phi(A) = UAU^* \qquad (A \in E(H)).$$

Corollary 2.7.7. *Let* $\phi : E(H) \rightarrow E(H)$ *be a sequential automorphism. Then* ϕ *preserves the order and zero product in both directions. In fact, there is an either unitary or antiunitary operator* U *on* H *such that*

$$\phi(A) = UAU^* \qquad (A \in E(H)).$$

2.7.3 Proofs

First we emphasize that the proof of our main result Theorem 2.7.1 has many common points with the proof of [190, Theorem]. In fact, the argument what we present below can be considered as an adaptation of the proof given in our paper [190] to another situation.

We begin with some notation and useful facts that we shall use in our arguments.

Firstly, we note that the concept of the strength of an effect along a ray plays important role in what follows. This notion was introduced by Busch and Gudder in [37] and reads as follows. If A is an effect on H, φ is a unit vector in H and P_φ is the rank-one projection onto the subspace generated by φ, then the quantity

$$\lambda(A, P_\varphi) = \sup\{\lambda \in [0, 1] : \lambda P_\varphi \leq A\}$$

is called the strength of A along the ray represented by φ. Due to [37, Theorem 4] there is a very useful formula to compute the strength. In fact, we have

$$\lambda(A, P_\varphi) = \begin{cases} \|A^{-1/2}\varphi\|^{-2}, & \text{if } \varphi \in \text{rng}(A^{1/2}); \\ 0, & \text{else.} \end{cases} \qquad (2.7.2)$$

(The symbol rng denotes the range of operators and $A^{-1/2}$ denotes the inverse of $A^{1/2}$ on its range.)

Let $\phi : E(H) \rightarrow E(H)$ be a bijective map which preserves the order in both directions. It was shown by Ludwig in the proof of [146, Theorem 5.8., p. 220] that ϕ necessarily preserves the projections in both directions. It is then trivial to see that ϕ also preserves the rank of the projections (cf. the proof of [175, Theorem 1]).

An easy fact follows what we shall use several times. Namely, if A, B are effects, B is of rank one and $A \leq B$, then A is a scalar multiple of B. This observation and the previous one have, among others, the following corollary. Let ϕ be as above, i.e., suppose that it is a bijection of $E(H)$ which preserves the order in both directions. Then for every rank-one projection P, there is a function $f_P : [0, 1] \rightarrow [0, 1]$ such that

$$\phi(tP) = f_P(t)\phi(P) \qquad (t \in [0,1]).$$

By the order preserving property of ϕ and ϕ^{-1} we see that f_P is strictly increasing and bijective. In fact, we have

$$\phi^{-1}(t\phi(P)) = f_P^{-1}(t)P \qquad (t \in [0,1]). \tag{2.7.3}$$

Now, we turn to the proofs. In the proof of our main result Theorem 2.7.1 we use the following proposition which presents the solution of a certain functional equation.

Proposition 2.7.8. *Let $f, g :]0, 1[\to]0, 1[$ be functions and suppose that f is a strictly monotone increasing bijection. Let*

$$f\left(\frac{x}{x + (1-x)y}\right) = \frac{f(x)}{f(x) + (1 - f(x))g(y)} \qquad (x, y \in]0, 1[). \tag{2.7.4}$$

Then there are positive real numbers a, b, c such that

$$f(x) = \frac{x^c}{x^c + a(1-x)^c} \qquad (x \in]0, 1[)$$

and

$$g(y) = by^c \qquad (y \in]0, 1[).$$

Proof. First note that since the function f is continuous, equation (2.7.4) implies the continuity of g.

Next observe that with the notation

$$\alpha(t) = \frac{1}{1 + e^t} \qquad (t \in \mathbb{R}),$$

$$\beta(x) = \ln \frac{1-x}{x} \qquad (x \in]0, 1[),$$

$$\gamma(y) = \ln y \qquad (y \in]0, 1[),$$

we have the identity

$$\frac{x}{x + (1-x)y} = \frac{1}{1 + \exp\left(\ln\frac{1-x}{x} + \ln y\right)} = \alpha\big(\beta(x) + \gamma(y)\big)$$

for all $x, y \in]0, 1[$. Therefore, equation (2.7.4) can be rewritten as

$$f \circ \alpha\big(\beta(x) + \gamma(y)\big) = \alpha\big(\beta \circ f(x) + \gamma \circ g(y)\big) \qquad (x, y \in]0, 1[). \tag{2.7.5}$$

Substituting $x = \beta^{-1}(u)$ and $y = \gamma^{-1}(v)$ into (2.7.5) and applying the inverse function of α to both sides of (2.7.5), we get

$$\alpha^{-1} \circ f \circ \alpha(u + v) = \beta \circ f \circ \beta^{-1}(u) + \gamma \circ g \circ \gamma^{-1}(v) \tag{2.7.6}$$

for all $u \in \mathbb{R}$ and $v \in] - \infty, 0[$. This means that the functions

$$F = \alpha^{-1} \circ f \circ \alpha, \quad G = \beta \circ f \circ \beta^{-1}, \quad \text{and} \quad H = \gamma \circ g \circ \gamma^{-1}$$

satisfy the following so-called Pexider equation

$$F(u + v) = G(u) + H(v) \qquad (u \in \mathbb{R}, \ v \in] - \infty, 0[).$$

Then, by known results of the theory of functional equations (cf. [1], or [131]) and by the continuity of F, G, H, it follows that there exist constants $a, b, c \in \mathbb{R}$ such that

$$
\begin{aligned}
F(w) &= cw + a + b & (w \in \mathbb{R}), \\
G(u) &= cu + a & (u \in \mathbb{R}), & \text{(2.7.7)} \\
H(v) &= cv + b & (v \in] - \infty, 0[). & \text{(2.7.8)}
\end{aligned}
$$

Using (2.7.7) and the definition of G, we get $\beta \circ f(x) = c\beta(x) + a$. Easy computation yields that

$$f(x) = \frac{x^c}{x^c + e^a (1 - x)^c} \qquad (x \in]0, 1[).$$

Similarly, the definition of H and γ together with equation (2.7.8) give

$$g(y) = e^b y^c \qquad (y \in]0, 1[).$$

The function f being strictly increasing, G is also increasing and hence we get $c > 0$. $\qquad \square$

Now, we are in a position to prove the main result of the section.

Proof of Theorem 2.7.1. The clue of the proof is to show that the functions f_P (see above) do not depend on the rank-one projections P. This will be done in what follows.

Let P, Q be arbitrary mutually orthogonal rank-one projections. By the order, rank and orthogonality preserving properties of ϕ on set of all projections we have

$$\phi(P + Q) = \phi(P) + \phi(Q).$$

Let $\lambda \in [0, 1]$. From the inequality

$$\phi(Q) \leq \phi(\lambda P + Q) \leq \phi(P + Q) = \phi(P) + \phi(Q)$$

we obtain

$$0 \leq \phi(\lambda P + Q) - \phi(Q) \leq \phi(P).$$

As $\phi(P)$ is of rank one, according to the first part of the present subsection 2.7.3. Proofs, this implies that there is a scalar $h_P(\lambda) \in [0, 1]$ such that

$$\phi(\lambda P + Q) - \phi(Q) = h_P(\lambda)\phi(P)$$

or, equivalently, such that

$$\phi(\lambda P + Q) = h_P(\lambda)\phi(P) + \phi(Q).$$

Since ϕ^{-1} has the same properties as ϕ, it can be seen that the function $h_P : [0,1] \rightarrow [0,1]$ is a strictly monotone increasing bijection. In fact, we have

$$\phi^{-1}(\lambda\phi(P) + \phi(Q)) = h_P^{-1}(\lambda)P + Q. \tag{2.7.9}$$

We assert that $h_P = f_P$. Indeed, since

$$f_P(\lambda)\phi(P) = \phi(\lambda P) \leq \phi(\lambda P + Q) = h_P(\lambda)\phi(P) + \phi(Q),$$

it follows that $f_P \leq h_P$. By (2.7.3) and (2.7.9), if one considers ϕ^{-1}, it follows that $f_P^{-1} \leq h_P^{-1}$. Since the functions $f_P, h_P : [0,1] \rightarrow [0,1]$ are monotone increasing, we conclude that $f_P = h_P$. From the inequality

$$f_P(\lambda)\phi(P) = \phi(\lambda P) \leq \phi(\lambda(P+Q)) \leq$$
$$\phi(\lambda P + Q) = h_P(\lambda)\phi(P) + \phi(Q) = f_P(\lambda)\phi(P) + \phi(Q)$$

we infer that

$$0 \leq \phi(\lambda(P+Q)) - f_P(\lambda)\phi(P) \leq \phi(Q).$$

As $\phi(Q)$ is of rank one, this implies that

$$\phi(\lambda(P+Q)) = f_P(\lambda)\phi(P) + k_P(\lambda)\phi(Q) \tag{2.7.10}$$

holds for some scalar $k_P(\lambda) \in [0,1]$.

With the notation $F = P+Q$ it follows from the equality (2.7.10) that the operator $\phi(\lambda F)$ is diagonizable with respect to any orthonormal basis in the range of the projection $\phi(F) = \phi(P) + \phi(Q)$. Consequently, we obtain that $\phi(\lambda F)$ is a constant multiple of $\phi(F)$. This gives us that $f_P = k_P$ and hence we have

$$\phi(\lambda(P+Q)) = f_P(\lambda)\phi(P) + f_P(\lambda)\phi(Q).$$

Let R be a rank-one projection whose range is included in the subspace generated by rng P and rng Q. Then we have

$$f_R(\lambda)\phi(R) = \phi(\lambda R) \leq \phi(\lambda F) = f_P(\lambda)(\phi(P) + \phi(Q)) = f_P(\lambda)\phi(F).$$

This gives us that

$$f_R \leq f_P \tag{2.7.11}$$

whenever P, R are rank-one projections. It is then trivial that there is in fact equality in (2.7.11) and this proves that f_P does not depend on P. Denote by f this common function.

Our next claim is to show that f satisfies a functional equation of the form (2.7.4). Fix mutually orthogonal rank-one projections P, Q on H. Pick

$\mu \in]0, 1[$ and let $E = \mu P + Q$. Take any rank-one projection R on H whose range is contained in the subspace generated by the ranges of P and Q and which is neither equal nor orthogonal to P. Using the formula (2.7.2), one can easily verify that

$$\lambda(E, R) = \frac{\mu}{\mu + (1 - \mu) \operatorname{tr} PR}.$$

By the definition of $\lambda(E, R)$ and the order preserving property of ϕ it is clear that

$$f(\lambda(E, R)) = \sup\{f(\lambda) : \lambda R \le E\} = \sup\{f(\lambda) : \phi(\lambda R) \le \phi(E)\} = \sup\{f(\lambda) : f(\lambda)\phi(R) \le \phi(E)\} = \lambda(\phi(E), \phi(R)).$$

Since $\phi(E) = \phi(\mu P + Q) = f(\mu)\phi(P) + \phi(Q)$, we obtain the equality

$$f\left(\frac{\mu}{\mu + (1 - \mu) \operatorname{tr} PR}\right) = \frac{f(\mu)}{f(\mu) + (1 - f(\mu)) \operatorname{tr} \phi(P)\phi(R)}.$$

As the quantities $\operatorname{tr} PR$ and $\operatorname{tr} \phi(P)\phi(R)$ do not depend on μ, it follows from this equality that $\operatorname{tr} \phi(P)\phi(R)$ can be uniquely expressed as a function of $\operatorname{tr} PR$. Denoting $g(\operatorname{tr} PR) = \operatorname{tr} \phi(P)\phi(R)$, we get a bijective function $g :]0, 1[\to]0, 1[$ for which

$$f\left(\frac{\mu}{\mu + (1 - \mu)\nu}\right) = \frac{f(\mu)}{f(\mu) + (1 - f(\mu))g(\nu)} \qquad (\mu, \nu \in]0, 1[).$$

This gives us the desired functional equation for f and g. By Proposition 2.7.8 we obtain that there are positive real numbers a, b, c such that

$$f(x) = \frac{x^c}{x^c + a(1 - x)^c} \qquad (x \in]0, 1[)$$

and

$$g(y) = by^c \qquad (y \in]0, 1[).$$

But our function g has the additional property that $g(1 - x) = 1 - g(x)$ $(x \in]0, 1[)$. Indeed, this follows from the equality

$$g(\operatorname{tr} PR) + g(1 - \operatorname{tr} PR) = g(\operatorname{tr} PR) + g(\operatorname{tr} QR) = \operatorname{tr} \phi(P)\phi(R) + \operatorname{tr} \phi(Q)\phi(R) = 1.$$

One can easily deduce that we necessarily have $b = 1, c = 1$, i.e., g is the identity on $]0, 1[$. This further implies that our function f is of the form

$$f(x) = \frac{x}{x + a(1 - x)} = \frac{x}{x(1 - a) + a} \qquad (x \in]0, 1[).$$

Because of continuity, the above equality holds also on the whole interval $[0, 1]$. Hence, we have that f is of the form $f = f_p$ where $p = 1 - a$.

Since the function g above is the identity, we have

$$\operatorname{tr} PQ = \operatorname{tr} \phi(P)\phi(Q)$$

for all rank-one projections P, Q on H. Hence, using Wigner's theorem on symmetry transformations we obtain that there exists an either unitary or antiunitary operator U on H such that

$$\phi(P) = UPU^*$$

holds for every rank-one projection P on H. Consider the transformation

$$\psi : A \longmapsto f^{-1}(U^*\phi(A)U)$$

on $E(H)$. It is not hard to see that this map is a bijection of $E(H)$ which preserves the order (as well as zero product) in both directions and it has the additional property that it fixes the so-called weak atoms, that is, the effects of the form λP where $\lambda \in [0,1]$ and P is a rank-one projection. As, according to [37, Corollary 3], every effect is equal to the supremum of the set of all weak atoms it majorizes, we have that ψ is the identity on $E(H)$. Transforming back, we see that

$$\phi(A) = Uf(A)U^* \qquad (A \in E(H)).$$

The proof is complete. $\qquad\qquad\qquad\qquad\qquad\qquad\qquad\qquad\qquad$ $\square$

Proof of Proposition 2.7.2. Let $\phi : E(H) \to E(H)$ be an order preserving bijection and λ, μ be a pair of nontrivial scalars such that $\phi(\lambda I) = \mu I$. Keeping the notation introduced in the first part of this subsection 2.7.3. Proofs, we claim that

$$f_P(\lambda) = \mu, \qquad\qquad\qquad\qquad (2.7.12)$$

i.e., that $\phi(\lambda P) = \mu\phi(P)$ holds for every rank-one projection P. Indeed, from

$$f_P(\lambda)\phi(P) = \phi(\lambda P) \le \phi(\lambda I) = \mu I$$

we deduce that $f_P(\lambda) \le \mu$. Now, considering ϕ^{-1} and $\phi(P)$ in the place of ϕ and P, respectively, we also have $f_P^{-1}(\mu) \le \lambda$. Since f_P is increasing, this implies $\mu \le f_P(\lambda)$ and hence we get (2.7.12).

We next assert that ϕ preserves the orthogonality between rank-one projections. To see this, let P, Q be mutually orthogonal rank-one projections. Denote by $P^\perp = I - P$ the orthogonal complement of P. Consider the effect $E = \lambda P + P^\perp$. Clearly, we have $\lambda I \le E \le I$, the strength of E along P is λ and along Q (which is a subprojection of $P^\perp$) is 1. It follows from the order preserving property of ϕ and from (2.7.12) that

- $\mu I \le \phi(E) \le I$,
- the strength of $\phi(E)$ along $\phi(P)$ is μ,

-    the strength of $\phi(E)$ along $\phi(Q)$ is 1.

Now, Lemma 3 in [175] tells us that in this case the ranges of $\phi(P)$ and $\phi(Q)$ are subspaces of the eigenspaces of $\phi(E)$ corresponding to the eigenvalues μ and 1, respectively. This yields that the ranges of $\phi(P)$ and $\phi(Q)$ are orthogonal to each other.

As every projection is equal to the supremum of the set of all rank-one projections it majorizes, it follows that ϕ preserves the orthogonality of projections of any rank. It is also easy to verify that ϕ preserves the range projections of effects. This means that if R is the range projection of A (i.e., the projection onto $\overline{\text{rng } A}$), then $\phi(R)$ is the range projection of $\phi(A)$. Indeed, this preserving property follows from the simple fact that the range projection of the effect A is equal to the infimum of the set of all projections which are greater than or equal to A. It is clear that for any $A, B \in E(H)$, we have $AB = 0$ if and only if the range projections of A and B are orthogonal. Using these observations we infer that

$$AB = 0 \iff \phi(A)\phi(B) = 0.$$

This completes the proof.   □

Proof of Corollary 2.7.5. By [175, Lemma 1] an effect is coexistent with every other effect if and only if it is a scalar multiple of the identity. This implies that our transformation ϕ maps scalar operators to scalar operators. Hence, by Corollary 2.7.3 we infer that, up to unitary-antiunitary equivalence, ϕ is of the form

$$\phi(A) = f_p(A) \qquad (A \in E(H))$$

for some $p < 1$. We claim that $p = 0$, i.e., f_p is the identity. To see this, let P, Q be different rank-one projections. It was proved in [175, Lemma 2] that two rank-one effects with different ranges are coexistent if and only if their sum is an effect. Since λP and $(1 - \lambda)Q$ are always coexistent (indeed, their sum is an effect), we obtain that $\phi(\lambda P)$ and $\phi((1 - \lambda)Q)$ must be also coexistent. By the just mentioned characterization we obtain that for any $\lambda \in]0, 1[$ we have

$$f_p(\lambda)P + f_p(1 - \lambda)Q = \phi(\lambda P) + \phi((1 - \lambda)Q) \leq I.$$

If we let Q tend to P, we infer from this inequality that

$$f_p(\lambda) + f_p(1 - \lambda) \leq 1 \qquad (\lambda \in [0, 1]). \tag{2.7.13}$$

Since ϕ^{-1} has the same properties as ϕ, it also follows that

$$f_p^{-1}(f_p(\lambda)) + f_p^{-1}(1 - f_p(\lambda)) \leq 1.$$

This further implies

$$f_p^{-1}(1 - f_p(\lambda)) \leq 1 - \lambda$$

and by the monotonicity of f_p we obtain

$$1 - f_p(\lambda) \le f_p(1 - \lambda).$$

Comparing this with (2.7.13), we see that

$$f_p(1 - \lambda) = 1 - f_p(\lambda)$$

holds for every $\lambda \in [0, 1]$. It is easy to show that this implies $p = 0$ which completes the proof.   □

Proof of Corollary 2.7.6. Since ϕ preserves the orthocomplementation, we can compute

$$\phi\left(\frac{1}{2}I\right) = \phi\left(\left(\frac{1}{2}I\right)^{\perp}\right) = \phi\left(\frac{1}{2}I\right)^{\perp}$$

and this implies that $\phi(\frac{1}{2}I) = \frac{1}{2}I$. One can apply Corollary 2.7.4 to complete the proof.   □

Proof of Corollary 2.7.7. It was proved in [80] that the order on $E(H)$ is completely determined by the sequential product. More precisely, [80, Theorem 5.1] tells us that for any $A, B \in E(H)$ we have

$$A \le B \Longleftrightarrow \exists C \in E(H) : A = B \circ C.$$

As ϕ is a sequential automorphism, using this characterization it follows that ϕ preserves the order in both directions. It is easy to see that

$$A \circ B = (B^{\frac{1}{2}}A^{\frac{1}{2}})^* B^{\frac{1}{2}}A^{\frac{1}{2}} = 0 \Longleftrightarrow AB = 0.$$

Therefore, ϕ also preserves the zero product in both directions. By Theorem 2.7.1 we obtain that, up to unitary-antiunitary equivalence, ϕ is of the form

$$\phi(A) = f_p(A) \qquad (A \in E(H)) \tag{2.7.14}$$

for some $p < 1$. We assert that $p = 0$, i.e., f_p is the identity. In fact, as ϕ is a sequential automorphism, we obtain from (2.7.14) that f_p is a multiplicative function on the unit interval. It is then obvious that $p = 0$ and the proof is complete.   □

2.7.4 Remarks

Recall that in Section 2.5 we described the general form of all bijective maps (without assuming any kind of linearity) of the positive cone $B(H)^+$ of the C^*-algebra $B(H)$ which preserve the order in both directions. In fact, in Theorem 2.5.1 we proved that every such transformation is of the form

$$A \longmapsto TAT^*$$

with some invertible bounded either linear or conjugate-linear operator T on H. One might think that a somewhat similar statement should hold also for the operator interval $[0, I] = E(H) \subset B(H)^+$. But this is not the case. Indeed, as we noted in [189], for any fixed invertible operator $T \in E(H)$, the transformation

$$A \longmapsto \left(\frac{T^2}{2I - T^2}\right)^{-\frac{1}{2}} \left((I - T^2 + T(I + A)^{-1}T)^{-1} - I\right)\left(\frac{T^2}{2I - T^2}\right)^{-\frac{1}{2}}$$

is bijective, preserves the order in both directions but it has nothing to do with the nice form appearing above. The reason of this somewhat surprising fact lies in the serious differences among the classes of operator monotone functions defined on different subintervals of $\mathbb{R}$.

Finally, according to our promise given in the summary of the present section, we make some remarks on the relation among the class of our preservers and the collections of ortho-order automorphisms, respectively, sequential automorphisms in the general setting of von Neumann algebras. So, let $\mathcal{A}$ be a von Neumann algebra of operators acting on the Hilbert space H. Denote by $E(\mathcal{A})$ the set of all effects which belong to $\mathcal{A}$, i.e., let $E(\mathcal{A}) = E(H) \cap \mathcal{A}$. The definitions of the order, orthocomplementation, sequential product are straightforward and so are the definitions of ortho-order automorphisms and sequential automorphisms.

Let ϕ be an ortho-order automorphism of $E(\mathcal{A})$. It is easy to see that the sharp elements in $E(\mathcal{A})$ (i.e., the elements A for which the infimum of A and $A^\perp = I - A$ is 0) are exactly the projections. Hence we obtain that ϕ preserves the projections in $E(\mathcal{A})$ as well as their orthogonality in both directions. Since, as it is well-known, the range projection of any element of a von Neumann algebra also belongs to the algebra, we see that ϕ preserves the range projections of the elements of $E(\mathcal{A})$ in the same sense as it was mentioned in the proof of Proposition 2.7.2. Then one can argue as in that proof to verify that ϕ preserves zero product in both directions. So, we obtain that every ortho-order automorphism of $E(\mathcal{A})$ belongs to our class of preservers, that is, those automorphisms preserve the order and zero product in both directions.

Now, let $\phi : E(\mathcal{A}) \to E(\mathcal{A})$ be a sequential automorphism. It is not hard to prove that the mentioned characterization of the order by means of the sequential product due to Gudder and Greechie (see the proof of Corollary 2.7.7) holds true also in the setting of von Neumann algebras. This means that for any $A, B \in E(\mathcal{A})$ we have

$$A \leq B \iff \exists C \in E(\mathcal{A}) : A = B \circ C. \tag{2.7.15}$$

Indeed, if $A, B \in E(\mathcal{A})$ and $A \leq B$, then by [80, Theorem 5.1] there is an operator $C \in E(H)$ such that $A = B \circ C$. Following the proof in [80], one can see that the existence of this operator C is a consequence of a well-known result of Douglas [67]. In fact, in the corresponding part of the proof of Douglas' result this C was constructed. Examining the construction, it is not hard to verify that C belongs to the von Neumann algebra $\mathcal{A}$, i.e., we have $C \in E(\mathcal{A})$. This gives one implication from the asserted equivalence in (2.7.15). The other implication is trivial. Then, just as in the proof of Corollary 2.7.7, one can show that ϕ preserves the order and zero product in both directions.

To sum up, it has turned out that the ortho-order automorphisms and the sequential automorphisms all belong to our class of preservers even in the setting of von Neumann algebras. This seems to be a worthwhile observation as in that generality it is quite hard to see any connection between those two kinds of automorphisms.

In closing the section, we refer to the recent paper [125] of Kim for additional results on the automorphisms of the Hilbert space effect algebra.

2.8 Sequential Isomorphisms Between the Sets of von Neumann Algebra Effects

2.8.1 Summary

In this section we describe the structure of all sequential isomorphisms between the sets of von Neumann algebra effects. It turns out that if the underlying algebras have no commutative direct summands, then every sequential isomorphism between the sets of their effects extends to the direct sum of a *-isomorphism and a *-antiisomorphism between the underlying von Neumann algebras. The results of this section appeared in the paper [186].

2.8.2 Formulation of the Results

In Corollary 2.7.7 of Section 2.7 we have obtained that if H is a Hilbert space with $\dim H \geq 2$, then every sequential automorphism $\phi : E(H) \to E(H)$ is implemented by a unitary or antiunitary operator U on H. This means that ϕ is of the form

$$\phi(A) = UAU^* \qquad (A \in E(H)).$$

(The result also appears in [79] under the condition that $\dim H \geq 3$. The 2-dimensional case was treated in the paper [152] independently) The aim of this section is to extend this result from the case of $E(H)$ of all Hilbert space effects to the case of the sets of general von Neumann algebra effects.

In fact, it is a natural step in any research on operator algebras that if one has a result concerning $B(H)$ (the algebra of all bounded linear operators on H), then one is tempted to try to extend it in some way for more general operator algebras. The first and in many cases probably the most important candidates for such a purpose are the von Neumann algebras. Therefore, after describing the sequential automorphisms of the set of all Hilbert space effects it is a natural problem to investigate the question for von Neumann algebra effects, that is, for the set $E(\mathcal{A})$ of all elements of a given von Neumann algebra $\mathcal{A}$ which are positive and majorized by the identity. (Observe that we have $E(\mathcal{A}) = E(H) \cap \mathcal{A}$, where H is the Hilbert space on which the elements of $\mathcal{A}$ act.) This is the aim of this section.

We begin with the notation and some definitions that we shall use throughout the section. If $\mathcal{A}$ is a unital C^*-algebra, then the effects in $\mathcal{A}$ are the positive elements of $\mathcal{A}$ which are less than or equal to the unit of $\mathcal{A}$. The set of all effects in $\mathcal{A}$ is denoted by $E(\mathcal{A})$. The sequential product $\circ$ on $E(\mathcal{A})$ is defined by

$$A \circ B = A^{1/2}BA^{1/2} \qquad (A, B \in E(\mathcal{A})).$$

(For some interesting properties of this operation on $E(H)$ concerning associativity and commutativity see [81].) If $\mathcal{A}, \mathcal{B}$ are unital C^*-algebras, then a bijective map $\phi : E(\mathcal{A}) \to E(\mathcal{B})$ which satisfies

$$\phi(A \circ B) = \phi(A) \circ \phi(B) \qquad (A, B \in E(\mathcal{A}))$$

is called a sequential isomorphism.

Let $\psi_1 : \mathcal{A} \to \mathcal{B}$ be a *-isomorphism, that is, a bijective linear map such that

$$\psi_1(AB) = \psi_1(A)\psi_1(B), \quad \psi_1(A^*) = \psi_1(A)^* \quad (A, B \in \mathcal{A}).$$

It is easy to see that the restriction of ψ_1 onto $E(\mathcal{A})$ is a sequential isomorphism from $E(\mathcal{A})$ onto $E(\mathcal{B})$. Similar assertion applies for any *-antiisomorphism $\psi_2 : \mathcal{A} \to \mathcal{B}$ as well. The bijective linear map $\psi_2 : \mathcal{A} \to \mathcal{B}$ is called a *-antiisomorphism if it satisfies

$$\psi_2(AB) = \psi_2(B)\psi_2(A), \quad \psi_2(A^*) = \psi_2(A)^* \quad (A, B \in \mathcal{A}).$$

Clearly, the direct sum of sequential isomorphisms is again a sequential isomorphism. Therefore, the direct sum of a *-isomorphism and a *-antiisomorphism is, when restricted onto the set of effects, a sequential isomorphism. In the main result of the section which follows we see that if the underlying algebras $\mathcal{A}, \mathcal{B}$ are von Neumann algebras, then every sequential isomorphism between the sets of their effects can be obtained in this way on the 'non-commutative parts' of $E(\mathcal{A})$ and $E(\mathcal{B})$.

Theorem 2.8.1. *Let $\mathcal{A}, \mathcal{B}$ be von Neumann algebras and let $\phi : E(\mathcal{A}) \to E(\mathcal{B})$ be a sequential isomorphism. Then there are direct decompositions*

$$\mathcal{A} = \mathcal{A}_1 \oplus \mathcal{A}_2 \oplus \mathcal{A}_3 \quad and \quad \mathcal{B} = \mathcal{B}_1 \oplus \mathcal{B}_2 \oplus \mathcal{B}_3$$

within the category of von Neumann algebras and there are bijective maps

$$\phi_1 : E(\mathcal{A}_1) \to E(\mathcal{B}_1), \quad \Phi_2 : \mathcal{A}_2 \to \mathcal{B}_2, \quad \Phi_3 : \mathcal{A}_3 \to \mathcal{B}_3$$

such that

(i) *$\mathcal{A}_1, \mathcal{B}_1$ are commutative von Neumann algebras and the algebras $\mathcal{A}_2 \oplus \mathcal{A}_3$, $\mathcal{B}_2 \oplus \mathcal{B}_3$ have no commutative direct summands;*
(ii) *ϕ_1 is a multiplicative bijection, Φ_2 is a *-isomorphism, Φ_3 is a *-antiisomorphism and $\phi = \phi_1 \oplus \Phi_2 \oplus \Phi_3$ holds on $E(\mathcal{A})$.*

(Of course, the above decomposition is meant in the sense that some of the direct summands can be missing.)

If there is a scalar $\lambda \in]0,1[$ such that $\phi(\lambda I) = \lambda I$, or $\mathcal{A}$ has no commutative direct summand, then ϕ extends to the direct sum of a *-isomorphism and a *-antiisomorphism between subalgebras of $\mathcal{A}$ and $\mathcal{B}$.

To see that between the 'commutative parts' of the underlying algebras the behavior of ϕ can be quite 'irregular', consider an arbitrary compact Hausdorff space X. Let $C(X)$ denote the algebra of all continuous complex valued functions on X. Take any strictly positive continuous function $p : X \to [0, \infty[$ and define $\phi : E(C(X)) \to E(C(X))$ by

$$\phi(f)(x) = f(x)^{p(x)} \qquad (x \in X, f \in E(C(X))).$$

It is easy to verify that ϕ is a sequential automorphism of $E(C(X))$ which, in general, does not extend to an automorphism of $C(X)$. (Observe that if $\mathcal{A}, \mathcal{B}$ are commutative, then the sequential product on $E(\mathcal{A}), E(\mathcal{B})$ coincides with the ordinary product and hence in that case the sequential isomorphisms between $E(\mathcal{A})$ and $E(\mathcal{B})$ are just the multiplicative bijections or, in other words, the semigroup isomorphisms.)

As for the case of factors, the main result of this section has the following immediate corollary which easily implies Corollary 2.7.7.

Corollary 2.8.2. *Let $\mathcal{A}, \mathcal{B} \neq \mathbb{C}I$ be factors and let $\phi : E(\mathcal{A}) \to E(\mathcal{B})$ be a sequential isomorphism. Then ϕ extends either to a *-isomorphism or to a *-antiisomorphism between the algebras $\mathcal{A}$ and $\mathcal{B}$.*

2.8.3 Proof

We introduce some further notation and definitions that we shall use in the proof of our main result. If $\mathcal{A}$ is a C^*-algebra, then $\mathcal{A}_s, \mathcal{A}^+$ denote the set of all self-adjoint elements of $\mathcal{A}$ and the set of all positive elements of $\mathcal{A}$, respectively. The set of all projections (i.e., self-adjoint idempotents) in $\mathcal{A}$ is denoted by $P(\mathcal{A})$ and $\mathcal{Z}(\mathcal{A})$ stands for the center of $\mathcal{A}$.

Our first auxiliary result concerns the so-called E-isomorphisms of the sets of C^*-algebra effects. Let $\mathcal{A}, \mathcal{B}$ be C^*-algebras. Let $\phi : E(\mathcal{A}) \to E(\mathcal{B})$ be a bijective map with the properties that

$$A + B \in E(\mathcal{A}) \Longleftrightarrow \phi(A) + \phi(B) \in E(\mathcal{B}) \qquad (A, B \in E(\mathcal{A}))$$

and

$$\phi(A + B) = \phi(A) + \phi(B) \qquad (A, B, A + B \in E(\mathcal{A})). \tag{2.8.1}$$

Then ϕ is called an E-isomorphism (cf. Section 0.3 of the Introduction).

To formulate our result on E-isomorphisms we recall the concept of Jordan *-isomorphisms. Let $\mathcal{A}, \mathcal{B}$ be *-algebras and let $\phi : \mathcal{A} \to \mathcal{B}$ be a bijective linear map with the properties that

$$\phi(A^2) = \phi(A)^2, \quad \phi(A^*) = \phi(A)^* \quad (A \in \mathcal{A}).$$

Then ϕ is called a Jordan *-isomorphism. It is easy to see that in the above definition the condition that $\phi(A^2) = \phi(A)^2$ holds for every $A \in \mathcal{A}$ can be replaced by the following one:

$$\phi(AB + BA) = \phi(A)\phi(B) + \phi(B)\phi(A) \quad (A, B \in \mathcal{A}).$$

One can readily verify that if $\mathcal{A}, \mathcal{B}$ are C^*-algebras and $\phi : \mathcal{A} \to \mathcal{B}$ is a Jordan *-isomorphism, then its restriction onto $E(\mathcal{A})$ is an E-isomorphism from $E(\mathcal{A})$ onto $E(\mathcal{B})$. Our first result says that the converse is also true, that is, every E-isomorphism from $E(\mathcal{A})$ onto $E(\mathcal{B})$ extends to a Jordan *-isomorphism from $\mathcal{A}$ onto $\mathcal{B}$.

Proposition 2.8.3. *Let $\mathcal{A}, \mathcal{B}$ be C^*-algebras and let $\phi : E(\mathcal{A}) \to E(\mathcal{B})$ be an E-isomorphism. Then there is a Jordan *-isomorphism $\Phi : \mathcal{A} \to \mathcal{B}$ such that*

$$\phi(A) = \Phi(A) \quad (A \in E(\mathcal{A})).$$

Proof. The idea of the proof is easy. Using a quite elementary argument, we can extend ϕ to a linear transformation $\Phi : \mathcal{A} \to \mathcal{B}$. We next show that Φ is bijective, unital and preserves the order in both directions. Finally, we apply the well-known result of Kadison on order isomorphisms between C^*-algebras to conclude that Φ is a Jordan *-isomorphism.

Turning to the details, first we show that ϕ preserves the order. Let $A, B \in E(\mathcal{A})$ be such that $A \leq B$. Then we have $B = A + (B - A)$, where $A, B - A, B \in E(\mathcal{A})$. As ϕ is an E-isomorphism, we obtain

$$\phi(B) = \phi(A) + \phi(B - A) \geq \phi(A).$$

Therefore, ϕ preserves the order. Since ϕ^{-1} has similar properties as ϕ, it follows that ϕ preserves the order in both directions. We easily deduce that $\phi(0) = 0$ and $\phi(I) = I$.

Let $A \in E(\mathcal{A})$ be arbitrary. By the partial additivity of ϕ (see (2.8.1)) we have

$$\phi(A) = \phi\left(n\frac{1}{n}A\right) = n\phi\left(\frac{1}{n}A\right)$$

which implies

$$\phi\left(\frac{1}{n}A\right) = \frac{1}{n}\phi(A)$$

for every $n \in \mathbb{N}$. If $n, k \in \mathbb{N}$ and $k \leq n$, using the partial additivity of ϕ again, we obtain that

$$\frac{k}{n}\phi(A) = k\phi\left(\frac{1}{n}A\right) = \phi\left(\frac{k}{n}A\right).$$

By the order preserving property and the just proved rational-homogeneity of ϕ one can easily verify that

$$\phi(\lambda A) = \lambda \phi(A)$$

holds for every $A \in E(\mathcal{A})$ and $\lambda \in [0,1]$.

Now it requires only elementary arguments to check that the transformation $\psi_1 : \mathcal{A}^+ \to \mathcal{B}^+$ defined by

$$\psi_1(A) = \begin{cases} \|A\| \phi\left(\frac{A}{\|A\|}\right), & 0 \neq A \in \mathcal{A}^+; \\ 0, & A = 0 \end{cases}$$

is bijective, additive, positive homogeneous, unital, preserves the order in both directions and extends ϕ. We omit the details. To proceed, we define the transformation $\psi_2 : \mathcal{A}_s \to \mathcal{B}_s$ by

$$\psi_2(A) = \psi_1(A^+) - \psi_1(A^-) \qquad (A \in \mathcal{A}_s)$$

where A^+, A^- denote the positive part and the negative part of A, respectively. It is easy to verify that ψ_2 is bijective, linear, unital, preserves the order in both directions and extends ψ_1. Finally, let $\Phi : \mathcal{A} \to \mathcal{B}$ be defined by

$$\Phi(A) = \psi_2(\operatorname{Re} A) + i\psi_2(\operatorname{Im} A) \qquad (A \in \mathcal{A})$$

where $\operatorname{Re} A$ and $\operatorname{Im} A$ denote the real part and the imaginary part of A, respectively. One can check that Φ is bijective, linear, unital, preserves the order in both directions, extends ψ_2 and hence extends ϕ as well. We now refer to the result Theorem A.3 of Kadison stating that any bijective unital linear transformation between C^*-algebras which preserves the order in both directions is necessarily a Jordan *-isomorphism. Applying this result to Φ we complete the proof. $\qquad\qquad\qquad\qquad\qquad\qquad\qquad\qquad\qquad\square$

Observe that Proposition 2.8.3 readily implies the fact that every E-automorphism of $E(H)$ is implemented by a unitary or antiunitary operator which was mentioned in the corresponding part of the Introduction (see Section 0.3). In fact, if $\phi : E(H) \to E(H)$ is an E-automorphism, then by Proposition 2.8.3 it extends to a Jordan *-automorphism Φ of $B(H)$. By the result Theorem A.7 of Herstein (also mentioned in Section 0.4 in relation with our new approach to the proof of Wigner's theorem) it follows that Φ is either a *-automorphism or a *-antiautomorphism. Applying Theorem A.8 we have an either unitary or antiunitary operator U on H such that

$$\Phi(A) = UAU^* \qquad (A \in E(H))$$

and this results in what we have asserted.

Furthermore, the simple result Proposition 2.8.3 will play important role in the proof of the main theorem of the section. In fact, the basic idea of that proof is the following. Let $\mathcal{A}, \mathcal{B}$ be von Neumann algebras and let $\phi : E(\mathcal{A}) \to E(\mathcal{B})$ be a sequential isomorphism. Consider the type decomposition of $\mathcal{A}$ and $\mathcal{B}$ (see, for example, [121, 6.5.2. Theorem]). We shall see that ϕ

maps the effects in the type I_n direct summand of $\mathcal{A}$ to the effects in the type I_n direct summand of $\mathcal{B}$. Therefore, roughly speaking, we can consider our problem separately on type I_n algebras and on algebras without type I direct summands and then take direct sums. It will turn out that on such algebras (with the exception of type I_1 algebras), the sequential isomorphisms are all E-isomorphisms and hence we can apply Proposition 2.8.3. This is the plan of the proof. The details follow.

First we present a result concerning type I_1 (i.e., commutative) algebras. As we have already mentioned, on such algebras the sequential isomorphisms coincide with the multiplicative bijections, i.e., the semigroup isomorphisms.

Proposition 2.8.4. *Let X, Y be compact Hausdorff spaces, $\phi : E(C(X)) \to E(C(Y))$ be a bijective multiplicative map. Suppose that there is a constant $\lambda \in]0, 1[$ such that $\phi(\lambda) = \lambda$. Then ϕ extends to a (linear) *-isomorphism $\Phi : C(X) \to C(Y)$.*

Proof. The basic idea of the proof that one should transfer multiplicativity to additivity is due to P. Šemrl.

Let $g, f \in E(C(X))$, $g \le f$ and suppose that f is nowhere vanishing. Then $\frac{g}{f} \in E(C(X))$, so there is a function $f' \in E(C(X))$ such that $g = f f'$. It follows that

$$\phi(g) = \phi(f)\phi(f') \le \phi(f).$$

This shows a certain order preserving property of ϕ.

Let $f \in E(C(X))$ be nowhere vanishing. Then there exists a positive integer n such that $\lambda^n \le f$. Since $\phi(\lambda) = \lambda$ and ϕ is multiplicative, it follows that

$$\lambda^n = \phi(\lambda^n) \le \phi(f).$$

This implies that $\phi(f)$ is also nowhere vanishing. Hence ϕ preserves the nowhere vanishing elements between $E(C(X))$ and $E(C(Y))$. Since ϕ^{-1} has similar properties as ϕ, we obtain that the above two preserving properties of ϕ hold in both directions.

It is now easy to see that the transformation $\psi : C(X)^+ \to C(Y)^+$ defined by

$$\psi(h) = -\ln \phi(\exp(-h)) \qquad (h \in C(X)^+)$$

is bijective, additive and preserves the order in both directions. In particular, we infer that ψ is positive homogeneous. Following the construction in the proof of Proposition 2.8.3 where we obtain Φ from ψ_1, we see that our present map ψ extends to a bijective unital linear transformation $\Phi : C(X) \to C(Y)$ which preserves the order in both directions. Hence, by the result Theorem A.3 of Kadison, Φ is a Jordan *-isomorphism. Because of commutativity, it means that Φ is an isomorphism between the function algebras $C(X)$ and $C(Y)$. The general form of such transformations is well-known. Namely, there is a homeomorphism $\varphi : Y \to X$ for which we have

$$\Phi(h) = h \circ \varphi \qquad (h \in C(X)).$$

Referring to the relation between Φ and our original transformation ϕ we easily obtain that $\phi(f) = f \circ \varphi$ holds for every nowhere vanishing function $f \in E(C(X))$. We assert that this formula is valid for every element of $E(C(X))$. In order to see it, consider the transformation

$$\phi' : f \longmapsto \phi(f) \circ \varphi^{-1}.$$

Clearly, this is a bijective multiplicative selfmap of $E(C(X))$ which acts as the identity on the nowhere vanishing elements of $E(C(X))$. We prove that the same holds on the whole set $E(C(X))$. In fact, suppose that $f \in E(C(X))$. It is trivial to see that there is a sequence (f_n) of nowhere vanishing elements of $E(C(X))$ such that $f \leq f_n$ and (f_n) converges (uniformly) to f. By the order preserving property of ϕ' obtained in the second paragraph of the proof for ϕ, we have $\phi'(f) \leq \phi'(f_n) = f_n$. Taking limit, we arrive at

$$\phi'(f) \leq f.$$

Since ϕ'^{-1} has similar properties as ϕ', we also have

$$f = \phi'^{-1}(\phi'(f)) \leq \phi'(f).$$

These result in $\phi'(f) = f$ ($f \in E(C(X))$). Therefore, we obtain

$$\phi(f) = f \circ \varphi \qquad (f \in E(C(X)))$$

and the proof is complete.   □

In the remaining part of the section $\mathcal{A}, \mathcal{B}$ denote von Neumann algebras and $\phi : E(\mathcal{A}) \to E(\mathcal{B})$ is a sequential isomorphism.

Lemma 2.8.5. We have $\phi(P(\mathcal{A})) = P(\mathcal{B})$. The restriction of ϕ onto $P(\mathcal{A})$ is a bijective map from $P(\mathcal{A})$ onto $P(\mathcal{B})$ which preserves the order and the orthogonality in both directions and hence it is completely orthoadditive.

Proof. It is trivial that for an arbitrary $A \in E(\mathcal{A})$ we have

$$A \text{ is a projection } \Longleftrightarrow A \circ A = A.$$

This implies that ϕ preserves the projections in both directions.

Let $P, Q \in P(\mathcal{A})$. Clearly, we have

$$P \leq Q \Longleftrightarrow Q \circ P = P.$$

This implies that ϕ preserves the order among projections in both directions. In particular, we have $\phi(0) = 0$ and $\phi(I) = I$. As for the orthogonality between projections, we have

$$PQ = 0 \Longleftrightarrow P \circ Q = 0.$$

This implies that ϕ preserves the orthogonality between projections in both directions.

Now, let (P_α) be an arbitrary collection of mutually orthogonal projections in $\mathcal{A}$ with sum P. Using the above verified properties of ϕ, we deduce that $(\phi(P_\alpha))$ is a collection of mutually orthogonal projections and we have $\phi(P_\alpha) \leq \phi(P)$. This implies that

$$\sum_\alpha \phi(P_\alpha) \leq \phi(P) = \phi(\sum_\alpha P_\alpha).$$

Since ϕ^{-1} has similar properties as ϕ, it follows that

$$P = \sum_\alpha \phi^{-1}(\phi(P_\alpha)) \leq \phi^{-1}(\sum_\alpha \phi(P_\alpha)).$$

By the order preserving property of ϕ we obtain that

$$\phi(P) \leq \sum_\alpha \phi(P_\alpha).$$

Therefore, we have

$$\sum_\alpha \phi(P_\alpha) = \phi(P)$$

which means that ϕ is completely orthoadditive on $P(\mathcal{A})$. □

Lemma 2.8.6. *Let $Z, Z' \in E(\mathcal{A})$ be central elements in $\mathcal{A}$ and let $P, P' \in \mathcal{A}$ be mutually orthogonal projections. Then we have*

$$\phi(ZP + Z'P') = \phi(ZP) + \phi(Z'P').$$

Proof. First observe that $\phi(ZP + Z'P')$ is defined. Indeed, for $ZP + Z'P' = PZP + P'Z'P'$ we have

$$0 \leq PZP + P'Z'P' \leq PIP + P'IP' = P + P' \leq I,$$

that is, $ZP + Z'P' \in E(\mathcal{A})$. Next, since $ZP + Z'P'$ commutes with $P, P', P + P'$, the same holds true for its square root. Hence, using the orthoadditivity of the sequential isomorphism ϕ on $P(\mathcal{A})$ (see Lemma 2.8.5) we can compute

$$\phi(ZP + Z'P') =$$
$$\phi((ZP + Z'P')^{\frac{1}{2}}(P + P')(ZP + Z'P')^{\frac{1}{2}}) =$$
$$\phi(ZP + Z'P')^{\frac{1}{2}}\phi(P + P')\phi(ZP + Z'P')^{\frac{1}{2}} =$$
$$\phi(ZP + Z'P')^{\frac{1}{2}}(\phi(P) + \phi(P'))\phi(ZP + Z'P')^{\frac{1}{2}} =$$
$$\phi(ZP + Z'P')^{\frac{1}{2}}\phi(P)\phi(ZP + Z'P')^{\frac{1}{2}} +$$
$$\phi(ZP + Z'P')^{\frac{1}{2}}\phi(P')\phi(ZP + Z'P')^{\frac{1}{2}} =$$
$$\phi((ZP + Z'P')^{\frac{1}{2}}P(ZP + Z'P')^{\frac{1}{2}}) + \phi((ZP + Z'P')^{\frac{1}{2}}P'(ZP + Z'P')^{\frac{1}{2}}) =$$
$$\phi(ZP) + \phi(Z'P').$$

The proof is complete. □

Lemma 2.8.7. *The sequential isomorphism ϕ preserves commutativity in both directions. Therefore, we have $\phi(E(\mathcal{Z}(\mathcal{A}))) = E(\mathcal{Z}(\mathcal{B}))$.*

Proof. The result [82, Corollary 3] says that for any Hilbert space effects A, B we have $A \circ B = B \circ A$ if and only if $AB = BA$. Since ϕ is a sequential isomorphism, this characterization of commutativity clearly implies that ϕ preserves commutativity in both directions. To the second statement observe that the set of all elements of $E(\mathcal{A})$ which commute with every effect in $\mathcal{A}$ equals to $E(\mathcal{Z}(\mathcal{A}))$. □

We say that the collection (E_{ij}) of operators in the von Neumann algebra $\mathcal{A}$ forms a self-adjoint system of $n \times n$ matrix units if n is any cardinal number, the index set in which i, j run has cardinality n, for every i, j, k, l we have

$$E_{ij} E_{kl} = \begin{cases} 0, & j \neq k; \\ E_{il}, & j = k, \end{cases}$$

$\sum_i E_{ii} = I$ in the strong operator topology and $E_{ij}^* = E_{ji}$ holds for all i, j.

Lemma 2.8.8. *Suppose that $\mathcal{A}$ has a self-adjoint system of $n \times n$ matrix units for some cardinal number $n \geq 2$. Then the restriction of ϕ onto $E(\mathcal{Z}(\mathcal{A}))$ is an E-isomorphism onto $E(\mathcal{Z}(\mathcal{B}))$. Moreover, ϕ is homogeneous in the sense that*

$$\phi(\lambda A) = \lambda \phi(A)$$

holds for every $A \in E(\mathcal{A})$ and $\lambda \in [0, 1]$.

Proof. From Lemma 2.8.7 we know that the restriction of ϕ onto $E(\mathcal{Z}(\mathcal{A}))$ is a sequential isomorphism onto $E(\mathcal{Z}(\mathcal{B}))$.

Let (E_{ij}) be a self-adjoint system of $n \times n$ matrix units in $\mathcal{A}$. Let $Z, Z' \in E(\mathcal{A})$ be central elements. Suppose that $Z, Z' \leq \frac{1}{2}I$. For temporary use, fix the matrix units $E = E_{ii}, V = E_{ij}\ i \neq j$. Define

$$P = \frac{1}{2}(E + V)^*(E + V), \quad P' = \frac{1}{2}(E - V)^*(E - V).$$

(The trick to use these operators to prove a kind of additivity of ϕ comes from the proof of Lemma 2.1 in [95] where the author studied the problem of the additivity of so-called Jordan *-maps between operator algebras.) It is obvious that P, P' are mutually orthogonal projections. It follows from $Z, Z' \leq \frac{1}{2}I$ that $Z + Z', 2Z, 2Z' \in E(\mathcal{A})$. We have

$$E(2ZP)E = 2Z(EPE) = ZE, \quad E(2Z'P')E = 2Z'(EP'E) = Z'E. \quad (2.8.2)$$

Since ϕ preserves commutativity in both directions, $\phi(Z), \phi(Z'), \phi(Z+Z')$ are central elements in $E(\mathcal{B})$. Using Lemma 2.8.6 and (2.8.2) we can compute

$$\phi(Z + Z')\phi(E) = \phi(Z + Z')^{\frac{1}{2}}\phi(E)\phi(Z + Z')^{\frac{1}{2}} =$$
$$\phi((Z + Z')E) = \phi(E(2ZP + 2Z'P')E) =$$
$$\phi(E)\phi(2ZP + 2Z'P')\phi(E) = \phi(E)(\phi(2ZP) + \phi(2Z'P'))\phi(E) =$$
$$\phi(E)\phi(2ZP)\phi(E) + \phi(E)\phi(2Z'P')\phi(E) =$$
$$\phi(2EZPE) + \phi(2EZ'P'E) = \phi(ZE) + \phi(Z'E) =$$
$$\phi(Z)\phi(E) + \phi(Z')\phi(E) = (\phi(Z) + \phi(Z'))\phi(E).$$

Therefore, we obtain that

$$\phi(Z + Z')\phi(E_{ii}) = (\phi(Z) + \phi(Z'))\phi(E_{ii}) \qquad (2.8.3)$$

holds for every i. We know that $\sum_i E_{ii} = I$ and by the complete orthoadditivity of ϕ (see Lemma 2.8.5) this yields that $\sum_i \phi(E_{ii}) = I$. Consequently, we deduce from (2.8.3) that

$$\phi(Z + Z') = \phi(Z) + \phi(Z')$$

holds for every central elements $Z, Z' \in E(\mathcal{A})$ with $Z, Z' \leq \frac{1}{2}I$. In particular, we obtain that

$$I = \phi(I) = \phi\left(\frac{1}{2}I + \frac{1}{2}I\right) = \phi\left(\frac{1}{2}I\right) + \phi\left(\frac{1}{2}I\right)$$

and this implies that

$$\phi\left(\frac{1}{2}I\right) = \frac{1}{2}I.$$

Restricting ϕ onto $E(\mathcal{Z}(\mathcal{A}))$, we have a sequential isomorphism between the sets of commutative von Neumann algebra effects which maps a nontrivial scalar to itself. Applying Proposition 2.8.4, it follows that this restriction of ϕ can be extended to a *-isomorphism from $\mathcal{Z}(\mathcal{A})$ onto $\mathcal{Z}(\mathcal{B})$. This implies the first assertion in our statement.

The homogeneity of ϕ is now easy to see. In fact, pick $\lambda \in [0,1]$ and $A \in E(\mathcal{A})$. As ϕ is homogeneous on $E(\mathcal{Z}(\mathcal{A}))$, we compute

$$\phi(\lambda A) = \phi(A)^{\frac{1}{2}}\phi(\lambda I)\phi(A)^{\frac{1}{2}} = \phi(A)^{\frac{1}{2}}\lambda\phi(I)\phi(A)^{\frac{1}{2}} = \lambda\phi(A).$$

$\square$

Lemma 2.8.9. *Suppose that $P \in \mathcal{A}$ is an abelian projection and $A \in \mathcal{A}^+$. Then there is a positive element Z in $\mathcal{Z}(\mathcal{A})$ such that $PAP = ZP$.*

Proof. The operator PAP is a positive element in the C^*-algebra $P\mathcal{A}P$. Therefore, there exists an element $B \in P\mathcal{A}P$ such that $PAP = B^*B$. But, as P is abelian, we know that

$$P\mathcal{A}P = \mathcal{Z}(\mathcal{A})P$$

(see [121, 6.4.2. Proposition]). So, there is a central element Z in $\mathcal{A}$ such that $B = ZP$. We have

$$PAP = B^*B = (ZP)^*(ZP) = Z^*ZP$$

and this verifies our statement. $\square$

Lemma 2.8.10. *Let $\mathcal{A}$ be a von Neumann algebra of type I_n with $n < \infty$. Let $A \in \mathcal{A}$ be such that $PAP = 0$ holds for every abelian projection $P \in \mathcal{A}$. Then we have $A = 0$.*

Proof. Considering the real and imaginary parts of A, we can clearly assume that A is self-adjoint.

By [121, 8.2.8. Theorem], on every finite von Neumann algebra there exists a center-valued trace. What concerns $\mathcal{A}$, denote it by τ. In what follows we assume that τ has the properties listed in that theorem. We have

$$0 = \tau(PAP) = \tau(APP) = \tau(AP)$$

for every abelian projection P in $\mathcal{A}$. It is known that any nonzero projection in a type I algebra is the sum of mutually orthogonal abelian projections (see [121, 6.4.8. Proposition and 6.5.1. Definition]). Hence, if we pick an arbitrary projection Q in $\mathcal{A}$, then there is a collection (P_α) of mutually orthogonal abelian projections such that $Q = \sum_\alpha P_\alpha$. As the weak and ultraweak topologies coincide on the unit ball of any von Neumann algebra (see the discussion in [121, 7.4.4. Remark]), it follows that $AQ = \sum_\alpha AP_\alpha$ holds in the ultraweak topology. By the ultraweak continuity of τ we infer that

$$\tau(AQ) = \sum_\alpha \tau(AP_\alpha) = 0. \tag{2.8.4}$$

Since the linear span of the set of all projections in a von Neumann algebra is dense with respect to the norm topology and τ is norm continuous, it follows from (2.8.4) that $\tau(A^2) = 0$. By the definiteness of the trace, we obtain that $A^2 = 0$. This implies $A = 0$ and the proof is complete. $\square$

Lemma 2.8.11. *The sequential isomorphism ϕ preserves the abelian projections and the equivalence among them in both directions.*

Proof. Clearly, $P \in \mathcal{A}$ is abelian if and only if the set $PE(\mathcal{A})P$ is commutative. As ϕ is a sequential isomorphism, it preserves commutativity in both directions (see Lemma 2.8.7). Therefore, the set $PE(\mathcal{A})P$ is commutative if and only if so is its image under ϕ. But this image equals

$$\phi(PE(\mathcal{A})P) = \phi(P)\phi(E(\mathcal{A}))\phi(P) = \phi(P)E(\mathcal{B})\phi(P).$$

Therefore, P is abelian if and only if $\phi(P)$ is abelian.

As for the preservation of the equivalence between abelian projections, we recall that two abelian projections are equivalent if and only if their central carriers coincide (see [121, 6.2.8. Proposition and 6.4.6 Proposition]). But the notion of the central carrier is expressed by order and commutativity both of them being preserved by ϕ in both directions. This implies the second assertion of our statement.   □

After this preparation, now we are in a position to prove that every sequential isomorphism between the sets of effects in type I_n algebras ($2 \leq n < \infty$) is an E-isomorphism.

Proposition 2.8.12. *Suppose that* $\mathcal{A}, \mathcal{B}$ *are type* I_n *algebras with* $2 \leq n < \infty$. *Then* ϕ *is an E-isomorphism.*

Proof. Let $A, A' \in E(\mathcal{A})$ be such that $A + A' \in E(\mathcal{A})$. Pick an arbitrary abelian projection $P \in \mathcal{A}$. By Lemma 2.8.9 we have positive central elements Z, Z' in $\mathcal{A}$ such that

$$PAP = ZP, \quad PA'P = Z'P. \tag{2.8.5}$$

Clearly, we can choose a scalar $\lambda \in]0, 1]$ such that $0 \leq \lambda(Z + Z') \leq I$. It is well-known that for $n \geq 2$, any type I_n algebra has a self-adjoint system of $n \times n$ matrix units (see [121, 6.6.3. Lemma]). Therefore, Lemma 2.8.8 applies and using (2.8.5) we can compute

$$\lambda\phi(P)\phi(A + A')\phi(P) =$$
$$\lambda\phi(P(A + A')P) = \phi(\lambda P(A + A')P) = \phi(\lambda(ZP + Z'P)) =$$
$$\phi(P)\phi(\lambda(Z + Z'))\phi(P) = \phi(P)(\phi(\lambda Z) + \phi(\lambda Z'))\phi(P) =$$
$$\phi(P)\phi(\lambda Z)\phi(P) + \phi(P)\phi(\lambda Z')\phi(P) = \phi(\lambda PZP) + \phi(\lambda PZ'P) =$$
$$\phi(\lambda PAP) + \phi(\lambda PA'P) = \lambda(\phi(PAP) + \phi(PA'P)) =$$
$$\lambda\phi(P)(\phi(A) + \phi(A'))\phi(P).$$

Consequently, we obtain that

$$\phi(P)\phi(A + A')\phi(P) = \phi(P)(\phi(A) + \phi(A'))\phi(P)$$

holds for every abelian projection $P \in \mathcal{A}$. Referring to Lemma 2.8.11 we infer that

$$Q(\phi(A + A') - (\phi(A) + \phi(A')))Q = 0$$

is valid for every abelian projection Q in $\mathcal{B}$. By Lemma 2.8.10 this implies that

$$\phi(A + A') = \phi(A) + \phi(A').$$

In particular, we also get that $\phi(A) + \phi(A') \in E(\mathcal{B})$. Since ϕ^{-1} has similar properties as ϕ, we obtain that ϕ is an E-isomorphism.   □

In the case of von Neumann algebras of the remaining types we can follow a different approach based on the solution of the so-called Mackey-Gleason problem due to Bunce and Wright (see Theorem A.12). We have the following result.

Proposition 2.8.13. *Suppose that $\mathcal{A}, \mathcal{B}$ are both of type I_n with $2 < n$ or $\mathcal{A}, \mathcal{B}$ have no type I direct summands. Then ϕ is an E-isomorphism.*

Proof. We know from Lemma 2.8.5 that ϕ, when restricted onto the set of all projections in $\mathcal{A}$, is orthoadditive. We now apply the deep result Theorem A.12 of Bunce and Wright stating that every bounded orthoadditive map from the set of all projections of a von Neumann algebra without type I_2 direct summand into a Banach space can be extended to a continuous linear transformation defined on the whole algebra. Therefore, we have a continuous linear transformation $L : \mathcal{A} \to \mathcal{B}$ such that

$$L(P) = \phi(P) \qquad (P \in P(\mathcal{A})).$$

We claim that L coincides with ϕ on the whole set $E(\mathcal{A})$. To see this, first observe that

$$L(\lambda P) = \phi(\lambda P)$$

whenever $\lambda \in [0,1], P \in P(\mathcal{A})$. Indeed, this follows from the homogeneity of ϕ which was asserted in Lemma 2.8.8 (notice that by [121, 6.5.6. Lemma and 6.6.4. Lemma], $\mathcal{A}$ has a self-adjoint system of $n \times n$ matrix units). Now, let (P_i) be a finite collection of mutually orthogonal projections in $\mathcal{A}$ with sum I and pick scalars $\lambda_i \in [0,1]$. Since $\sum_i \phi(P_i) = I$, we can compute

$$\phi(\sum_i \lambda_i P_i) = \phi(\sum_i \lambda_i P_i)^{\frac{1}{2}} \sum_k \phi(P_k)\phi(\sum_i \lambda_i P_i)^{\frac{1}{2}} =$$

$$\sum_k \phi(\sum_i \lambda_i P_i)^{\frac{1}{2}} \phi(P_k)\phi(\sum_i \lambda_i P_i)^{\frac{1}{2}} = \sum_k \phi((\sum_i \lambda_i P_i)^{\frac{1}{2}} P_k (\sum_i \lambda_i P_i)^{\frac{1}{2}}) =$$

$$\sum_k \phi(\lambda_k P_k) = \sum_k L(\lambda_k P_k) = L(\sum_k \lambda_k P_k).$$

Let $A \in E(\mathcal{A})$ be arbitrary. It follows easily from the spectral theorem of normal operators and the properties of the spectral integral that for any $\epsilon > 0$ there are operators $A_l, A^u \in E(\mathcal{A})$ of the form $\sum_i \lambda_i P_i$ (where λ_i, P_i are like above) and operators $A_l', A^{u'} \in E(\mathcal{A})$ such that the whole set $\{A, A_l, A^u, A_l', A^{u'}\}$ is commutative, $\|A_l - A\|, \|A^u - A\| < \epsilon$, and

$$A_l = AA_l', \quad A = A_u A_u'.$$

We compute

$$L(A_l) = \phi(A_l) = \phi(A^{\frac{1}{2}} A_l' A^{\frac{1}{2}}) = \phi(A)^{\frac{1}{2}} \phi(A_l')\phi(A)^{\frac{1}{2}} \leq \phi(A)^{\frac{1}{2}} I \phi(A)^{\frac{1}{2}} = \phi(A)$$

and one can prove in a similar manner that

$$\phi(A) \le \phi(A_u) = L(A_u).$$

Therefore, we have $L(A_l) \le \phi(A) \le L(A^u)$ and by the continuity of L we infer that

$$L(A) \le \phi(A) \le L(A).$$

Consequently, we have $\phi(A) = L(A)$ for any $A \in E(\mathcal{A})$.

It follows that whenever $A, A' \in E(\mathcal{A})$ are such that $A + A' \in E(\mathcal{A})$, we have

$$\phi(A + A') = L(A + A') = L(A) + L(A') = \phi(A) + \phi(A').$$

In particular, we have $\phi(A) + \phi(A') \in E(\mathcal{B})$. Just as in the proof of Proposition 2.8.12, referring to the fact that ϕ^{-1} has similar properties as ϕ, we obtain that ϕ is an E-isomorphism. $\qquad\square$

Putting together all the information that we have collected so far, it is now an easy task to prove the main result of this section.

Proof of Theorem 2.8.1. Let $P \in \mathcal{A}$ be a nonzero projection. Then $\phi(P)$ is also a nonzero projection and the restriction of ϕ onto $PE(\mathcal{A})P = E(P\mathcal{A}P)$ gives rise to a sequential isomorphism from $E(P\mathcal{A}P)$ onto $E(\phi(P)\mathcal{B}\phi(P))$. By the properties of the sequential isomorphisms formulated in Lemmas 2.8.5, 2.8.7, 2.8.11, it follows that $P\mathcal{A}P$ is of type I, or of type I_n, or has no direct summand of type I, or has no commutative direct summand if and only if the same holds for $\phi(P)\mathcal{B}\phi(P)$.

Consider the type decomposition of $\mathcal{A}$ (see, for example, [121, 6.5.2. Theorem]). For any cardinal number n (not exceeding the dimension of the Hilbert space on which the elements of $\mathcal{A}$ act) let $P_n \in \mathcal{A}$ be a central projection such that the algebra $\mathcal{A}P_n$ is of type I_n (or $P_n = 0$) and in case $Q = I - \sum_n P_n \ne 0$ the algebra $\mathcal{A}Q$ has no type I direct summand. It follows from the first paragraph of the proof that the collection $(\phi(P_n))$ of central projections in $\mathcal{B}$ has the same properties (relating the algebra $\mathcal{B}$ in the place of $\mathcal{A}$, of course).

Clearly, the algebras $\mathcal{A}_1 = \mathcal{A}P_1$ and $\mathcal{B}_1 = \mathcal{B}\phi(P_1)$ are commutative and their direct complements in $\mathcal{A}$, respectively in $\mathcal{B}$ have no commutative direct summands.

Apply our auxiliary result Proposition 2.8.12 for the direct summands $\mathcal{A}P_n$ and $\mathcal{B}\phi(P_n)$ whenever $2 \le n < \infty$. Moreover, apply Proposition 2.8.13 for the direct summands $\mathcal{A}P_n$ and $\mathcal{B}\phi(P_n)$ whenever n is an infinite cardinal and do the same for $\mathcal{A}Q$ and $\mathcal{B}\phi(Q)$. Taking direct sums, we obtain that the restriction of ϕ onto $E(\mathcal{A})(I - P_1) = E(\mathcal{A}(I - P_1))$ is an E-isomorphism onto $E(\mathcal{B}(I - \phi(P_1)))$. By Proposition 2.8.3 this can be extended to a Jordan *-isomorphism from $\mathcal{A}(I - P_1)$ onto $\mathcal{B}(I - \phi(P_1))$. It is well-known that every Jordan *-isomorphism between two von Neumann algebras induces a direct sum decomposition of the underlying algebras according to which the Jordan *-isomorphism under consideration is the direct sum of a *-isomorphism and a *-antiisomorphism (see [121, 10.5.26. Exercise]). This verifies the existence of a decomposition having the properties (i)-(ii) in our theorem. As for the last assertion, observe that if $P_1 \neq 0$ (i.e., when $\mathcal{A}_1, \mathcal{B}_1$ 'really appear') and ϕ maps a nontrivial scalar to itself, then by Proposition 2.8.4, the sequential isomorphism ϕ_1 from $E(\mathcal{A}_1)$ onto $E(\mathcal{B}_1)$ can also be extended to a *-isomorphism from $\mathcal{A}_1$ onto $\mathcal{B}_1$. The proof is complete.   □

2.8.4 Remarks

As we have seen above, sequential isomorphisms have a nice structure on the 'non-commutative parts' of the underlying algebras. What concerns the 'commutative parts' we have shown that these isomorphisms can behave on them quite 'irregularly' (see the discussion after the formulation of Theorem 2.8.1). In the recent paper [280] Marovt presented the complete description of the sequential automorphisms of $E(C(X))$ for a first countable compact Hausdorff space X. He proved that every such map ϕ is of the form

$$\phi(f)(x) = f(\varphi(x))^{p(x)} \qquad (x \in X),$$

where $\varphi : X \to X$ is a homeomorphism and $p : X \to]0, \infty[$ is a continuous function.

In Sections 2.7 and 2.8 we have presented some of our results concerning different automorphisms of the sets of effects as well as certain preservers on them. For further results we refer to the papers [170, 175].

3

Local Automorphisms and Local Isometries of Operator Algebras and Function Algebras

3.1 The Automorphism and Isometry Groups of $B(H)$ Are Topologically Reflexive

3.1.1 Summary

We prove that for an infinite dimensional separable Hilbert space H the automorphism and isometry groups of $B(H)$ are topologically reflexive. The content of this section appeared in the paper [161].

3.1.2 Formulation of the Results

Let $\mathcal{X}$ be a Banach space. From Section 0.5 in the Introduction we recall the concept of algebraically or topologically reflexive subsets of $B(\mathcal{X})$. A subset $\mathcal{E} \subset B(\mathcal{X})$ is called topologically [algebraically] reflexive if $T \in B(\mathcal{X})$ belongs to $\mathcal{E}$ whenever T has the property that $Tx \in \overline{\mathcal{E}x}$ [$Tx \in \mathcal{E}x$] for every $x \in \mathcal{X}$.

As noted in the mentioned part of the Introduction, the study of reflexive linear subspaces of the whole operator algebra on a Hilbert space is one of the most extensive research areas in operator theory. In fact, it is in a close connection with the famous problem of invariant subspaces.

Investigations on reflexivity questions in the case when $\mathcal{X}$ above is an operator algebra (rather than a Hilbert space) and $\mathcal{E}$ is some important set of transformations on $\mathcal{X}$ (for example, the Lie algebra of all derivation or the group of all automorphisms) were begun by Kadison, Larson and Sourour. For the most fundamental results in this direction we refer to Section 0.5 of the Introduction. What concerns the present section we recall an important result of Brešar and Šemrl. They proved in [27] that if H is an infinite dimensional separable Hilbert space then the automorphism group of the whole operator algebra $B(H)$ is algebraically reflexive.

A general result of Shulman [231] says that the linear space of all derivations of a C^*-algebra is topologically reflexive. Having this in mind, it is a natural question whether something similar holds also for the automorphism group of C^*-algebras (reflexivity problems concerning the automorphism group were initiated by Larson in [133]). Unfortunately, the easy answer to this question is negative. As an example, consider the function algebra $C[a, b]$. Choosing a sequence (φ_n) of homeomorphisms of $[a, b]$ onto itself which converges uniformly to a non-injective function $\varphi : [a, b] \rightarrow [a, b]$, we obtain a continuous linear transformation (in fact, a unital homomorphism) ϕ of $C[a, b]$ defined by $\phi(f) = f \circ \varphi$ ($f \in C[a, b]$) which is the pointwise limit of a sequence of automorphisms but ϕ itself is not an automorphism. Therefore, a Shulman-type general result does not hold for the automorphism groups of C^*-algebras. As a possible next step, it is natural to study the problem for particular but important C^*-algebras. In the present section we consider the probably most fundamental (non-commutative) C^*-algebra $B(H)$ and obtain two positive reflexivity results.

Let us turn to the formulation of our statements. We begin with the following key result which gives us the basement to prove our reflexivity theorems. It can be considered as an automatic surjectivity result for the Jordan homomorphisms of $B(H)$. Namely, surprisingly enough, it tells us that the inclusion of merely two extremal operators, one with rank one and one with dense range, in the range of a Jordan homomorphism ϕ of $B(H)$ assures that ϕ is automatically bijective.

Theorem 3.1.1. *Let H be a separable infinite dimensional Hilbert space and $\phi : B(H) \rightarrow B(H)$ be a linear mapping. If ϕ is a Jordan homomorphism whose range contains a rank-one operator and an operator with dense range, then ϕ is either an automorphism or an antiautomorphism.*

Using this theorem we can prove our results on the topological reflexivity of the automorphism and isometry groups of $B(H)$ which follow.

Theorem 3.1.2. *Let H be a separable infinite dimensional Hilbert space. Then the group of all automorphisms of $B(H)$ is topologically reflexive in $B(B(H))$.*

Theorem 3.1.3. *Let H be a separable infinite dimensional Hilbert space. Then the group of all surjective linear isometries of $B(H)$ is topologically reflexive in $B(B(H))$.*

3.1.3 Proofs

We begin with the following two lemmas. The first statement is an automatic continuity result on Jordan homomorphisms. In the proof we use the fact that any Jordan homomorphism $\phi : \mathcal{A} \rightarrow \mathcal{B}$ between algebras $\mathcal{A}$ and $\mathcal{B}$ satisfies

$$\phi(ABA) = \phi(A)\phi(B)\phi(A) \tag{3.1.1}$$

for every $A, B \in \mathcal{A}$ (see [206, 6.3.2 Lemma]).

Lemma 3.1.4. *Let H be as above. Then any Jordan homomorphism ϕ : $B(H) \to B(H)$ is continuous.*

Proof. We first show that there exists a projection (i.e., self-adjoint idempotent) $P \in B(H)$ with infinite rank and infinite corank for which the mapping $A \mapsto \phi(PAP)$ is continuous. Suppose that $\phi \neq 0$. According to [72, Theorem 3], every Jordan ideal of $B(H)$ is an associative ideal. Therefore, the same holds for the kernel of ϕ.

Let P be an infinite dimensional projection. If $\phi(P) = 0$, then using the ideal property of $\ker \phi$, we easily obtain that $I \in \ker \phi$ yielding $\ker \phi = B(H)$ which contradicts $\phi \neq 0$. So, we have $\phi(P) \neq 0$.

Now, let (P_n) be a sequence of pairwise orthogonal infinite dimensional projections. We assert that there exists an $n \in \mathbb{N}$ for which the linear transformation $A \mapsto \phi(P_n A P_n)$ is bounded. To see this, assume on the contrary that for every $n \in \mathbb{N}$ there is an operator $A_n \in B(H)$ such that $\|A_n\| = 1$ and

$$\|\phi(P_n A_n P_n)\| \geq n 2^n \|\phi(P_n)\|^2.$$

Defining $A = \sum_n \frac{1}{2^n} P_n A_n P_n$, we obtain that $A \in B(H)$ and

$$\|\phi(P_n)\|^2 \|\phi(A)\| \geq \|\phi(P_n)\phi(A)\phi(P_n)\| = \|\phi(P_n A P_n)\| =$$
$$\frac{1}{2^n}\|\phi(P_n A_n P_n)\| \geq n\|\phi(P_n)\|^2.$$

Since $\|\phi(P_n)\| \neq 0$ and the inequality above holds for every $n \in \mathbb{N}$, we arrive at a contradiction.

Therefore, we obtain that there is a projection $P \in B(H)$ with infinite rank and infinite corank for which the mapping $A \mapsto \phi(PAP)$ is continuous. Write $P = \sum_{n=1}^{\infty} e_n \otimes e_n$, where (e_n) is an orthonormal sequence. Let (f_n) be a complete orthonormal sequence in H which extends (e_n). Consider the operators

$$T = \sum_n f_n \otimes e_n, \qquad S = \sum_n e_n \otimes f_n.$$

It follows that $TPS = I$ and $SPT = Q$ is a projection with infinite dimensional range. Since

$$\phi(AQ) = \phi(TPSASPT) = \phi(T)\phi(P(SAS)P)\phi(T)$$

and the mapping $A \mapsto \phi(P(SAS)P)$ is obviously continuous, we obtain the continuity of the transformation $A \mapsto \phi(AQ)$ and a similar argument gives the same property for the mapping $A \mapsto \phi(QA)$. Therefore, with the notation $Q^\perp = I - Q$ we have the continuity of the linear mappings

$$A \longmapsto \phi((QA)Q) = \phi(QAQ),$$
$$A \longmapsto \phi((Q^\perp A)Q) = \phi(Q^\perp AQ),$$
$$A \longmapsto \phi(Q(AQ^\perp)) = \phi(QAQ^\perp).$$

Let $Q = \sum_{n=1}^{\infty} e'_n \otimes e'_n$ be with some orthonormal sequence (e'_n). Extend (e'_n) by (f'_n) to a complete orthonormal sequence in H and define

$$R = \sum_n f'_n \otimes e'_n + \sum_n e'_n \otimes f'_n$$

Plainly, $RQR = Q^\perp$ and hence the mapping

$$A \longmapsto \phi(Q^\perp A Q^\perp) = \phi(RQRARQR) = \phi(R)\phi(Q(RAR)Q)\phi(R)$$

is continuous. Finally, since

$$\phi(A) = \phi(QAQ) + \phi(QAQ^\perp) + \phi(Q^\perp A Q) + \phi(Q^\perp A Q^\perp) \qquad (A \in B(H)),$$

we obtain the continuity of ϕ.   □

Lemma 3.1.5. *Let H be as above. If $\phi : B(H) \to B(H)$ is a Jordan homomorphism, then there exists an idempotent $E \in B(H)$ such that for any maximal family (P_n) of pairwise orthogonal rank-one projections, the sequence $(\sum_{k=1}^{n} \phi(P_k))$ converges strongly to E. Moreover, E commutes with the range of ϕ.*

Proof. Some substeps of the argument below have been motivated by ideas in the proof of [209, 2.2 Lemma]. In what follows 'span' denotes generated subspace while '$\overline{\text{span}}$' stands for generated closed subspace.

We call two idempotents $P, Q \in B(H)$ (algebraically) orthogonal if $PQ = QP = 0$ holds true. Observe that, as ϕ is a Jordan homomorphism, it maps idempotents into idempotents and preserves the orthogonality between them. Indeed, if P, Q are mutually orthogonal idempotents, then we have

$$0 = \phi(PQ + QP) = \phi(P)\phi(Q) + \phi(Q)\phi(P)$$

which readily implies the orthogonality between the idempotents $\phi(P)$ and $\phi(Q)$.

Now, let (P_n) be a maximal family of pairwise orthogonal rank-one projections in $B(H)$. Let $S_n = \sum_{k=1}^{n} \phi(P_k)$ and define $E(P)$ as the idempotent having range $R = \overline{\text{span}}\{\text{rng } S_n : n \in \mathbb{N}\}$ and kernel $K = \cap_n \ker S_n$. To verify that $E(P)$ is well-defined we have to prove that $R \oplus K = H$. We first show that $R \cap K = \{0\}$. Let (x_n) be a sequence in $\text{span}\{\text{rng } S_n : n \in \mathbb{N}\}$ which converges to an $r \in K$. From Lemma 3.1.4 we learn that ϕ is bounded. Let M denote the norm of ϕ. For every $\epsilon > 0$ there exists an index $n_0 \in \mathbb{N}$ such that $\|r - x_n\| < \epsilon/M$ $(n \geq n_0)$. It is easy to see that $(\text{rng } S_n)$ is a monotone increasing sequence of subspaces. Therefore, x_n is in the range of an idempotent S_k while r is in its kernel which imply that

$$\|0 - x_n\| = \|S_k r - S_k x_n\| \leq \|S_k\|\|r - x_n\| =$$

$$\|\phi(\sum_{i=1}^{k} P_i)\|\|r - x_n\| \leq M\|r - x_n\| < \epsilon$$

for every $n \geq n_0$. Thus $x_n \longrightarrow 0$ and we have $r = 0$. This gives us that $R \cap K = \{0\}$. We next prove that $R + K = H$. Let $h \in H$. For every $n \in \mathbb{N}$ we know that

$$h = S_n h + (I - S_n)h.$$

Since $(S_n h)$ is a bounded sequence in a Hilbert space (indeed, we have $\|S_n\| \leq M$ for every $n \in \mathbb{N}$), it has a weakly convergent subsequence. For the sake of simplicity, assume that this is $(S_n h)$ itself. Since the closed subspaces of H are weakly closed, we infer that in the sum

$$h = \text{w-}\lim S_n h + \text{w-}\lim (I - S_n)h,$$

the first term belongs to R. Moreover, for every $n \leq m$ we have $S_n S_m = S_m S_n = S_n$. Hence, using the fact that the bounded operators on H are weakly continuous, we obtain that the second term above is in K. Therefore, we have proved that $R \oplus K = H$. So, $E(P) \in B(H)$ is the idempotent corresponding to this decomposition of H.

It is easy to see that $S_n h \longrightarrow E(P)h$ whenever $h \in \text{span}\{\text{rng } S_n : n \in \mathbb{N}\}$ or $h \in K$. As (S_n) is uniformly bounded, using the Banach-Steinhaus theorem, we get that (S_n) converges strongly to $E(P)$. We next show that $E(P)$ is independent of the choice of (P_n). Let (Q_n) be another maximal family of pairwise orthogonal rank-one projections and denote by T_n the nth partial sum of the series $\sum_n \phi(Q_n)$. Clearly, for any $i \in \mathbb{N}$ we have

$$\sum_{k=1}^{n} P_k Q_i + Q_i \sum_{k=1}^{n} P_k \xrightarrow{n \to \infty} 2Q_i$$

in the operator norm. By the continuity of ϕ, we infer that

$$S_n T_k + T_k S_n \xrightarrow{n \to \infty} 2T_k$$

also in the operator norm. Since $S_n \longrightarrow E(P)$ strongly, it follows that $E(P)T_k + T_k E(P) = 2T_k$ ($k \in \mathbb{N}$). We then conclude that $E(P)E(Q) + E(Q)E(P) = 2E(Q)$. Changing the role of P and Q, we have $E(Q)E(P) + E(P)E(Q) = 2E(P)$. Hence, $E(P) = E(Q)$ as we have claimed. In what follows denote $E = E(P)$.

We prove that $E\phi(A) = \phi(A)E$ holds for every $A \in B(H)$. Let Q be a projection of arbitrary rank. Choose a maximal family (P_n) of pairwise orthogonal rank-one projections in such a way that for every $n \in \mathbb{N}$ either $P_n Q = QP_n = 0$ or $P_n Q = QP_n = P_n$ holds true. In the first case we have $P_n Q + QP_n = 0$ and hence

$$\phi(P_n)\phi(Q) + \phi(Q)\phi(P_n) = 0,$$

while in the second one we obtain

$$\phi(P_n)\phi(Q) + \phi(Q)\phi(P_n) = 2\phi(P_n).$$

Since the involved operators $\phi(P_n), \phi(Q)$ are idempotents, an easy algebraic argument proves that

$$\phi(P_n)\phi(Q) = \phi(Q)\phi(P_n)$$

holds in both cases ($n \in \mathbb{N}$). We now have $E\phi(Q) = \phi(Q)E$. Since Q was an arbitrary projection, using the spectral theorem of self-adjoint operators and the continuity of ϕ, we deduce that E commutes with the range of ϕ. This completes the proof of the lemma. $\square$

Now, we are in a position to prove the first theorem of this section.

Proof of Theorem 3.1.1. By Lemma 3.1.4, using the fact that every Jordan ideal of $B(H)$ is an associative ideal (see [72, Theorem 3]), we deduce that $\ker \phi$ is a closed ideal. We intend to prove that $\ker \phi = \{0\}$, that is, ϕ is injective. By a classical theorem of Calkin the only nontrivial closed ideal of $B(H)$ is the ideal $K(H)$ of all compact operators. Hence, supposing $\ker \phi \neq \{0\}$, we have $\ker \phi = K(H)$. We have already learnt that ϕ maps idempotents into idempotents and preserves their orthogonality. It is well-known that there exists an uncountable family I of infinite subsets of $\mathbb{N}$ with the property that every two different members of I have finite intersection. Therefore, one can construct an uncountable family of projections in $B(H)$ such that the product of any two of them is a finite rank projection. Taking images under ϕ, we obtain that $B(H)$ contains an uncountable family (P_α) of pairwise orthogonal nonzero idempotents. But this is a contradiction. To see this, for every α, pick a vector x_α from the range of P_α for which $\|x_\alpha\| \geq \|P_\alpha\|$. We then have

$$\|P_\alpha\|\|x_\alpha - x_\beta\| \geq \|P_\alpha x_\alpha - P_\alpha x_\beta\| = \|x_\alpha - 0\| \geq \|P_\alpha\|$$

and thus $\|x_\alpha - x_\beta\| \geq 1$ holds true whenever $\alpha \neq \beta$. This gives us that the set of all x_α's is a non-separable subspace of a separable metric space. But this is untenable and we get the injectivity of ϕ.

We next assert that there is a rank-one idempotent whose image under ϕ is also rank-one. Let $T = x \otimes y$ be a rank-one operator in the range of ϕ. Suppose that $\langle x, y \rangle \neq 0$. In this case multiplying T by an appropriate scalar, we have a rank-one idempotent P in $\operatorname{rng} \phi$ and one can easily verify that the idempotent $\phi^{-1}(P)$ is also rank-one. Now, suppose that $\langle x, y \rangle = 0$. Then $T^2 = 0$. Let A be such that $\phi(A) = T$. Consider a rank-two projection P with the property that $PAP \neq 0$. We infer $0 \neq \phi(PAP) = \phi(P)T\phi(P)$. Since the operator $\phi(P)T\phi(P)$ is also rank-one, we can assume that it is square-zero (otherwise we are done as above). Consequently, we have a rank-one, square-zero operator S and an operator B with rank not greater than 2 for which $\phi(B) = S$. As ϕ is an injective Jordan homomorphism, B is also square-zero. Suppose that B has rank exactly 2. Then there are two pairs x, x' and y, y' of linearly independent vectors such that

$$B = x \otimes y + x' \otimes y'.$$

Using the property that $B^2 = 0$, it is easy to shows that $\{x, x'\} \perp \{y, y'\}$. Let $\lambda, \mu \in \mathbb{C}$ be such that

$$x' - \lambda x = x_0 \perp x \quad \text{and} \quad y' - \mu y = y_0 \perp y.$$

Let z, z' be orthogonal unit vectors such that $\{z, z'\} \perp \{x, x', y, y'\}$. Consider the operator

$$C = \frac{1}{\|x\|^2} z \otimes x + \frac{1}{\|x_0\|^2}(z' - \lambda z) \otimes x_0 + \frac{1}{\|y\|^2} y \otimes z + \frac{1}{\|y_0\|^2} y_0 \otimes (z' - \mu z).$$

One can easily check that $CBC = z \otimes z + z' \otimes z'$ and hence $\phi(C)S\phi(C) = \phi(CBC) \neq 0$ is a rank-one operator which is the sum of two orthogonal nonzero idempotents. But this is a contradiction. Therefore, we obtain that B is rank-one which means that it is of the form $B = x \otimes y$. Defining

$$C = \frac{1}{\|x\|^2} z \otimes x + \frac{1}{\|y\|^2} y \otimes z$$

with a unit vector $z \perp \{x, y\}$, it follows just as above that $\phi(z \otimes z) = \phi(CBC)$ is a rank-one idempotent. Consequently, we have proved that ϕ maps some rank-one idempotent to a rank-one idempotent.

We now claim that $\phi_{|F(H)}$ is either a homomorphism or an antihomomorphism. In fact, this follows from the classical result Theorem A.5 of Jacobson and Rickart stating that every Jordan homomorphism on a locally matrix algebra is the sum of a homomorphism and an antihomomorphism. So we have a homomorphism $\psi_1 : F(H) \to B(H)$ and an antihomomorphism $\psi_2 : F(H) \to B(H)$ such that $\phi_{|F(H)} = \psi_1 + \psi_2$. Let $P \in F(H)$ be an idempotent for which $\phi(P)$ is rank-one. Since $\phi(P) = \psi_1(P) + \psi_2(P)$ and the terms of this sum are also idempotents, we obtain that either $\psi_1(P) = 0$ or $\psi_2(P) = 0$. By the simplicity of the ring $F(H)$ this gives that either $\psi_1 = 0$ or $\psi_2 = 0$ holds true. Plainly, this is equivalent to our claim. In what follows, with no loss of generality we can assume that $\phi_{|F(H)}$ is a homomorphism.

Now, let $y, z \in H$ be such that $\phi(y \otimes y)z \neq 0$. Define a linear operator T on H by

$$Tx = \phi(x \otimes y)z \quad (x \in H).$$

Then T is bounded and, by the multiplicativity of ϕ on $F(H)$, it is very easy to see that T satisfies $TA = \phi(A)T$ ($A \in F(H)$). If $Tx = 0$, then $TAx = \phi(A)Tx = 0$ holds for every $A \in F(H)$. Obviously, this implies $x = 0$ and hence T is injective.

We claim that T is surjective as well. To show this, we first prove that in our present case the idempotent E given in Lemma 3.1.5 is the identity on H. In fact, since E commutes with the range of ϕ, the mapping

$$\psi : A \longmapsto \phi(A)(I - E)$$

is a Jordan homomorphism of $B(H)$. Moreover, as one can easily verify it, ψ vanishes on every finite rank projection and hence on the whole $F(H)$. Now, by the first part of the proof it follows that $\psi = 0$, i.e., we have $(I - E)\phi(A) = \phi(A)(I - E) = 0$ for every $A \in B(H)$. Considering an operator A with the property that $\phi(A)$ has dense range, we infer that $E = I$. We now are in a position to show that the range of T is dense. Let (P_n) be a maximal family of pairwise orthogonal rank-one projections. The transformation ϕ maps rank-one operators into rank-one operators. Indeed, this follows easily from the fact that there exists a rank-one operator whose image under ϕ is also rank-one and from the assumption that $\phi_{|F(H)}$ is a homomorphism. Therefore, every $\phi(P_n)$ has rank one and the series $\sum_n \phi(P_n)$ converges strongly to $E = I$. Consequently, we have sequences (e_n), (f_n) in H for which $f_n \otimes e_n = \phi(P_n)$ and

$$\sum_n \langle x, e_n \rangle \langle f_n, y \rangle = \langle x, y \rangle \qquad (x, y \in H)$$

holds true. This immediately implies that the subspace generated by $\{f_n : n \in \mathbb{N}\}$ is dense. But every $f_n \in \operatorname{rng} T$. Indeed, since $TP_n = \phi(P_n)T = f_n \otimes e_n T = f_n \otimes T^* e_n$ and, by the injectivity of T, we have $TP_n \neq 0$, hence it follows that $f_n \in \operatorname{rng} T$. Therefore, $\operatorname{rng} T$ is dense. We next prove that T is in fact surjective. Let $y \in H$ be arbitrary and let (x_n) be a sequence in H such that $Tx_n \longrightarrow y$. Since $TA = \Phi(A)T$ $(A \in F(H))$, we obtain that (TAx_n) is convergent for every finite rank operator A. Thus, for every $u \in H$ and for a fixed nonzero $v \in H$ we infer that the sequences $(\langle x_n, u \rangle Tv) = ((Tv \otimes u)(x_n))$ and, consequently, $(\langle x_n, u \rangle)$ are convergent. Plainly, this gives us that (x_n) converges weakly to some $x \in H$. Using the weak continuity of T, we have $Tx = y$ and this proves the surjectivity of T.

It is now apparent that $\phi(A) = TAT^{-1}$ $(A \in F(H))$. In particular, we obtain that the range of ϕ contains every finite rank operator. Following the proof of Herstein's theorem (see Theorem A.7) given in [206, 6.3.2 Lemma, 6.3.6 Lemma and 6.3.7 Theorem] one can verify that this inclusion of $F(H)$ in the range of ϕ implies that ϕ is either a homomorphism or an antihomomorphism. If ϕ is an antihomomorphism, then for every $A, B \in F(H)$ we have

$$TAT^{-1}TBT^{-1} = \phi(A)\phi(B) = \phi(BA) = T(BA)T^{-1},$$

i.e., we have $AB = BA$. Since this is a contradiction, it follows that ϕ is a homomorphism and one can check that $TA = \phi(A)T$ and hence $\phi(A) = TAT^{-1}$ holds for every $A \in B(H)$. This completes the proof of the theorem. $\square$

Remark 3.1.6. As a matter of curiosity we mention that one could ask whether only one operator in the range of a Jordan homomorphism of $B(H)$ can be enough to imply a similar automatic surjectivity result as in Theorem 3.1.1. The answer is easily seen to be affirmative. Indeed, consider a rank-one operator $x \otimes y$ with the property that $\langle x, y \rangle = 2$. If $A = I - x \otimes y$ is in the range of a Jordan homomorphism of $B(H)$, then we obtain that its square, $A^2 = I$ also belongs to it and hence the same is true for $x \otimes y$. Theorem 3.1.1 now applies.

Proof of Theorem 3.1.2. Let $\phi : B(H) \to B(H)$ be a continuous linear transformation with the property that for every $A \in B(H)$ there is a sequence (ϕ_n) of automorphisms of $B(H)$ (depending on A) such that $\phi(A) = \lim_n \phi_n(A)$. Clearly, ϕ maps idempotents into idempotents. By Theorem A.4 it follows that ϕ is a Jordan homomorphism.

Let A be a rank-one operator whose spectrum consists of two points. Since $\phi(A)$ is the norm limit of a sequence of operators all of them being similar to A, it follows from the continuity properties of the spectrum that the operator $\phi(A)$ of rank at most one has the same spectrum as A. Consequently, the rank of $\phi(A)$ is exactly one. As the relation $\phi(I) = I$ obviously holds, the conditions in Theorem 3.1.1 are fulfilled. Consequently, we obtain that ϕ is either an automorphism or an antiautomorphism. If ϕ is an antiautomorphism, then for a unilateral shift $U \in B(H)$ we infer that $I = \phi(I) = \phi(U^*U) = \phi(U)\phi(U^*)$ which means that $\phi(U)$ has a right inverse. But $\phi(U)$ is the limit of a sequence of operators which are all similar to U. Therefore, neither the elements of this sequence nor its limit $\phi(U)$ have right inverses. Since this is a contradiction, we conclude that ϕ is an automorphism and the proof is complete. □

Remark 3.1.7. It seems natural to ask what happens in the finite dimensional case. The argument given in the proof of Theorem 3.1.2 shows that if H is finite dimensional and ϕ is a linear mapping on $B(H)$ which can be approximated at every operator by a sequence of automorphisms, then ϕ is either an automorphism or an antiautomorphism. However, nothing more can be asserted. To see this, consider the mapping $\psi : A \mapsto A^{tr}$ (recall that A^{tr} stands for transpose of A with respect to a fixed complete orthogonal system in H). This is an antiautomorphism which has the property that for every A, $\psi(A) = A^{tr}$ is similar to A (a known fact in matrix theory). In other words, ψ locally belongs to the automorphism group of $B(H)$, but it fails to be an automorphism (if $\dim H \geq 2$). So, in the finite dimensional case the automorphism group of $B(H)$ is not reflexive even algebraically.

We next prove the topological reflexivity of the isometry group.

Proof of Theorem 3.1.3. According to Theorem A.9, every surjective linear isometry of $B(H)$ is either of the form $A \mapsto UAV$ or of the form $A \mapsto UA^{tr}V$ with some unitaries U, V.

Let $\phi : B(H) \to B(H)$ be a linear mapping with the property that for every $A \in B(H)$ there exists a sequence (ϕ_n) of surjective linear isometries such that $\phi(A) = \lim_n \phi_n(A)$. Plainly, ϕ is a linear isometry. Our aim is to show that ϕ is surjective as well. Since the surjective linear isometries of $B(H)$ map the unitary operators into unitary operators, we obtain that the same holds for ϕ. Of course, we may suppose that $\phi(I) = I$. Then by the result Theorem A.2 of Russo and Dye on the structure of unitary group preserving mappings we obtain that ϕ is a Jordan *-homomorphism. Since ϕ is easily seen to preserve the rank-one operators, hence Theorem 3.1.1 can be applied again to obtain the surjectivity of ϕ. This completes the proof. □

Remark 3.1.8. Here we give an example in order to emphasize how the topological reflexivity of the groups of all automorphisms as well as surjective isometries of $B(H)$ should be considered exceptional even among such 'nice' C^*-subalgebras of $B(H)$ as the standard ones (those which contain all finite rank operators).

So, let us consider the C^*-algebra $K(H)$ of all compact operators on H. Let (e_n) be a fixed complete orthonormal sequence in H. Choose unitary operators U_n such that

$$U_n e_k = e_{k+1} \qquad (n \in \mathbb{N}, \, 1 \leq k \leq n)$$

hold true. If U denotes the unilateral shift corresponding to the sequence (e_n), then it is obvious that $U_n e_k \xrightarrow{n \to \infty} U e_k$ ($k \in \mathbb{N}$). The Banach-Steinhaus theorem gives us that (U_n) converges strongly to U. Let

$$\phi(A) = UAU^*, \quad \phi_n(A) = U_n A U_n^* \quad (n \in \mathbb{N}).$$

Clearly, ϕ_n is a *-automorphism and a surjective linear isometry of $K(H)$ ($n \in \mathbb{N}$). By the strong convergence of (U_n) to U we infer that $\phi_n(A) \longrightarrow \phi(A)$ holds for every rank-one operator A and consequently, by the Banach-Steinhaus theorem again, for every $A \in K(H)$ as well. Therefore, ϕ is in the topological reflexive closures of the automorphism and isometry groups, but it is not surjective. Consequently, the automorphism group and the isometry group of $K(H)$ are not topologically reflexive.

3.1.4 Remarks

In relation with the theorems of the present section and Remark 3.1.8 above we note that the topological reflexivity of the automorphism and isometry groups fails to be true even in the case of more 'full' standard C^*-subalgebras of $B(H)$ than $K(H)$. To see this, we refer to the results of our paper [168]. It was proved there that on the one hand, the automorphism and isometry groups of the extension of $K(H)$ by any separable commutative C^*-algebra are algebraically reflexive. (These algebras are the basic objects in the famous Brown-Douglas-Fillmore theory which was elaborated to solve the classification problem of essentially normal operators, one of the main achievements in operator theory.) On the other hand, we showed there that those groups are not topologically reflexive in the case of the probably most important extensions by the algebra $C(\mathbb{T})$ of all continuous complex valued functions on the perimeter of the unit disc. In particular, this is the case with the C^*-algebra generated by the unilateral shift (Toeplitz algebra).

Keeping these 'negative' results in mind one might be interested in the question whether there exists at all a proper standard C^*-subalgebra of $B(H)$ whose automorphism and isometry groups are topologically reflexive. The answer is affirmative. In our paper [162] we presented such an algebra. Namely, we proved the following assertion. Let H be a separable infinite dimensional

Hilbert space. Let $H_1, \ldots, H_n$ be a finite sequence of pairwise orthogonal closed subspaces of H which generate H. Then the automorphism and isometry groups of the C^*-algebra $K(H) + (B(H_1) \oplus \ldots \oplus B(H_n))$ are topologically reflexive.

There are even von Neumann algebras whose automorphism and isometry groups are not topologically reflexive. In fact, in [16] we showed that any infinite dimensional commutative von Neumann algebra acting on a separable Hilbert space has this property.

More or less this is what we know about the topological reflexivity of the automorphism and isometry groups of (non-commutative) operator algebras. Therefore, this area of research is still open and a lot of work could be done. A natural direction of further investigations would be, for example, to consider the question for certain classes of von Neumann algebras. But as it can be suspected from the proofs of our results above, the problem is probably quite difficult.

3.2 Reflexivity of the Automorphism and Isometry Groups of $C(X)$

3.2.1 Summary

We prove that if X is a first countable compact Hausdorff space then the automorphism and isometry groups of the function algebra $C(X)$ are algebraically reflexive. The content of this section represents a small fragment of the papers [196] and [44].

3.2.2 Formulation of the Result

In Section 3.1 we proved that the automorphism and isometry groups of the C^*-algebra $B(H)$ (H being an infinite dimensional separable Hilbert space) are topologically reflexive. In this section we study the reflexivity of those groups in the case of another fundamental class of C^*-algebras, namely, the commutative ones. Any such algebra is well-known to be isomorphic to the algebra $C(X)$ of all continuous complex valued functions on a compact Hausdorff space X.

First observe that unlike in the case of $B(H)$, the topological reflexivity of the automorphism and isometry groups of $C(X)$ is out of question as can be seen from the example given in the subsection 3.1.2 of Section 3.1. So, we can expect at most algebraic reflexivity. However, even this does not hold for every compact Hausdorff space X. In fact, in Remark 3.2.2 below we present a counterexample. Therefore, some restriction should be made on the underlying topological space in order to have a positive result. In what follows we show that the first countability is a suitable such condition. (Recall that a topological space is called first countable if each of its points has a countable basis of neighbourhoods.) More precisely, we have the following result.

Theorem 3.2.1. *Let X be a first countable compact Hausdorff space. Then the group of all automorphisms and the group of all surjective linear isometries of $C(X)$ are algebraically reflexive.*

3.2.3 Proof

In what follows we shall use the concept of local isometries. Let $\mathcal{X}$ be a Banach space. The linear map $\phi : \mathcal{X} \to \mathcal{X}$ is called a local isometry if for every $x \in \mathcal{X}$ there exists a surjective linear isometry ϕ_x of $\mathcal{X}$ such that $\phi(x) = \phi_x(x)$.

Proof of Theorem 3.2.1. We begin with the reflexivity of the isometry group. By the Banach-Stone theorem (see Theorem A.10) every surjective linear isometry of $C(X)$ is of the form

$$f \longmapsto \tau \cdot f \circ \varphi, \tag{3.2.1}$$

where τ is a continuous complex valued function of modulus 1 and $\varphi : X \to X$ is a homeomorphism.

Now, let $\phi : C(X) \to C(X)$ be a local isometry. It follows from the local form of ϕ (see (3.2.1)) that it maps continuous functions of modulus 1 into functions of the same kind. But in the commutative C^*-algebra $C(X)$ these functions are exactly the unitary elements. Therefore, ϕ preserves the unitary group and we can apply the result Theorem A.2 of Russo and Dye. It then follows that ϕ is a homomorphism multiplied by a fixed continuous function of modulus 1. Without any loss of generality we may assume that this function is identically 1, i.e., $\phi(1) = 1$. Therefore, ϕ is a unital endomorphism of the algebra $C(X)$. It is a folk result that every such transformation is of the form

$$f \longmapsto f \circ \varphi$$

with some continuous mapping $\varphi : X \to X$. Since ϕ is an isometry, by Urysohn's lemma the mapping φ must be surjective. It remains to prove that φ is injective as well. To this end, let $x, y \in X$ be points such that $\varphi(x) = \varphi(y) = z$. We construct a continuous function $f : X \to \mathbb{C}$ in the following way. Let (U_n) be a monotone decreasing sequence of open sets in X such that $\cap_n U_n = \{z\}$ (this is the place where we use the first countability of X). By Urysohn's lemma, for every n we have a continuous function $f_n : X \to [0,1]$ such that

$$f_n(z) = 1 \text{ and } f_n(t) = 0 \ (t \in X \setminus U_n).$$

Let

$$f = 1 - \sum_{n=1}^{\infty} \frac{1}{2^n} f_n.$$

It is easy to see that f is a continuous function and $f(t) = 0$ if and only if $t = z$. Since ϕ is a local isometry, it follows that to this f there exist a homeomorphism $\varphi_f : X \to X$ and a continuous function $\tau_f : X \to \mathbb{C}$ of modulus 1 such that

$$f \circ \varphi = \phi(f) = \tau_f \cdot f \circ \varphi_f.$$

It follows that

$$\{x, y\} \subset (f \circ \varphi)^{-1}(0) = (f \circ \varphi_f)^{-1}(0) = \varphi_f^{-1}(z).$$

Since φ_f is bijective, we obtain $x = y$. This means that φ is injective which implies the surjectivity of ϕ. This verifies the statement of our theorem for the isometry group of $C(X)$.

As for the automorphism group, the result follows easily from the fact that every automorphism of $C(X)$ is of the form

$$f \longmapsto f \circ \varphi$$

where $\varphi : X \to X$ is a homeomorphism and from our statement concerning the isometry group.   □

Remark 3.2.2. As we have promised above, we show that in Theorem 3.2.1 the condition on first countability is essential. Consider the compact Hausdorff space $\mathbb{N}^* = \beta\mathbb{N} \setminus \mathbb{N}$, where $\beta\mathbb{N}$ stands for the Stone-Čech compactification of the positive integers. This is a very exotic space which contains two disjoint copies of itself which are clopen in $\mathbb{N}^*$ (see [244, 3.10, 3.14, 3.15]). This gives us the existence of a surjective non-injective continuous mapping $\varphi : \mathbb{N}^* \to \mathbb{N}^*$. Now consider the transformation $\phi : f \mapsto f \circ \varphi$. It is a strange property of $\mathbb{N}^*$ that if two functions $f, g \in C(\mathbb{N}^*)$ have the same range, then there is a homeomorphism $h : \mathbb{N}^* \to \mathbb{N}^*$ such that $f = g \circ h$ (see [244, first section on p. 83]). It then follows that for every $f \in C(\mathbb{N}^*)$ there exists a homeomorphism $h : \mathbb{N}^* \to \mathbb{N}^*$ such that $f \circ \varphi = f \circ h$. Consequently, ϕ is a local automorphism and a local isometry of $C(\mathbb{N}^*)$ which is not surjective.

3.2.4 Remarks

In this section we have presented the most simple and basic result on the reflexivity of the automorphism and isometry groups of function algebras. Our deeper results can be found in [44] (concerning spaces of measurable functions, Hardy spaces, Banach algebras of holomorphic functions and Fréchet algebras of holomorphic functions), and in [197] (concerning group algebras of compact metric groups).

Our results motivated further investigations by other researchers on the considered reflexivity problem concerning spaces of scalar-valued functions. For corresponding results we refer to the papers [41, 109, 214]. For example, the main result in [41] states that the automorphism and isometry groups of $L_\infty[0, 1]$ are algebraically reflexive. In [109] Jarosz and Rao proved that the local isometries (for the definition see the subsection 3.2.3 above) of any finite-codimensional subspace of $C(X)$ (X being first countable) are all surjective linear isometries.

3.3 Reflexivity of the Automorphism and Isometry Groups of the Suspension of $B(H)$

3.3.1 Summary

In this section we show that the automorphism and isometry groups of the suspension of $B(H)$ (H being a separable infinite dimensional Hilbert space) are algebraically reflexive. This means that every local automorphism, respectively local isometry of the tensor product $C_0(\mathbb{R}) \otimes B(H)$ is an automorphism, respectively a surjective linear isometry. The results of this section appeared in the paper [188].

3.3.2 Formulation of the Results

In Section 3.1 we proved that the automorphism and isometry groups of the operator algebra $B(H)$, H being an infinite dimensional separable Hilbert space, are topologically reflexive. The result in Section 3.2 says that those groups are algebraically reflexive in the case of the function algebra $C(X)$, X being a first countable compact Hausdorff space. In the present section we investigate the reflexivity of the automorphism and isometry groups of the tensor product of $C(X)$ and $B(H)$. In fact, we consider not only compact Hausdorff spaces X but also locally compact ones and the algebra $C_0(X)$ of all continuous complex valued functions on them which vanish at infinity.

Let $\mathcal{A}$ be a C^*-algebra. The tensor product $C_0(X) \otimes \mathcal{A}$ is well-known to be isomorphic to the algebra $C_0(X, \mathcal{A})$ of all continuous functions from X to $\mathcal{A}$ which vanish at infinity. Among the C^*-algebras $C_0(X) \otimes \mathcal{A}$ the probably most important one is $S\mathcal{A} = C_0(\mathbb{R}) \otimes \mathcal{A}$. This is called the suspension of $\mathcal{A}$ and it plays very important role in the K-theory of C^*-algebras. The reason is that the K_1-group of $\mathcal{A}$ is isomorphic to the K_0-group of $S\mathcal{A}$. The results of the present section culminate in Corollary 3.3.5 which states that the automorphism and isometry groups of the suspension of $B(H)$ are algebraically reflexive.

Let us turn to the precise formulations of the results of the section. For a C^*-algebra $\mathcal{A}$, denote by $\mathrm{Aut}(\mathcal{A})$ and $\mathrm{Iso}(\mathcal{A})$ the group of automorphisms (i.e. multiplicative linear bijections) and the group of surjective linear isometries of $\mathcal{A}$, respectively. In what follows let H be a separable infinite dimensional Hilbert space. In our first theorem we describe the general form of the elements of $\mathrm{Aut}(C_0(X, B(H)))$ and $\mathrm{Iso}(C_0(X, B(H)))$.

Theorem 3.3.1. *Let X be a locally compact Hausdorff space. A linear map $\Phi : C_0(X, B(H)) \to C_0(X, B(H))$ is an automorphism if and only if there exist a function $\tau : X \to \mathrm{Aut}(B(H))$ and a bijection $\varphi : X \to X$ so that*

$$\Phi(f)(x) = [\tau(x)](f(\varphi(x))) \qquad (f \in C_0(X, B(H)), x \in X). \qquad (3.3.1)$$

Similarly, a linear map $\Phi : C_0(X, B(H)) \to C_0(X, B(H))$ is a surjective linear isometry if and only if there exist a function $\tau : X \to \mathrm{Iso}(B(H))$ and a bijection $\varphi : X \to X$ so that Φ is of the form (3.3.1).

Moreover, if the linear map $\Phi : C_0(X, B(H)) \to C_0(X, B(H))$ is an automorphism, respectively a surjective linear isometry, then for the maps τ, φ appearing in (3.3.1) we obtain that $x \mapsto \tau(x), x \mapsto \tau(x)^{-1}$ are strongly continuous and that $\varphi : X \to X$ is a homeomorphism.

The following two results show how the algebraic reflexivity of our groups in the case of $C_0(X)$ implies the algebraic reflexivity of $\mathrm{Aut}(C_0(X) \otimes B(H))$ and $\mathrm{Iso}(C_0(X) \otimes B(H))$. Note that in general $\mathrm{Aut}(C_0(X))$ and $\mathrm{Iso}(C_0(X))$ are not algebraically reflexive. In fact, as one can easily verify, this is the case with any uncountable discrete space. (If one considers only compact spaces the situation is not so trivial, see Remark 3.2.2.)

Theorem 3.3.2. *Let X be a locally compact Hausdorff space. If the automorphism group of $C_0(X)$ is algebraically reflexive, then so is the automorphism group of $C_0(X, B(H))$.*

Theorem 3.3.3. *Let X be a σ-compact locally compact Hausdorff space. If the isometry group of $C_0(X)$ is algebraically reflexive, then so is the isometry group of $C_0(X, B(H))$.*

To obtain the algebraic reflexivity of the automorphism and isometry groups of the suspension of $B(H)$ we prove the following assertion.

Theorem 3.3.4. *Let $\Omega \subset \mathbb{R}^n$ be an open convex set. The automorphism and isometry groups of $C_0(\Omega)$ are algebraically reflexive.*

We remark that the proof of this result will show how difficult it might be to treat our reflexivity problem for the tensor product of general C^*-algebras or even for the suspension of any C^*-algebra with algebraically reflexive automorphism and isometry groups. For the details see the subsection 3.3.4 Remarks.

Finally, we arrive at the statement announced in the summary.

Corollary 3.3.5. *The automorphism and isometry groups of the suspension of $B(H)$ are algebraically reflexive.*

3.3.3 Proofs

We begin with the following lemma which characterizes certain closed ideals in $C_0(X, B(H))$.

Lemma 3.3.6. *Let X be a locally compact Hausdorff space. A closed ideal $\mathfrak{J}$ in $C_0(X, B(H))$ is of the form*

$$\mathfrak{J} = \mathfrak{J}_{x_0} = \{f \in C_0(X, B(H)) : f(x_0) = 0\}$$

for some point $x_0 \in X$ if and only if $\mathfrak{I}$ is a proper subset of a maximal ideal $\mathfrak{I}_m$ in $C_0(X, B(H))$, there is no closed ideal properly between $\mathfrak{I}$ and $\mathfrak{I}_m$, and $\mathfrak{I}$ is not the intersection of two different maximal ideals in $C_0(X, B(H))$.

Proof. The structure of closed ideals in Banach algebras of vector valued functions is known. See, for example, [199, Remark on p. 342]. Using this result, $\mathfrak{I}$ is a closed ideal in $C_0(X, B(H))$ if and only if it is of the form

$$\mathfrak{I} = \{f \in C_0(X, B(H)) : f(x) \in \mathcal{I}_x\},$$

where every $\mathcal{I}_x$ is a closed ideal of $B(H)$. By the separability of H it follows that every $\mathcal{I}_x$ is either $\{0\}$ or $K(H)$ (the ideal of all compact operators) or $B(H)$. By the help of Uryson's lemma on the construction of continuous functions on X with compact support, one can readily verify that the maximal ideals in $C_0(X, B(H))$ are exactly those ideals which are of the form

$$\mathfrak{I} = \{f \in C_0(X, B(H)) : f(x_0) \in K(H)\}$$

for some point $x_0 \in X$. Now, the statement of the lemma follows quite easily. □

Proof of Theorem 3.3.1. We begin with the proof of the statement concerning isometries. Let Φ be a surjective linear isometry of $C_0(X, B(H))$. As a consequence of a deep result due to Kaup (see, for example, [61]) we obtain that every surjective linear isometry ϕ between not necessarily unital C^*-algebras $\mathcal{A}$ and $\mathcal{B}$ has a certain algebraic property, namely ϕ is a Jordan *-triple isomorphism. This means that ϕ satisfies the equality

$$\phi(ab^*c) + \phi(cb^*a) = \phi(a)\phi(b)^*\phi(c) + \phi(c)\phi(b)^*\phi(a)$$

for every $a, b, c \in \mathcal{A}$. We assert that ϕ preserves the closed ideals in both directions. Indeed, if $\mathcal{I} \subset \mathcal{A}$ is a closed ideal, then, since $\mathcal{I} = \mathcal{I}^*$, we have

$$\phi(a)\phi(b)^*\phi(c) + \phi(c)\phi(b)^*\phi(a) \in \phi(\mathcal{I}) \qquad (a, c \in \mathcal{A}, b \in \mathcal{I}).$$

Let $\mathcal{I}' = \phi(\mathcal{I})$. We obtain that $a'\mathcal{I}'^*c' + c'\mathcal{I}'^*a' \in \mathcal{I}'$ ($a', c' \in \mathcal{B}$). Since $\mathcal{I}'$ is a closed linear subspace of $\mathcal{B}$, if c' runs through an approximate identity, we deduce

$$a'\mathcal{I}'^* + \mathcal{I}'^*a' \in \mathcal{I}' \qquad (a' \in \mathcal{B}). \tag{3.3.2}$$

If now a' runs through an approximate identity, then we obtain

$$\mathcal{I}'^* \subset \mathcal{I}'. \tag{3.3.3}$$

This easily implies that in fact we have $\mathcal{I}'^* = \mathcal{I}'$. We next infer from (3.3.2) that $a'\mathcal{I}' + \mathcal{I}'a' \subset \mathcal{I}'$ ($a' \in \mathcal{B}$), i.e., $\mathcal{I}'$ is a closed Jordan ideal of $\mathcal{B}$. It is known that in the case of C^*-algebras, every closed Jordan ideal is an associative ideal (see, for example, [53, 5.3. Theorem]) and hence the same is true for $\mathcal{I}'$.

By Lemma 3.3.6 we infer that our map $\varPhi$ preserves the ideals

$$\mathfrak{I}_x = \{f \in C_0(X, B(H)) \ : \ f(x) = 0\} \qquad (x \in X)$$

in both directions. This gives us that there exists a bijection $\varphi : X \rightarrow X$ for which

$$\varPhi(f)(x) = 0 \Longleftrightarrow f(\varphi(x)) = 0 \tag{3.3.4}$$

holds true for every $f \in C_0(X, B(H))$ and $x \in X$. For any $x \in X$, let us define $\tau(x)$ by the formula

$$[\tau(x)](f(\varphi(x))) = \varPhi(f)(x) \qquad (f \in C_0(X, B(H))). \tag{3.3.5}$$

Because of (3.3.4) we obtain that $\tau(x)$ is a well-defined injective linear map on $B(H)$. Since $\varPhi$ is surjective, we have the surjectivity of $\tau(x)$. Now, we compute

$$[\tau(x)](f(\varphi(x))g(\varphi(x))^* f(\varphi(x))) =$$
$$\varPhi(fg^* f)(x) = \varPhi(f)(x)\varPhi(g)(x)^*\varPhi(f)(x) =$$
$$[\tau(x)](f(\varphi(x)))([\tau(x)](g(\varphi(x))))^*[\tau(x)](f(\varphi(x)))$$

for every $f, g \in C_0(X, B(H))$. This implies that $\tau(x)$ is a Jordan *-triple automorphism of $B(H)$. The Jordan *-triple homomorphisms clearly preserve the partial isometries. As every operator with norm less than 1 is the average of unitaries, it now follows that $\tau(x)$ is a contraction. Applying the same argument to the inverse of $\tau(x)$, we obtain that $\tau(x) \in \mathrm{Iso}(B(H))$. This proves that $\varPhi$ is of the form (3.3.1) given in the statement of our theorem.

Let now $\varPhi : C_0(X, B(H)) \rightarrow C_0(X, B(H))$ be a linear map of the form

$$\varPhi(f)(x) = [\tau(x)](f(\varphi(x))) \qquad (f \in C_0(X, B(H)), x \in X), \tag{3.3.6}$$

where $\tau : X \rightarrow \mathrm{Iso}(B(H))$ and $\varphi : X \rightarrow X$ is a bijection. The function φ is continuous. Indeed, this follows easily from the equality $\|f(\varphi(x))\| = \|\varPhi(f)(x)\|$ and from Uryson's lemma. To see the strong continuity of $\tau :$ $X \rightarrow \mathrm{Iso}(B(H))$, let (x_α) be a net in X converging to $x \in X$. Let $y_\alpha = \varphi(x_\alpha), y = \varphi(x)$. We may suppose that every y_α belongs to a fixed compact neighbourhood of y. If $f \in C_0(X)$ is identically 1 on this neighbourhood, then for every operator $A \in B(H)$ we have

$$[\tau(x_\alpha)](A) = [\tau(x_\alpha)](f(\varphi(x_\alpha))A) = \varPhi(fA)(x_\alpha) \longrightarrow \varPhi(fA)(x) =$$
$$[\tau(x)](f(\varphi(x))A) = [\tau(x)](A).$$

Next, from the equality

$$\|[\tau(x_\alpha)^{-1}](A) - [\tau(x)^{-1}](A)\| = \|[\tau(x_\alpha)^{-1}\tau(x)\tau(x)^{-1}](A) - [\tau(x)^{-1}](A)\| =$$
$$\|[\tau(x)]([\tau(x)^{-1}](A)) - [\tau(x_\alpha)]([\tau(x)^{-1}](A))\|$$

we get the strong continuity of the map $x \mapsto \tau(x)^{-1}$. We prove that φ^{-1} is also continuous. Since Φ maps into $C_0(X, B(H))$, it is quite easy to see from (3.3.6) that $f \circ \varphi \in C_0(X)$ holds true for every $f \in C_0(X)$. If $K \subset X$ is an arbitrary compact set and $f \in C_0(X)$ is a function which is identically 1 on K, then it follows from $f \circ \varphi \in C_0(X)$ that there exists a compact set $K' \subset X$ for which $\varphi(x) \in K^c$ holds true for all $x \in K'^c$. Thus, we have $K \subset \varphi(K')$. Let (x_α) be a net in X such that $(\varphi(x_\alpha))$ converges to some $\varphi(x)$. Obviously, we may suppose that every $\varphi(x_\alpha)$ belongs to a compact neighbourhood K of $\varphi(x)$. By what we have just seen, there exists a compact set $K' \subset X$ which contains the net (x_α) and the point x as well. Since K' is compact, the net (x_α) has a convergent subnet. Because of the continuity of the bijection φ, it is easy to see that the limit of this subnet is x. The continuity of φ^{-1} is now apparent. Finally, one can verify quite readily that Φ is a surjective linear isometry of $C_0(X, B(H))$.

Let us turn to the proof of our statement concerning automorphisms. So, let Φ be an automorphism of $C_0(X, B(H))$. Every automorphism Ψ of a C^*-algebra $\mathcal{A}$ is continuous and its norm equals the norm of its inverse. This follows, for example, from [220, 4.1.12. Lemma] and from the proof of [220, 4.1.13. Proposition] where it is proved that $\|a\|/\|\Psi\| \leq \|\Psi(a)\|$ $(a \in \mathcal{A})$ which implies that $\|\Psi^{-1}\| \leq \|\Psi\|$. Using these facts, one can get the form (3.3.1) in a way very similar to what was followed in the case of isometries. Let now $\Phi : C_0(X, B(H)) \to C_0(X, B(H))$ be a linear map of the form

$$\Phi(f)(x) = [\tau(x)](f(\varphi(x))) \qquad (f \in C_0(X, B(H)), x \in X), \qquad (3.3.7)$$

where $\tau : X \to \mathrm{Aut}(B(H))$ and $\varphi : X \to X$ is a bijection. We show that φ is continuous. Let (x_α) be a net in X converging to $x \in X$. By (3.3.7) we have

$$f(\varphi(x_\alpha))I = [\tau(x_\alpha)](f(\varphi(x_\alpha))I) \longrightarrow [\tau(x)](f(\varphi(x))I) = f(\varphi(x))I$$

for every $f \in C_0(X)$. Referring to Uryson's lemma again, we infer that $\varphi(x_\alpha) \to \varphi(x)$. This verifies the continuity of φ. We claim that the function τ is bounded. In fact, by the principle of uniform boundedness, in the opposite case we would obtain that there exists an operator $A \in B(H)$ for which $[\tau(.)](A)$ is not bounded. Then there is a sequence (x_n) in X with the property that $\|[\tau(x_n)](A)\| > n^3$ $(n \in \mathbb{N})$. Using Uryson's lemma, it is an easy task to construct a nonnegative function $f \in C_0(X)$ for which $f(\varphi(x_n)) \geq 1/n^2$. Indeed, for every $n \in \mathbb{N}$ let $f_n : X \to [0,1]$ be a continuous function with compact support such that $f_n(\varphi(x_n)) = 1$ and define $f = \sum_n (1/n^2)f_n$. We have $\|\Phi(fA)(x_n)\| = \|f(\varphi(x_n))[\tau(x_n)](A)\| > n$ $(n \in \mathbb{N})$ which contradicts the boundedness of the function $\Phi(fA)$. The strong continuity of τ can be proved as it was done in the case of isometries. Using the inequality

$$\|[\tau(x_\alpha)^{-1}](A) - [\tau(x)^{-1}](A)\| = \|[\tau(x_\alpha)^{-1}\tau(x)\tau(x)^{-1}](A) - [\tau(x)^{-1}](A)\| \leq$$
$$\|\tau(x_\alpha)^{-1}\|\|[\tau(x)]([\tau(x)^{-1}](A)) - [\tau(x_\alpha)]([\tau(x)^{-1}](A))\| =$$
$$\|\tau(x_\alpha)\|\|[\tau(x)]([\tau(x)^{-1}](A)) - [\tau(x_\alpha)]([\tau(x)^{-1}](A))\|$$

and the boundedness of τ, we get the strong continuity of the map $x \mapsto \tau(x)^{-1}$. The proof can be completed as in the case of isometries. $\qquad\square$

The following two lemmas are needed for the proof of Theorem 3.3.2.

Lemma 3.3.7. *Let* τ, τ_1, τ_2 *be automorphisms of* $B(H)$ *and let* $\lambda \in \mathbb{C}$, $0 \neq \lambda_1, \lambda_2 \in \mathbb{C}$ *be scalars so that*

$$\lambda\tau(A) = \lambda_1\tau_1(A) + \lambda_2\tau_2(A) \qquad (A \in B(H)).$$

Then we have $\tau_1 = \tau_2$.

Proof. Since the automorphisms of $B(H)$ are all inner (see Theorem A.8), hence there exist invertible operators $T, T_1, T_2 \in B(H)$ such that

$$\lambda T A T^{-1} = \lambda_1 T_1 A T_1^{-1} + \lambda_2 T_2 A T_2^{-1} \qquad (A \in B(H)). \tag{3.3.8}$$

It is apparent that if $a, b, x, y, u, v \in X$ and

$$a \otimes b = x \otimes y + u \otimes v,$$

then either $\{x, u\}$ or $\{y, v\}$ is linearly dependent. Using this elementary observation and putting $A = x \otimes y$ into (3.3.8), we infer that either $\{T_1 x, T_2 x\}$ is linearly dependent for all $x \in H$ or $\{T_1^{-1*}y, T_2^{-1*}y\}$ is linearly dependent for all $y \in H$. By Theorem A.14, in both cases we have the linear dependence of $\{T_1, T_2\}$ which results in $\tau_1 = \tau_2$. $\qquad\square$

The next lemma states that the set of all scalar multiples of all automorphisms of $B(H)$ is algebraically reflexive.

Lemma 3.3.8. *Let* $\Phi : B(H) \to B(H)$ *be a bounded linear map with the property that for every* $A \in B(H)$ *there exist a number* $\lambda_A \in \mathbb{C}$ *and an automorphism* $\tau_A \in \mathrm{Aut}(B(H))$ *so that* $\Phi(A) = \lambda_A \tau_A(A)$. *Then there exist a number* $\lambda \in \mathbb{C}$ *and an automorphism* $\tau \in \mathrm{Aut}(B(H))$ *such that* $\Phi(A) = \lambda\tau(A)$ *holds for every* $A \in B(H)$.

Proof. First suppose that $\Phi(I) = 0$. Assume that there exists a projection $0, I \neq P \in B(H)$ for which $\Phi(P) \neq 0$. Applying an appropriate transformation, we may suppose that $\Phi(P) = P$. Then we have $\Phi(I - P) = -P$. If ϵ, δ are different nonzero numbers, then by our assumption we infer that $\Phi(\epsilon P + \delta(I - P))$ is a scalar multiple of an invertible operator which, on the other hand, equals $(\epsilon - \delta)P$. This clearly implies that $\epsilon = \delta$, which is a contradiction. Hence, we obtain that $\Phi(P) = 0$ holds true for every projection $P \in B(H)$. Using the spectral theorem and the continuity of Φ, we conclude that $\Phi = 0$.

Next suppose that $\Phi(I) \neq 0$. Apparently, we may assume that $\Phi(I) = I$. By the linearity of Φ, for an arbitrary projection $0, I \neq P \in B(H)$ we obtain

$$I = \Phi(I) = \Phi(P) + \Phi(I - P) = \lambda_P Q + \lambda_{I-P} R,$$

where Q, R are idempotents different from $0, I$. Taking squares on both sides in the equality

$$I - \lambda_{I-P} R = \lambda_P Q,$$

we have

$$I + \lambda_{I-P}^2 R - 2\lambda_{I-P} R = \lambda_P^2 Q.$$

But we also have

$$\lambda_P (I - \lambda_{I-P} R) = \lambda_P^2 Q.$$

Comparing these equalities and using $R \neq 0, I$, we deduce that $\lambda_P = 1$. This means that $\Phi(P)$ is an idempotent. Therefore, Φ sends projections to idempotents. By Theorem A.4 it follows that ϕ is a Jordan homomorphism. Clearly, the range of Φ contains a rank-one operator (e.g. a rank-one idempotent) and an operator with dense range (e.g. the identity). Using our former result Theorem 3.1.1, we infer that Φ is either an automorphism or an antiautomorphism of $B(H)$. But Φ cannot be an antiautomorphism. In fact, in this case we would obtain that the image $\Phi(S)$ of a unilateral shift S has a right inverse. But, on the other hand, since Φ is locally a scalar multiple of an automorphism of $B(H)$, it follows that $\Phi(S)$ is not right invertible. This contradiction justifies our assertion. □

Before proving Theorem 3.3.2 we recall that the automorphisms of the function algebra $C_0(X)$ are of the form $f \mapsto f \circ \varphi$, where $\varphi : X \to X$ is a homeomorphism.

Proof of Theorem 3.3.2. Let $\Phi : C_0(X, B(H)) \to C_0(X, B(H))$ be a local automorphism of $C_0(X, B(H))$. We mean that Φ is a bounded linear map which agrees with some automorphism at each element of $C_0(X, B(H))$. By Theorem 3.3.1, for every $f \in C_0(X, B(H))$ there exist a homeomorphism $\varphi_f : X \to X$ and a function $\tau_f : X \to \text{Aut}(B(H))$ such that

$$\Phi(f)(x) = [\tau_f(x)](f(\varphi_f(x))) \qquad (x \in X).$$

It follows that for every $f \in C_0(X)$ there exists a homeomorphism $\psi_f : X \to X$ for which $\Phi(fI) = (f \circ \psi_f)I$. Since, by assumption, the automorphism group of $C_0(X)$ is reflexive, we obtain that there is a homeomorphism $\varphi : X \to X$ for which

$$\Phi(fI) = (f \circ \varphi)I \qquad (f \in C_0(X)). \tag{3.3.9}$$

Let $f \in C_0(X)$ and $x \in X$. Consider the linear transformation $\Psi : A \mapsto \Phi(fA)(x)$ on $B(H)$. By the form (3.3.1) of the automorphisms of $C_0(X, B(H))$ it follows that Ψ has the property that for every $A \in B(H)$ there exist a number λ_A and an automorphism $\tau_A \in \text{Aut}(B(H))$ such that

$$\Psi(A) = \lambda_A \tau_A(A).$$

Now, Lemma 3.3.8 tells us that there exist functions $\tau_f : X \to \text{Aut}(B(H))$ and $\lambda_f : X \to \mathbb{C}$ such that

$$\Phi(fA)(x) = [\tau_f(x)](\lambda_f(x)A) \qquad (f \in C_0(X), A \in B(H), x \in X).$$

From (3.3.9) we obtain that $\lambda_f = f \circ \varphi$ and hence we have

$$\Phi(fA)(x) = [\tau_f(x)](f(\varphi(x))A) \quad (f \in C_0(X), A \in B(H), x \in X). \qquad (3.3.10)$$

Let $x \in X$ be fixed for a moment. Pick functions $f, g \in C_0(X)$ with the property that $f(\varphi(x)), g(\varphi(x)) \neq 0$. Because of linearity we get

$$\{[\tau_f(x)]\}(f(\varphi(x))A) + [\tau_g(x)](g(\varphi(x))A) =$$
$$\Phi(fA)(x) + \Phi(gA)(x) = \Phi((f + g)A)(x) =$$
$$[\tau_{f+g}(x)](f(\varphi(x))A + g(\varphi(x))A)$$

for every $A \in B(H)$. Using Lemma 3.3.7 we infer that $\tau_f(x) = \tau_g(x)$. By the formula (3.3.10) it follows readily that there is a function $\tau : X \to \mathrm{Aut}(B(H))$ for which

$$\Phi(fA)(x) = [\tau(x)](f(\varphi(x))A) \quad (f \in C_0(X), A \in B(H), x \in X). \qquad (3.3.11)$$

Since the linear span of the set of functions fA ($f \in C_0(X), A \in B(H)$) is dense in $C_0(X, B(H))$ (see, for example, [198, 6.4.16. Lemma]), the equality in (3.3.11) gives us that

$$\Phi(f)(x) = [\tau(x)](f(\varphi(x))) \qquad (x \in X)$$

holds true for every $f \in C_0(X, B(H))$. By Theorem 3.3.1, the proof is complete. $\qquad \square$

The next lemma states that the set of all scalar multiples of all surjective linear isometries of $B(H)$ is algebraically reflexive. This is a very similar result to Lemma 3.3.8. Recall that by Theorem A.9, every surjective linear isometry of $B(H)$ is either of the form

$$A \longmapsto UAV$$

or of the form

$$A \longmapsto UA^{tr}V,$$

with some unitary operators U, V on H. In what follows $P(H)$ and $U(H)$ denote the sets of all projections and all unitaries on H, respectively.

Lemma 3.3.9. *Let $\Phi : B(H) \to B(H)$ be a bounded linear map with the property that for every $A \in B(H)$ there exist a number $\lambda_A \in \mathbb{C}$ and a surjective linear isometry $\tau_A \in \mathrm{Iso}(B(H))$ so that $\Phi(A) = \lambda_A \tau_A(A)$. Then there exist a number $\lambda \in \mathbb{C}$ and a surjective linear isometry $\tau \in \mathrm{Iso}(B(H))$ for which $\Phi(A) = \lambda\tau(A)$ ($A \in B(H)$).*

Proof. Just as in the proof of Lemma 3.3.8, first suppose that $\Phi(I) = 0$. Assume that there exists a projection $0, I \neq P \in B(H)$ for which $\Phi(P) \neq 0$. Apparently, we may suppose that $\Phi(P) = P$. Then we have $\Phi(I - P) = -P$. Since for any different nonzero numbers $\epsilon, \delta \in \mathbb{C}$, the operator $\epsilon P + \delta(I - P)$ is invertible, we obtain that $(\epsilon - \delta)P = \Phi(\epsilon P + \delta(I - P))$ is a scalar multiple of an invertible operator. But this is a contradiction and hence we have $\Phi(P) = 0$ for every projection P. Like in the proof of Lemma 3.3.8 this gives us that $\Phi = 0$.

So, let us suppose that $\Phi(I) \neq 0$. Clearly, we may assume that $\Phi(I) = I$ and that the constants λ_A are all nonnegative. Let $P \neq 0, I$ be a projection. Let λ, μ be nonnegative numbers and let U, V be partial isometries for which $\Phi(P) = \lambda U, \Phi(I - P) = \mu V$. We have

$$\lambda U + \mu V = I \quad \text{and} \quad \epsilon \lambda U + \delta \mu V \in \mathbb{C}U(H) \quad (|\epsilon| = |\delta| = 1). \qquad (3.3.12)$$

Since $P \neq 0, I$, it follows that $\lambda, \mu > 0$. Choose different complex numbers ϵ and δ with $|\epsilon| = |\delta| = 1$. Since by (3.3.12) it follows that the operator

$$\delta I + (\epsilon - \delta)\lambda U = \epsilon \lambda U + \delta(I - \lambda U) = \epsilon \lambda U + \delta \mu V$$

is normal, we obtain that U and then that V are both normal partial isometries. Therefore, U has a matrix representation

$$U = \begin{bmatrix} U_0 & 0 \\ 0 & 0 \end{bmatrix}$$

where U_0 is unitary on a proper closed linear subspace H_0 of H. In accordance with (3.3.12), we have the following matrix representation of V

$$V = \begin{bmatrix} (I - \lambda U_0)/\mu & 0 \\ 0 & I/\mu \end{bmatrix}.$$

Using the characteristic property $VV^*V = V$ of partial isometries, we get that $\mu = 1$ and, by symmetry, that $\lambda = 1$. Taking the matrix representations above into account, it is easy to see that $I - U_0$ is a normal partial isometry and that $\epsilon U_0 + \delta(I - U_0)$ is a scalar multiple of a unitary operator for every $\epsilon, \delta \in \mathbb{C}$ with $|\epsilon| = |\delta| = 1$. Since $I - U_0$ is a normal partial isometry, the spectrum of U_0 must consist of such numbers c of modulus 1 for which either $1 - c$ has modulus 1 or $1 - c = 0$. This gives us that $\sigma(U_0) \subset \{1, e^{i\pi/3}, e^{-i\pi/3}\}$. Let P_1, P_2, P_3 denote the projections onto the subspaces $\ker(U_0 - I), \ker(U_0 - e^{i\pi/3}I), \ker(U_0 - e^{-i\pi/3}I)$ of H_0, respectively. We assert that two of the operators P_1, P_2, P_3 are necessarily zero. In fact, if for example, $P_2, P_3 \neq 0$, then it follows from the second property in (3.3.12) that

$$|\epsilon e^{i\pi/3} + \delta e^{-i\pi/3}| = |\epsilon e^{-i\pi/3} + \delta e^{i\pi/3}|$$

holds for every ϵ, δ of modulus 1. But this is an obvious contradiction. The other cases can be treated in a similar way. Therefore, for any projection P we have $\Phi(P) = U \in \{1, e^{i\pi/3}, e^{-i\pi/3}\}P(H)$.

Now, let P be a projection having infinite rank and infinite corank. Since in this case P is unitarily equivalent to $I-P$, it follows that P and $I-P$ can be connected by a continuous curve within the set of projections. Consequently, we obtain that $\Phi(P)$ and $\Phi(I-P)$ have the same nonzero eigenvalue. Since $\Phi(I-P) = I - \Phi(P)$, it follows that this eigenvalue is 1. Thus we obtain that $\Phi(P)$ is a projection. If P is a finite rank projection, then P is the difference of two projections having infinite rank and corank. Then we obtain that $\Phi(P)$ is the difference of two projections and consequently $\Phi(P)$ is self-adjoint. On the other hand, we have $\Phi(P) \in \{1, e^{i\pi/3}, e^{-i\pi/3}\}P(H)$. These result in $\Phi(P) \in P(H)$ and we deduce that Φ sends every projection to a projection. It follows from Theorem A.4 that Φ is a Jordan *-endomorphism of $B(H)$. Since, by our condition, the range of Φ contains a rank-one operator and an operator with dense range, using Theorem 3.1.1 again, we infer that Φ is either a *-automorphism or a *-antiautomorphism of $B(H)$. In both cases we obtain that Φ is a surjective linear isometry of $B(H)$ and this completes the proof. □

Lemma 3.3.10. *If* $\mathcal{M} \subset \mathbb{C}U(H)$ *is a linear subspace, then* $\mathcal{M}$ *is either 1-dimensional or 0-dimensional.*

Proof. Let $A, B \in \mathcal{M}$. For every $\lambda \in \mathbb{C}$ we have $(A+\lambda B)^*(A+\lambda B) \in \mathbb{C}I$. Since A^*A, B^*B are scalar operators, choosing $\lambda = 1$ and then $\lambda = i$, it follows easily that A^*B is also scalar. This clearly gives us that A, B are linearly dependent. □

Lemma 3.3.11. *Let* X *be a locally compact Hausdorff space. Let* $\mathcal{M} \subset \mathbb{C}^X$ *be a linear subspace containing a nowhere vanishing function* $f_0 \in \mathcal{M}$ *and having the property that* $|f| \in C_0(X)$ *for every* $f \in \mathcal{M}$. *Then there is a function* $t : X \to \mathbb{C}$ *of modulus 1 such that* $t\mathcal{M} \subset C_0(X)$.

Proof. We know that the function $|f + f_0|^2 - |f|^2 - |f_0|^2$ is continuous for every $f \in \mathcal{M}$. This gives us that $f\bar{f_0}$ is continuous for every $f \in \mathcal{M}$. Let $t = |f_0|/f_0$. Then we have $|t| = 1$ and the function $(tf)|f_0| = (tf)(\overline{tf_0}) = f\bar{f_0}$ is continuous. Consequently, we obtain $tf \in C_0(X)$. □

For the proof of Theorem 3.3.3 we recall the Banach-Stone theorem (see Theorem A.10) stating that the surjective linear isometries of the function algebra $C_0(X)$ are all of the form $f \mapsto \tau \cdot f \circ \varphi$, where $\tau : X \to \mathbb{C}$ is a continuous function of modulus 1 and $\varphi : X \to X$ is a homeomorphism. In what follows the notion of local isometries is meant in the same sense as in the first paragraph of the subsection 3.2.3. Proof.

Proof of Theorem 3.3.3. Let $\Phi : C_0(X, B(H)) \to C_0(X, B(H))$ be a local isometry. Pick a function $f \in C_0(X)$ and a point $x \in X$, and consider the linear map $\Psi : B(H) \to B(H)$

$$\Psi : A \longmapsto \Phi(fA)(x).$$

It follows from Theorem 3.3.1 that for every $A \in B(H)$ there exist a number λ_A and a surjective linear isometry $\tau_A \in \mathrm{Iso}(B(H))$ such that $\Psi(A) = \lambda_A \tau_A(A)$. By Lemma 3.3.9 we infer that there exist a nonnegative number $\lambda_{f,x}$ and a surjective linear isometry $\tau_{f,x} \in \mathrm{Iso}(B(H))$ for which

$$\Phi(fA)(x) = \lambda_{f,x}\tau_{f,x}(A) \qquad (3.3.13)$$

holds true for every $f \in C_0(X)$, $A \in B(H)$ and $x \in X$. Now, let $U \in B(H)$ be a unitary operator and $x \in X$. The linear map

$$f \longmapsto \Phi(fU)(x)$$

maps $C_0(X)$ into $\mathbb{C}U(H)$. Since the range of this map is a linear subspace, by Lemma 3.3.10 we infer that it is either 1-dimensional or 0-dimensional. Thus there is a linear functional $F_{U,x} : C_0(X) \to \mathbb{C}$ and a unitary operator $[\tau(x)](U)$ such that

$$\Phi(fU)(x) = F_{U,x}(f)[\tau(x)](U) \qquad (f \in C_0(X), U \in U(H), x \in X).$$

Clearly, the map $F_U : C_0(X) \to \mathbb{C}^X$ defined by $F_U(f)(x) = F_{U,x}(f)$ is linear and we have

$$\Phi(fU)(x) = F_U(f)(x)[\tau(x)](U) \quad (f \in C_0(X), U \in U(H), x \in X). \quad (3.3.14)$$

Since Φ is a local isometry of $C_0(X, B(H))$, it follows from Theorem 3.3.1 that for every $f \in C_0(X)$ there exist a strongly continuous function $\tau_{f,U} : X \to \mathrm{Iso}(B(H))$ and a homeomorphism $\varphi_{f,U} : X \to X$ such that

$$\Phi(fU)(x) = f(\varphi_{f,U}(x))[\tau_{f,U}(x)](U) \qquad (x \in X). \qquad (3.3.15)$$

Apparently, we have $|F_U(f)| = |f| \circ \varphi_{f,U}$. Because of the σ-compactness of X, it is a quite easy consequence of Uryson's lemma that there exists a strictly positive function in $C_0(X)$. Therefore, the range of F_U contains a nowhere vanishing function and has the property that the absolute value of every function belonging to this range is continuous. By Lemma 3.3.11, there exists a function $t : X \to \mathbb{C}$ of modulus 1 such that the functions $tF_U(f)$ are all continuous ($f \in C_0(X)$). Consequently, we may suppose that the map F_U in (3.3.14) maps $C_0(X)$ into itself. Comparing (3.3.14) and (3.3.15) we have

$$F_U(f)(x)[\tau(x)](U) = f(\varphi_{f,U}(x))[\tau_{f,U}(x)](U) \qquad (x \in X). \qquad (3.3.16)$$

If $f \in C_0(X)$ is a nowhere vanishing function, then by the continuity of the functions $F_U(f), f \circ \varphi_{f,U}$ and $[\tau_{f,U}(.)](U)$, it follows that $[\tau(.)](U)$ is also continuous. From (3.3.16) we obtain that

$$F_U(f) = f(\varphi_{f,U}(x))[\tau_{f,U}(x)](U)[\tau(x)](U)^* \qquad (x \in X).$$

In particular, this implies that the function

$$x \longmapsto [\tau_{f,U}(x)](U)[\tau(x)](U)^*$$

can be considered as a continuous scalar valued function of modulus 1. Hence, F_U is a local isometry of $C_0(X)$. By our assumption this means that F_U is a surjective linear isometry, i.e., there exist a continuous function $t_U : X \to \mathbb{C}$ of modulus 1 and a homeomorphism $\varphi_U : X \to X$ such that $F_U(f) = t_U \cdot f \circ \varphi_U$ ($f \in C_0(X), U \in U(H)$). Having a look at (3.3.14), it is obvious that we may suppose that Φ satisfies

$$\Phi(fU)(x) = f(\varphi_U(x))[\tau(x)](U) \quad (f \in C_0(X), U \in U(H), x \in X), \quad (3.3.17)$$

where $[\tau(x)](U)$ is unitary. If $f \in C_0(X)$ is nonnegative, we see from (3.3.17) and (3.3.13) that

$$f(\varphi_U(x)) = \lambda_{f,x} = f(\varphi_I(x))$$

and

$$[\tau(x)](U) = \tau_{f,x}(U) \quad (U \in U(H), x \in X).$$

This verifies the existence of a homeomorphism φ of X and a function $\tau : X \to \mathrm{Iso}(B(H))$ for which

$$\Phi(fU)(x) = f(\varphi(x))[\tau(x)](U) \quad (U \in U(H), x \in X)$$

holds true for every nonnegative function $f \in C_0(X)$. Since every function in $C_0(X)$ is the linear combination of nonnegative functions in $C_0(X)$ and every operator in $B(H)$ is a linear combination of unitaries, we finally obtain that

$$\Phi(fA)(x) = f(\varphi(x))[\tau(x)](A) \quad (f \in C_0(X), A \in B(H), x \in X).$$

Referring to the fact once again that the linear span of the elementary tensors fA ($f \in C_0(X), A \in B(H)$) is dense in $C_0(X, B(H))$, we arrive at the form

$$\Phi(f)(x) = [\tau(x)](f(\varphi(x))) \quad (f \in C_0(X, B(H)), x \in X).$$

By Theorem 3.3.1, the proof is complete. $\qquad\square$

We now turn to the proof of Theorem 3.3.4. The next result describes the form of the local isometries of $C_0(X)$.

Lemma 3.3.12. *Let X be a first countable locally compact Hausdorff space. Let $F : C_0(X) \to C_0(X)$ be a local isometry. Then there exist a continuous function $t : X \to \mathbb{C}$ of modulus 1 and a homeomorphism g of X onto a subspace of X so that*

$$F(f) \circ g = t \cdot f \quad (f \in C_0(X)). \quad (3.3.18)$$

Proof. By the Banach-Stone theorem (see Theorem A.10) for every $f \in C_0(X)$ there exist a homeomorphism $\varphi_f : X \to X$ and a continuous function $\tau_f : X \to \mathbb{C}$ of modulus 1 such that

$$F(f) = \tau_f \cdot f \circ \varphi_f. \tag{3.3.19}$$

For any $x \in X$ let $\mathcal{S}_x$ denote the set of all functions $p \in C_0(X)$ which map into the interval $[0,1]$, $p(x) = 1$ and $p(y) < 1$ for every $x \neq y \in X$. By Uryson's lemma and the first countability of X, it is easy to verify that $\mathcal{S}_x$ is nonempty. Let $p, p' \in \mathcal{S}_x$. By (3.3.19) there exist $y, y' \in X$ for which $|F(p)| \in \mathcal{S}_y, |F(p')| \in \mathcal{S}_{y'}$. Similarly, since $(p + p')/2 \in \mathcal{S}_x$, there is a point $y'' \in X$ for which $|F((p+p')/2)| \in \mathcal{S}_{y''}$. Apparently, we have $y = y'$ and $F(p)(y) = F(p')(y')$. This shows that there are functions $t : X \to \mathbb{C}$ and $g : X \to X$ such that

$$t(x) = F(p)(g(x)) \tag{3.3.20}$$

holds true for every $x \in X$ and $p \in \mathcal{S}_x$. Clearly, $|t(x)| = 1$. Pick $x \in X$. It is easy to see that for any everywhere positive function $f \in C_0(X)$ with $f(x) = 1$ we have a function $p \in \mathcal{S}_x$ such that $p(y) < f(y)$ $(x \neq y \in X)$. Now, let $f \in C_0(X)$ be an arbitrary nonnegative function. Then there is a positive constant c for which the function $y \mapsto c + f(x) - f(y)$ is everywhere positive. Hence, we can choose a function $p \in \mathcal{S}_x$ such that $cp(y) < cp(x) + f(x) - f(y)$ $(x \neq y \in X)$. This means that the nonnegative function $cp + f$ takes its maximum only at x. By (3.3.20) we infer

$$t(x)(cp(x) + f(x)) = F(cp + f)(g(x)).$$

Clearly, we have

$$t(x)(cp(x)) = F(cp)(g(x)),$$

too. Therefore, we obtain

$$t \cdot f = F(f) \circ g \tag{3.3.21}$$

for every nonnegative f and then for every function in $C_0(X)$. We assert that g is a homeomorphism of X onto its range. To see this, first observe that for every function $p \in \mathcal{S}_y$ and net (y_α) in X, the condition that $p(y_\alpha) \to 1$ implies that $y_\alpha \to y$. Let (x_α) be a net in X converging to $x \in X$. Pick a function $p \in \mathcal{S}_x$. Since F is a local isometry, we have a homeomorphism φ_p of X for which

$$p = |t \cdot p| = |F(p) \circ g| = p \circ \varphi_p \circ g.$$

Since this implies that $p(\varphi_p(g(x_\alpha))) \to 1$, we obtain $\varphi_p(g(x_\alpha)) \to x = \varphi_p(g(x))$ and hence we have $g(x_\alpha) \to g(x)$. So, g is continuous. The injectivity of g immediately follows from (3.3.21) using the fact that the nonnegative elements of $C_0(X)$ separate the points of X. As for the continuity of g^{-1} and t, these follow from (3.3.21) again and from Uryson's lemma. $\quad\square$

Now, we are in a position to prove our last theorem.

Proof of Theorem 3.3.4. It is well-known that every open convex subset of $\mathbb{R}^n$ is homeomorphic to the open unit ball G of $\mathbb{R}^n$. Hence, it is sufficient to show that the automorphism and isometry groups of $C_0(G)$ are algebraically reflexive. Furthermore, by the forms of the automorphisms and surjective linear isometries of the function algebra $C_0(X)$ we are certainly done if we prove the statement for the isometry group. So, let $F : C_0(G) \to C_0(G)$ be a local isometry. Then F is of the form (3.3.18). The only thing we have to verify is that the function g appearing in this form is surjective. Consider the function $f \in C_0(G)$ defined by $f(x) = 1/(1 + \|x\|)$. Clearly, we may assume that $F(f) = f$. From (3.3.18) we infer that

$$\frac{1}{1 + \|g(x)\|} = \frac{1}{1 + \|x\|} \qquad (x \in G).$$

Therefore, the continuous function g maps the surface S_r of the closed ball $r\overline{G}$ $(0 \le r < 1)$ into itself. Suppose that $g(S_r) \subsetneq S_r$. It is obvious that every proper closed subset of S_r is homeomorphic to a subset of $\mathbb{R}^{n-1}$. Thus, by the famous Borsuk-Ulam theorem we would get that g takes the same value at two antipodal points of S_r. But this contradicts the injectivity of g. Consequently, we have $g(S_r) = S_r$ $(0 \le r < 1)$ which implies that g is surjective. This completes the proof. $\qquad \square$

3.3.4 Remarks

Observe that the proof of Theorem 3.3.4 shows how difficult it might be to treat the reflexivity problem of the isometry group of the suspension of general C^*-algebras. In fact, this is the situation even in the commutative case. The reason is the following. If X is an arbitrary compact Hausdorff space, then $C_0(\mathbb{R}) \otimes C(X)$ is isomorphic to $C_0(\mathbb{R} \times X)$. Therefore, we have to consider the local isometries of $C_0(\mathbb{R} \times X)$. If X is first countable, then Lemma 3.3.12 gives some information about the form of those transformations. The point would be to show that the function g appearing in (3.3.18) is surjective. In the proof of Theorem 3.3.4 we have faced a similar problem concerning the space $\mathbb{R} \times \mathbb{R}^{n-1}$. There we have used the deep result called Borsuk-Ulam theorem to prove the surjectivity of g. As for the general case of the spaces $\mathbb{R} \times X$ (X being an arbitrary first countable compact Hausdorff space), we have no idea how to attack the problem.

    After this remark it should not be surprising that we do not have such a general result which would state that the algebraic reflexivity of the groups Aut and Iso remains valid when taking tensor products of (unital) C^*-algebras. Just as above, to see this, it is enough to examine commutative algebras. If X_1, X_2 are compact Hausdorff spaces, then $C(X_1) \otimes C(X_2)$ is isomorphic to $C(X_1 \times X_2)$. Now, it is obvious from the forms of the automorphisms and surjective linear isometries of function algebras that the automorphisms (resp. surjective linear isometries) of $C(X_1 \times X_2)$ have nothing to do with the automorphisms (resp. surjective linear isometries) of $C(X_1)$ and $C(X_2)$. This causes the difficulty in the corresponding reflexivity problem.

    Fortunately, the situation is quite different with the tensor product of a function algebra and $B(H)$. In fact, as we have seen in Theorem 3.3.1, every automorphism as well as surjective linear isometry of $C_0(X) \otimes B(H)$ is an identifiable mixture of a 'function algebraic' and an 'operator algebraic' part. This made possible for us to achieve the reflexivity results of this section.

Next, we note that the automorphism and isometry groups of $C_0(\mathbb{R}) \otimes B(H)$ are not topologically reflexive. To verify this simple fact, one can produce a counterexample like in Subsection 3.1.2 relating the algebra $C[a, b]$. It is then a natural problem to determine the topological reflexive closures of $\mathrm{Aut}(C_0(\mathbb{R}) \otimes B(H))$ and $\mathrm{Iso}(C_0(\mathbb{R}) \otimes B(H))$ (see Section 0.5 of the Introduction). As for the isometry group, this has been recently done in the paper [88] of Győry (we also refer to his PhD thesis [86]).

In the present section we have investigated the reflexivity of the automorphism and isometry groups of the tensor product of a function algebra and an operator algebra. The tensor product is one possibility to produce a new C^*-algebra from those algebras. Beside this, there is another very important construction which is called extension. This concerns the algebra $K(H)$ of all compact operators and the function algebra $C(X)$ on a given compact metric space X. An extension of $K(H)$ by $C(X)$ is a C^*-subalgebra $\mathcal{A}$ of $B(H)$ together with a surjective *-homomorphism $\phi : \mathcal{A} \to C(X)$ whose kernel is $K(H)$. As we mentioned in Subsection 3.1.4, these C^*-algebras are the main objects of the Brown-Douglas-Fillmore (BDF) theory developed to solve the famous classification problem of essentially normal operators (see [62, Chapter IX] and the original source [34]). Our reflexivity results concerning those C^*-algebras appeared in [168] and they have already been summarized in the concluding remarks of Section 3.1.

In closing we refer to the paper [109] of Jarosz and Rao where the algebraic reflexivity of the isometry group of the Banach space of all continuous functions from a first countable compact Hausdorff space into a fixed Banach space has been investigated when this target space is finite dimensional or uniformly convex with algebraically reflexive isometry group. The papers [215] and [216] of Rao contain interesting results concerning the local isometries of the Banach space of all bounded linear operators from a uniformly smooth or uniformly convex Banach space $\mathcal{X}$ into the function space $C(X)$ over a first countable compact Hausdorff space X.

3.4 2-Local Automorphisms of Operator Algebras on Banach Spaces

3.4.1 Summary

In this section we present a significant extension of a result of Šemrl stating that every 2-local automorphism of the full operator algebra on a separable infinite dimensional Hilbert space is an automorphism. Namely, we prove that if X is a real or complex Banach space with a Schauder basis and $\mathcal{A}$ is a subalgebra of $B(X)$ which contains all compact operators, then every 2-local automorphism of $\mathcal{A}$ is an automorphism. The proof rests on an analogous statement concerning 2-local automorphisms of matrix algebras over general fields. The results of this section appeared in the paper [184].

3.4.2 Formulation of the Results

As we mentioned in Section 0.5 of the Introduction, we were motivated to study 2-local automorphisms because they are the natural non-linear substitutes of linear local automorphisms in the setting of non-linear structures. Using this concept one can investigate 2-local automorphisms of arbitrary algebraic structures (groups, lattices, etc.) and can get noteworthy new information about their automorphism groups.

Originally, the 2-local automorphisms were defined by P. Šemrl in [224] as follows. If $\mathcal{A}$ is an algebra, then a transformation (no linearity is assumed) $\phi : \mathcal{A} \to \mathcal{A}$ is called a 2-local automorphism of $\mathcal{A}$ if for every $A, B \in \mathcal{A}$ there exists an (algebra) automorphism $\phi_{A,B}$ of $\mathcal{A}$ such that $\phi(A) = \phi_{A,B}(A)$ and $\phi(B) = \phi_{A,B}(B)$. Just as with linear local maps, it is a remarkable fact on the algebra $\mathcal{A}$ if every 2-local automorphism of $\mathcal{A}$ is an automorphism. The first result on 2-local automorphisms is due to Šemrl [224] who proved that if H is a separable infinite dimensional Hilbert space, then every 2-local automorphism of the algebra $B(H)$ is an automorphism. (A similar result were also presented in [224] concerning 2-local derivations. The definition of such maps should be clear.) We remark that Šemrl's motivation to introduce this concept of 2-local automorphisms was a non-linear characterization of the characters of commutative Banach algebras due to Kowalski and Slodkowski [129]. (That result can be viewed as a non-linear version of the famous Gleason-Kahane-Żelazko theorem.)

In his paper [224], Šemrl mentioned that he could prove that the conclusion in his result holds also in the finite dimensional case, but his proof was rather long and involved tedious computations. Therefore, he raised the problem of presenting a short argument. In what follows we give such a proof. Furthermore, we extend Šemrl's result significantly, namely, we get the same conclusion as Šemrl for the case of any Banach space with a Schauder basis. Recall that most classical separable Banach spaces have Schauder bases (see the first chapter in [139]).

We turn to the formulation of our results. If $\mathbb{F}$ is a field, then $M_n(\mathbb{F})$ stands for the algebra of all $n \times n$ matrices with entries in $\mathbb{F}$. It is well-known that the automorphisms of the algebra $M_n(\mathbb{F})$ are exactly the transformations $A \mapsto TAT^{-1}$ where $T \in M_n(\mathbb{F})$ is nonsingular (see, for example, [207, Lemma, p. 230]). In what follows, if $A \in M_n(\mathbb{F})$, we denote by $\mathrm{eigv}(A)$ the system of all eigenvalues of A listed according to multiplicity. We emphasize that $\mathrm{eigv}(A)$ generally differs from the spectrum $\sigma(A)$ of A as a linear operator.

Our key result which follows gives a non-linear characterization of the automorphisms of $M_n(\mathbb{F})$ via a certain preserving property. The proofs of our results concerning 2-local automorphisms presented in this section rest on this observation.

Theorem 3.4.1. *Suppose that $\mathbb{F}$ is an algebraically closed field of characteristic 0. Let $\phi : M_n(\mathbb{F}) \to M_n(\mathbb{F})$ be a transformation (linearity is not assumed) such that*

$$\mathrm{eigv}(\phi(A)\phi(B)) = \mathrm{eigv}(AB) \qquad (3.4.1)$$

holds for every $A, B \in M_n(\mathbb{F})$. Then there exists a nonsingular matrix $T \in M_n(\mathbb{F})$ and $\lambda \in \{-1, 1\}$ such that ϕ is either of the form

$$\phi(A) = \lambda T A T^{-1} \qquad (A \in M_n(\mathbb{F}))$$

or of the form

$$\phi(A) = \lambda T A^{tr} T^{-1} \qquad (A \in M_n(\mathbb{F})).$$

We mention that similar characterizations of the automorphisms of operator algebras and function algebras can be found in our paper [180] but there we had to assume that the transformations in question are all surjective. The main point of the present theorem is that here there is no need to suppose surjectivity and this makes possible for us to get the following corollary on 2-local automorphisms of matrix algebras over general fields.

Corollary 3.4.2. *Let $\mathbb{F}$ be an algebraically closed field of characteristic 0. Then every 2-local automorphism of $M_n(\mathbb{F})$ is an automorphism.*

Since the field $\mathbb{R}$ of real numbers is not algebraically closed, we formulate the analogous statement for $M_n(\mathbb{R})$ as a separate proposition.

Proposition 3.4.3. *Every 2-local automorphism of $M_n(\mathbb{R})$ is an automorphism.*

The arguments applied in the proofs of our results above lead us to an extension of Šemrl's theorem for the case of certain subalgebras of the algebra $B(X)$ of all bounded linear operators on a Banach space X with a Schauder basis. This is the main result of the section and it reads as follows.

Theorem 3.4.4. *Let X be a real or complex Banach space with a Schauder basis. Suppose that $\mathcal{A}$ is a subalgebra of $B(X)$ which contains the ideal of all compact operators on X. If ϕ is a 2-local automorphism of $\mathcal{A}$, then ϕ is an automorphism of $\mathcal{A}$.*

3.4.3 Proofs

Proof of Theorem 3.4.1. Since the statement is obvious for $n = 1$, we suppose that $n \geq 2$. Denote by tr the usual trace functional on $M_n(\mathbb{F})$. It follows from the property (3.4.1), i.e., from the equality

$$\operatorname{eigv}(\phi(A)\phi(B)) = \operatorname{eigv}(AB)$$

that

$$\operatorname{tr}\phi(A)\phi(B) = \operatorname{tr}AB \qquad (3.4.2)$$

holds for every $A, B \in M_n(\mathbb{F})$.

As usual, denote by $E_{ij} \in M_n(\mathbb{F})$ the matrix whose ij entry is 1 and its all other entries are 0. We assert that the $\phi(E_{ij})$'s are linearly independent. Suppose that

$$\sum_{i,j} \lambda_{ij}\phi(E_{ij}) = 0$$

holds for some scalars $\lambda_{ij} \in \mathbb{F}$. Fix indices $k, l \in \{1, \ldots, n\}$. We have

$$\sum_{i,j} \lambda_{ij}\phi(E_{ij})\phi(E_{kl}) = 0.$$

Taking trace, we obtain

$$\sum_{i,j} \lambda_{ij} \operatorname{tr}\phi(E_{ij})\phi(E_{kl}) = 0.$$

By the property (3.4.2) of ϕ, it follows that

$$\sum_{i,j} \lambda_{ij} \operatorname{tr}E_{ij}E_{kl} = 0.$$

Since $E_{ij}E_{kl} = \delta_{jk}E_{il}$ (δ_{jk} is the Kronecker symbol), from this equality we easily deduce that $\lambda_{lk} = 0$. As k, l were arbitrary, it follows that the matrices $\phi(E_{ij})$, $i, j \in \{1, \ldots, n\}$ form a linearly independent set in $M_n(\mathbb{F})$. This implies that the range of ϕ linearly generates $M_n(\mathbb{F})$.

We are now in a position to prove that ϕ is linear. If $A, B \in M_n(\mathbb{F})$, we compute

$$\operatorname{tr}\phi(A+B)\phi(C) = \operatorname{tr}(A+B)C = \operatorname{tr}AC + \operatorname{tr}BC =$$
$$\operatorname{tr}(\phi(A)\phi(C) + \phi(B)\phi(C)) = \operatorname{tr}(\phi(A) + \phi(B))\phi(C)$$

for any $C \in M_n(\mathbb{F})$. Since the linear span of the $\phi(C)$'s is the whole space $M_n(\mathbb{F})$, it follows that $\phi(A+B) = \phi(A) + \phi(B)$. The homogenity of ϕ can be proved in a very similar way. So, ϕ is a surjective linear transformation on $M_n(\mathbb{F})$ which is hence bijective.

Let $A \in M_n(\mathbb{F})$ be of rank one. Then, using (3.4.1) and the surjectivity of ϕ, it follows that $\phi(A)B, B\phi(A)$ both have at most one nonzero eigenvalue and its multiplicity is one. This implies that $\phi(A)$ is of rank one. Consequently, ϕ is a rank-one preserving linear transformation of $M_n(\mathbb{F})$. The form of all such maps were determined by Marcus and Moyls. Namely, according to [150, Corollary] there are nonsingular matrices T, S in $M_n(\mathbb{F})$ such that ϕ is either of the form

$$\phi(A) = TAS \qquad (A \in M_n(\mathbb{F})) \qquad\qquad (3.4.3)$$

or of the form

$$\phi(A) = TA^{tr}S \qquad (A \in M_n(\mathbb{F})).$$

Without loss of generality we can assume that ϕ is of the first form. Composing our transformation ϕ with the automorphism $A \mapsto SAS^{-1}$, it follows that we can assume that the matrix S appearing in (3.4.3) is the identity I. We have

$$\operatorname{tr} TATB = \operatorname{tr} AB \qquad\qquad (3.4.4)$$

for every $A, B \in M_n(\mathbb{F})$. If we fix $A \in M_n(\mathbb{F})$ for a moment and let B run through $M_n(\mathbb{F})$, it follows from (3.4.4) that $TAT = A$. Now, if A runs through the set of all rank-one matrices in $M_n(\mathbb{F})$, we deduce from the equality $TAT = A$ that T, as a linear operator on a linear space X, has the property that for every $x \in X$, the vector Tx is a scalar multiple of x. By Theorem A.14 this gives us that $T = \lambda I$ for some scalar $\lambda \in \mathbb{F}$. Clearly, we have $\lambda^2 = 1$ and this completes the proof of our statement. $\qquad\qquad\square$

Proof of Corollary 3.4.2. First we recall again that the automorphisms of $M_n(\mathbb{F})$ are exactly the transformations of the form $A \mapsto TAT^{-1}$ with some nonsingular matrix $T \in M_n(\mathbb{F})$. Let now $\phi : M_n(\mathbb{F}) \to M_n(\mathbb{F})$ be a 2-local automorphism. Clearly, ϕ satisfies (3.4.1). Since $\phi(I) = I$, we infer from Theorem 3.4.1 that there exists a nonsingular matrix $T \in M_n(\mathbb{F})$ such that ϕ is either of the form

$$\phi(A) = TAT^{-1} \qquad (A \in M_n(\mathbb{F}))$$

or of the form

$$\phi(A) = TA^{tr}T^{-1} \qquad (A \in M_n(\mathbb{F})).$$

Choosing $A, B \in M_n(\mathbb{F})$ such that $AB = 0, BA \neq 0$, it follows from the 2-local property of ϕ that this second possibility above is excluded and we obtain the desired conclusion. $\qquad\qquad\square$

Proof of Proposition 3.4.3. The statement follows from obvious modifications of the proofs of Corollary 3.4.2 and Theorem 3.4.1. The only difference is that instead of referring to the result of Marcus and Moyls in [150] concerning the form of all linear transformations on $M_n(\mathbb{F})$ preserving rank-one matrices, here we refer, for example, to an analogous statement [203, Main Theorem] due to Omladič and Šemrl which covers the case of rank-one preserving surjective linear mappings on real matrices. $\qquad\qquad\square$

For the proof of Theorem 3.4.4 we need the following notation and definitions (they were already used in Section 2.2). Let X be a real or complex Banach space. As usual, the algebra of all bounded linear operators on X is denoted by $B(X)$ and $F(X)$ stands for the ideal of all finite rank operators in $B(X)$. An operator $P \in B(X)$ is called an idempotent if $P^2 = P$.

Two idempotents P, Q in $B(X)$ are said to be (algebraically) orthogonal if $PQ = QP = 0$. The dual space of X (that is the set of all bounded linear functionals on X) is denoted by $X^\sharp$. The Banach space adjoint of an operator $A \in B(X)$ is denoted by $A^\sharp$. If $x \in X$ and $f \in X^\sharp$, then $x \otimes f$ stands for the operator (of rank at most 1) defined by

$$(x \otimes f)(z) = f(z)x \qquad (z \in X).$$

It requires only elementary computations to show that

$$A \cdot x \otimes f = (Ax) \otimes f, \quad x \otimes f \cdot A = x \otimes (A^\sharp f), \quad x \otimes f \cdot y \otimes g = f(y) x \otimes g$$

hold for every $x, y \in X$, $f, g \in X^\sharp$ and $A \in B(X)$. It is easy to see that the elements of $F(X)$ are exactly the operators $A \in B(X)$ which can be written as a finite sum

$$A = \sum_i x_i \otimes f_i \tag{3.4.5}$$

for some $x_1, \ldots, x_n \in X$ and $f_1, \ldots, f_n \in X^\sharp$. Using this representation, the trace of A is defined by

$$\operatorname{tr} A = \sum_i f_i(x_i).$$

It is known that $\operatorname{tr} A$ is well-defined, that is, it does not depend on the particular representation (3.4.5) of A. It is easy to see that tr is a linear functional on $F(X)$ with the property that

$$\operatorname{tr} TA = \operatorname{tr} AT$$

holds for every $A \in F(X)$ and $T \in B(X)$. Finally, we recall that, similarly to the case of matrix algebras, every automorphism of any subalgebra of $B(X)$ which contains $F(X)$ is of the form $A \mapsto TAT^{-1}$ where $T \in B(X)$ is invertible (see Theorem A.8 where this was formulated for Hilbert spaces but the corresponding assertion holds for any Banach space as well).

Proof of Theorem 3.4.4. In view of our previous results in this section we can assume that X is infinite dimensional. Let ϕ be a 2-local automorphism of $\mathcal{A}$. If $P \in \mathcal{A}$ is a finite rank idempotent, then $\phi(P) = \tilde{P}$ is also a finite rank idempotent and the ranks of P and $\tilde{P}$ are the same. Consider the subalgebra $\mathcal{A}_P$ of $\mathcal{A}$ which consists of all operators $A \in \mathcal{A}$ for which $PAP = A$. It follows from the 2-local property of ϕ that $\phi(P)\phi(A)\phi(P) = \phi(A)$. Consequently, ϕ maps $\mathcal{A}_P$ into $\mathcal{A}_{\tilde{P}}$. Clearly, both algebras $\mathcal{A}_P$ and $\mathcal{A}_{\tilde{P}}$ are isomorphic to $M_n(\mathbb{K})$ ($\mathbb{K}$ stands for the real or complex field) where $n = \operatorname{rank} P = \operatorname{rank} \tilde{P}$. It is easy to see that ϕ has the property (3.4.2). Now, one can follow the arguments given in the proofs of the previous results of this section to show that ϕ is linear and multiplicative on $\mathcal{A}_P$. Since $P \in \mathcal{A}$ was an arbitrary finite rank idempotent, we can infer that the restriction ψ of ϕ onto $F(X)$ is an algebra endomorphism of $F(X)$ which preserves the rank.

Since ψ is an algebra homomorphism of $F(X)$, we have an injective linear operator $T : X \to X$ such that

$$TA = \psi(A)T \qquad (A \in F(X)). \tag{3.4.6}$$

Indeed, similarly as in [223] (also see the proof of Theorem 3.1.1) we define

$$Tx = \psi(x \otimes f_0)y_0 \qquad (x \in X)$$

where $y_0 \in X$ and $f_0 \in X^\sharp$ are fixed such that $\psi(x \otimes f_0)y_0 \neq 0$ holds for some $x \in X$. It is evident that T is a nonzero linear operator on X. It follows from the multiplicativity of ψ that $TA = \psi(A)T$ $(A \in F(X))$. To see the injectivity of T, let $Tx = 0$ and $x \neq 0$. Then we have $TAx = \psi(A)Tx = 0$ for every $A \in F(X)$. This gives us that $Ty = 0$ holds for every $y \in X$ which is an obvious contradiction.

We show that T is continuous. To see this, we apply the closed graph theorem. Let (x_n) be a sequence in X and $y \in X$ be such that $x_n \to 0$ and $Tx_n \to y$. We have to show that $y = 0$. Since TA is continuous for every $A \in F(X)$, from (3.4.6) we infer that

$$\psi(A)y = 0 \qquad (A \in F(X)). \tag{3.4.7}$$

Here comes the point where we use that X has a Schauder basis. Namely, this condition implies that we have a sequence (P_n) of pairwise orthogonal rank-one idempotents in $B(X)$ whose sum strongly converges to I. In particular, (P_n) is uniformly bounded. Clearly, we can write $P_n = x_n \otimes f_n$ where $x_n \in X, f_n \in X^\sharp$ are such that $f_i(x_j) = \delta_{ij}$ and $\|x_i\| < m, \|f_j\| < M$ $(i, j \in \mathbb{N})$ for some positive real numbers m, M. Choose a sequence (λ_n) of positive real numbers with the property that $\sum_{n=k+1}^{\infty} \lambda_n < \lambda_k$ holds for every $k \in \mathbb{N}$. For example, one can define $\lambda_n = (1/3)^n$ $(n \in \mathbb{N})$. In particular, it follows that $\sum_n \lambda_n P_n$ converges in the operator norm topology and hence its sum is a compact operator. Pick another sequence (μ_n) of positive real numbers for which $\sum_n \mu_n < \infty$. The operator $\sum_n \mu_n x_n \otimes f_{n+1}$ is also compact.

By the 2-local property of ϕ, composing ϕ with an automorphism of $\mathcal{A}$ if necessary, we can and do assume that for the particular operators $\sum_n \lambda_n P_n$ and $\sum_n \mu_n x_n \otimes f_{n+1}$ we have

$$\phi\left(\sum_n \lambda_n P_n\right) = \sum_n \lambda_n P_n \text{ and } \phi\left(\sum_n \mu_n x_n \otimes f_{n+1}\right) = \sum_n \mu_n x_n \otimes f_{n+1}. \tag{3.4.8}$$

Let $n_0 \in \mathbb{N}$ be arbitrary. By the 2-local property of ϕ we have an invertible bounded linear operator $U \in B(X)$ such that

$$\phi\left(\sum_n \lambda_n P_n\right) = U\left(\sum_n \lambda_n P_n\right)U^{-1} \quad \text{and} \quad \phi(P_{n_0}) = UP_{n_0}U^{-1}.$$

From

$$\sum_n \lambda_n U P_n U^{-1} = U(\sum_n \lambda_n P_n)U^{-1} = \phi(\sum_n \lambda_n P_n) = \sum_n \lambda_n P_n$$

we can infer that $U P_n U^{-1} = P_n$ holds for every $n \in \mathbb{N}$. Indeed, let $U P_n U^{-1} = Q_n$ ($n \in \mathbb{N}$). Dividing both sides of the equality

$$\sum_n \lambda_n Q_n = \sum_n \lambda_n P_n$$

by λ_1 and then taking the kth powers of the operators on both sides and, finally, letting k tend to infinity, by the property of the sequence (λ_n) we easily obtain that $Q_1 = P_1$. Therefore, we have

$$\sum_{n=2}^{\infty} \lambda_n Q_n = \sum_{n=2}^{\infty} \lambda_n P_n$$

and one can proceed in the same way to show that $Q_n = P_n$ holds for every $n \in \mathbb{N}$. In particular, we have $\phi(P_{n_0}) = Q_{n_0} = P_{n_0}$. But n_0 was arbitrary and hence we obtain that

$$\phi(P_n) = P_n \qquad (n \in \mathbb{N}). \tag{3.4.9}$$

We now go back to the proof that T is continuous. By (3.4.7) we have $\psi(P_n)y = 0$ ($n \in \mathbb{N}$) and this, together with (3.4.9), yields that $P_n y = 0$ holds for every $n \in \mathbb{N}$. This implies that $y = 0$ verifying the continuity of T.

Beside the fact that ψ is an algebra endomorphism of $F(X)$, we know that ψ is rank preserving. The form of ψ can now be described. It follows from Theorem A.1 that we have two possibilities: either there are linear operators $S : X \to X$ and $R : X^\sharp \to X^\sharp$ such that

$$\psi(x \otimes f) = Sx \otimes Rf \qquad (x \in X, f \in X^\sharp)$$

or there are linear operators $R : X \to X^\sharp$ and $S : X^\sharp \to X$ such that

$$\psi(x \otimes f) = Sf \otimes Rx \qquad (x \in X, f \in X^\sharp).$$

This second possibility can be excluded easily after referring to the multiplicativity of ψ. So, we have linear operators $S : X \to X$ and $R : X^\sharp \to X^\sharp$ such that

$$\psi(x \otimes f) = Sx \otimes Rf \qquad (x \in X, f \in X^\sharp). \tag{3.4.10}$$

It follows from (3.4.6) that

$$Tx \otimes f = T \cdot x \otimes f = Sx \otimes Rf \cdot T = Sx \otimes T^\sharp Rf.$$

This implies that Tx, Sx are linearly dependent for every $x \in X$. By Theorem A.14, S is a scalar multiple of T. Therefore, in what follows we can and do assume that the linear operator S appearing in (3.4.10) is equal to T.

By the local form of ϕ it follows that ψ is trace preserving. Consequently, we have

$$(Rf)(Tx) = f(x) \qquad (x \in X, f \in X^\sharp).$$

This means that $T^\sharp R$ equals the identity on $X^\sharp$. In particular, the range of $T^\sharp$ is closed which is well-known to imply that the range of T is also closed. On the other hand, T has dense range which follows from (3.4.6) and (3.4.9). Therefore, we can infer that $T, T^\sharp$ are invertible and $R = T^{\sharp^{-1}} = T^{-1\sharp}$. This gives us that

$$\psi(x \otimes f) = T \cdot x \otimes f \cdot T^{-1} \qquad (x \in X, f \in X^\sharp) \qquad (3.4.11)$$

and hence we have $\phi(A) = \psi(A) = TAT^{-1}$ for every $A \in F(X)$. We show that $\phi(A) = TAT^{-1}$ holds also for every $A \in \mathcal{A}$. In order to do this, let $A \in \mathcal{A}$ be arbitrary. Pick any $x \in X, f \in X^\sharp$ and let $B = x \otimes f$. By the 2-local property of ϕ we have an invertible element U of $B(X)$ for which

$$\phi(A) = UAU^{-1} \quad \text{and} \quad \phi(B) = UBU^{-1}.$$

Since $B \in F(X)$, we can compute

$$TBT^{-1}\phi(A)TBT^{-1} = \phi(B)\phi(A)\phi(B) =$$
$$UBU^{-1}UAU^{-1}UBU^{-1} = UBABU^{-1}.$$

Taking traces in this equality, we get

$$f(T^{-1}\phi(A)Tx)f(x) = f(Ax)f(x).$$

This implies that for any $x \in X, f \in X^\sharp$ with $f(x) \neq 0$ we have

$$f(T^{-1}\phi(A)Tx) = f(Ax). \qquad (3.4.12)$$

Moreover, if $x \in X, 0 \neq f \in X^\sharp$ and $f(x) = 0$, then we can find a sequence (x_n) in X such that $x_n \to x$ and $f(x_n) \neq 0$ for every $n \in \mathbb{N}$. Hence, we deduce that the equality (3.4.12) holds without any assumption on x and f. This gives us that

$$\phi(A) = TAT^{-1} \qquad (A \in \mathcal{A}). \qquad (3.4.13)$$

To complete the proof it remains to show that in fact we have $\phi(A) = A$ for every $A \in \mathcal{A}$ (this is to prove that ϕ is surjective). From (3.4.9) we learn that $\phi(x_n \otimes f_n) = x_n \otimes f_n$ for every $n \in \mathbb{N}$. Due to (3.4.11) this readily imples that $Tx_n = \epsilon_n x_n$ and $T^{-1\sharp}f_n = \delta_n f_n$ hold for some scalars ϵ_n, δ_n. Moreover, we have $\epsilon_n \delta_n = 1$ for every $n \in \mathbb{N}$. Obviously, we can assume that $\epsilon_1 = \delta_1 = 1$. We have supposed in (3.4.8) that

$$\phi\left(\sum_n \mu_n x_n \otimes f_{n+1}\right) = \sum_n \mu_n x_n \otimes f_{n+1}.$$

By (3.4.13) this yields

$$\sum_n \mu_n T x_n \otimes T^{-1^\sharp} f_{n+1} = \sum_n \mu_n x_n \otimes f_{n+1}.$$

Therefore, we have

$$\sum_n \epsilon_n \delta_{n+1} \mu_n x_n \otimes f_{n+1} = \sum_n \mu_n x_n \otimes f_{n+1}.$$

Considering the values of the operators on both sides at $x_2, x_3, \ldots$ one after the other, we get in turn that $\delta_2 = 1, \epsilon_2 = 1,\ \delta_3 = 1, \epsilon_3 = 1,\ \ldots$. Therefore, T is the identity on X and we have

$$\phi(A) = A \qquad (A \in \mathcal{A}).$$

This completes the proof. $\qquad\qquad\qquad\qquad\qquad\qquad\qquad\qquad\qquad\quad$ $\square$

3.4.4 Remarks

To conclude the section we make the following remarks. Firstly, in the recent paper [126] Kim and Kim have proved that every 2-local *-automorphism of the matrix algebra $M_n(\mathbb{C})$ is a *-automorphism and that similar assertion holds for the 2-local derivations of $M_n(\mathbb{C})$.

We mention that the method of the proof of Theorem 3.4.1 has been utilized in the recent paper [253] of Xie and Lu to show that every 2-local automorphism of a symmetric digraph algebra is an automorphism. For some results on 2-local *-automorphisms and derivations on AF C^*-algebras we refer to the paper [127]. Similar results concerning unital C^*-algebras of real rank zero and nest algebras can be found in [92].

Beside 2-local automorphisms and 2-local derivations one can also investigate 2-local isometries. (Trivially, by a 2-local isometry we mean a map which is equal to a surjective linear isometry on the two-point subsets of its domain.) Such study was done in the paper [182] where we investigated 2-local isometries of standard C^*-subalgebras of $B(H)$. We proved that if the isometry group of such a subalgebra $\mathcal{A}$ of $B(H)$ is algebraically reflexive, then every 2-local isometry of $\mathcal{A}$ is a surjective linear isometry. In particular, we obtained that every 2-local isometry of $B(H)$ is a surjective linear isometry. Furthermore, in [182] we raised the problem to study similar questions for function algebras. In his nice paper [85] Győry presented such a result concerning the function algebra $C_0(X)$.

3.5 2-Local Automorphisms of Some Quantum Structures

3.5.1 Summary

Let H be a separable infinite dimensional Hilbert space. We prove that every continuous 2-local automorphism of the poset (partially ordered set) of all idempotents on H is an automorphism. Similar results are presented for the 2-local automorphisms of the orthomodular lattice of all projections and for those of the Jordan algebra of all self-adjoint operators on H even without assuming continuity. The content of this section appeared in the paper [179].

3.5.2 Formulation of the Results

As we mentioned in the Introduction (see Section 0.3), the orthomodular lattice or quantum logic $P(H)$ of all projections on a Hilbert space H plays important role in the mathematical foundations of quantum mechanics. Because of some theoretical problems the interest in the poset $I(H)$ of all idempotents or, in other words, skew projections has been also aroused (see [204], its references and its review MR 95a:46093).

In this section we investigate the local behaviour of the automorphism groups of $P(H)$ and $I(H)$. We emphasize that these are highly non-linear structures, so here we cannot consider linear local automorphisms. Instead, our main objects are the 2-local automorphisms which can be defined in relation with any algebraic structure. We recall that if $\mathcal{R}$ is an arbitrary algebraic structure, then a transformation $\phi : \mathcal{R} \to \mathcal{R}$ is called a 2-local automorphism of $\mathcal{R}$ if for every $x, y \in \mathcal{R}$ there exists an automorphism $\phi_{x,y}$ of $\mathcal{R}$ such that $\phi(x) = \phi_{x,y}(x)$ and $\phi(y) = \phi_{x,y}(y)$.

In what follows we prove that every continuous 2-local automorphism of $I(H)$ is an automorphism and that every 2-local automorphism of $P(H)$ is also an automorphism. Furthermore, we obtain a similar result concerning the Jordan algebra $B_s(H)$ of all self-adjoint bounded linear operators on H. As we already mentioned several times, the elements of this set represent observable physical quantities and therefore they form one of the most fundamental quantum structures.

We now turn to the precise formulation of our results. First we recall that the partial order $\leq$ on $I(H)$ is defined in the following way. If $P, Q \in I(H)$, then we write $P \leq Q$ if $PQ = QP = P$. Moreover, the operations what make $P(H)$ an orthomodular lattice are the partial order $\leq$ defined above and the orthocomplementation $\perp : P \mapsto I - P$.

The main result of the section is the following theorem. Examining its proof, we shall easily get our other two results.

Theorem 3.5.1. *Let H be a separable infinite dimensional Hilbert space. Every continuous 2-local automorphism of the poset $I(H)$ is an automorphism.*

Theorem 3.5.2. *Every 2-local automorphism of the orthomodular poset $P(H)$ is an automorphism.*

Theorem 3.5.3. *Every 2-local automorphism of the Jordan algebra $B_s(H)$ is an automorphism.*

3.5.3 Proofs

We begin with the forms of the automorphisms of the structures $I(H)$, $P(H)$ and $B_s(H)$. It is an important result of Ovchinnikov [204] (see Theorem A.13) that in the case of an infinite dimensional Hilbert space H, the automorphisms of $I(H)$ with respect to the partial order $\leq$ are exactly the maps

$$P \longmapsto TPT^{-1}, \qquad P \longmapsto TP^*T^{-1}$$

where T is an invertible bounded either linear or conjugate-linear operator on H. What concerns the automorphisms of our other two structures we recall from the Introduction (see Section 0.3) that the so-called ortho-order automorphisms of $P(H)$ are the transformations of the form

$$P \longmapsto UPU^*$$

where U is an either unitary or antiunitary operator on H and the same form applies to the Jordan automorphisms of $B_s(H)$.

In the proof of the main result of the section we shall use the concept of orthogonality between idempotents. For any $P, Q \in I(H)$ we say that P, Q are orthogonal to each other if $PQ = QP = 0$.

We shall also need the following easy lemma.

Lemma 3.5.4. *Let $A, B \in B(H)$. Suppose that for every $x \in H$ we have either $\langle Ax, x \rangle = 0$ or $\langle Bx, x \rangle = 0$. Then either $A = 0$ or $B = 0$.*

Proof. The sets $\{x \in H : \langle Ax, x \rangle = 0\}$ and $\{x \in H : \langle Bx, x \rangle = 0\}$ are closed and their union is H. By Baire's category theorem we obtain that one of these sets has nonempty interior. So, we can suppose that there exist $x_0 \in H$ and $\epsilon_0 > 0$ such that

$$\langle A(x_0 + \epsilon x), x_0 + \epsilon x \rangle = 0$$

holds for every unit vector $x \in H$ and $0 \leq \epsilon < \epsilon_0$. This gives us that

$$\langle Ax_0, \epsilon x \rangle + \langle A(\epsilon x), x_0 \rangle + \langle A(\epsilon x), \epsilon x \rangle = 0,$$

that is,

$$\epsilon \langle Ax_0, x \rangle + \epsilon \langle Ax, x_0 \rangle + \epsilon^2 \langle Ax, x \rangle = 0.$$

By the arbitrariness of ϵ we obtain that $\langle Ax, x \rangle = 0$ for every unit vector $x \in H$. This results in $A = 0$. $\qquad\square$

Proof of Theorem 3.5.1. Let $\phi : I(H) \to I(H)$ be a continuous 2-local automorphism. By the local form of ϕ it is obvious that ϕ preserves the partial order $\leq$ and the orthogonality between the elements of $I(H)$. Furthermore, it is clear that ϕ preserves the rank of finite rank idempotents.

In what follows we describe the form of ϕ on the set $I_f(H)$ of all finite rank elements of $I(H)$. We first show that ϕ is finitely orthoadditive on $I_f(H)$. To see this, let $P, Q \in I_f(H)$ be orthogonal. By the order preserving property of ϕ we have $\phi(P), \phi(Q) \leq \phi(P + Q)$. Since $\phi(P)\phi(Q) = \phi(Q)\phi(P) = 0$, it follows that $\phi(P) + \phi(Q) \leq \phi(P + Q)$. From the rank preserving property of ϕ we obtain that

$$\operatorname{rank}(\phi(P) + \phi(Q)) = \operatorname{rank}\phi(P) + \operatorname{rank}\phi(Q) =$$

$$\operatorname{rank} P + \operatorname{rank} Q = \operatorname{rank}(P + Q) = \operatorname{rank}\phi(P + Q)$$

which yields that

$$\phi(P) + \phi(Q) = \phi(P + Q). \tag{3.5.1}$$

This means that ϕ is finitely orthoadditive on $I_f(H)$.

We now extend ϕ from the set $P_f(H)$ of all finite rank projections to a Jordan homomorphism of the ideal $F(H)$ of all finite rank operators in $B(H)$. Let H_d denote an arbitrary d-dimensional (d is finite) subspace of H. Consider the natural embedding $B(H_d) \hookrightarrow B(H)$ and for any $h \in H$ let ϕ_h be defined by $\phi_h(P) = \langle \phi(P)h, h \rangle$ for every projection P on H_d. Since ϕ is continuous, thus ϕ_h is a bounded orthoadditive function on the set $P(H_d)$ of all projections on H_d (observe that $P(H_d)$ is compact in the norm-topology). If $d \geq 3$, then by Gleason's theorem [68, Theorem 3.2.16] there exists an operator T_h on H_d such that

$$\phi_h(P) = \operatorname{tr} T_h P \qquad (P \in P(H_d)). \tag{3.5.2}$$

Let $P_1, \ldots, P_n \in P(H)$ be finite rank projections (their pairwise orthogonality is not assumed) and let $\lambda_1, \ldots, \lambda_n$ be complex numbers. Define

$$\psi(\sum_k \lambda_k P_k) = \sum_k \lambda_k \phi(P_k). \tag{3.5.3}$$

We have to check that ψ is well-defined. To see this, let $P_1', \ldots, P_n' \in P(H)$ be finite rank projections and $\mu_1, \ldots, \mu_n \in \mathbb{C}$ be such that

$$\sum_k \lambda_k P_k = \sum_k \mu_k P_k'.$$

Take a finite dimensional subspace H_d of H with dimension $d \geq 3$ which includes the ranges of the projections P_k, P_k' ($k = 1, \ldots, n$). Let T_h denote the linear operator on H_d which corresponds to ϕ_h as in (3.5.2). We compute

$$\langle \sum_k \lambda_k \phi(P_k)h, h \rangle = \sum_k \lambda_k \phi_h(P_k) = \operatorname{tr} T_h(\sum_k \lambda_k P_k) =$$

$$\operatorname{tr} T_h(\sum_k \mu_k P_k') = \sum_k \mu_k \phi_h(P_k') = \langle \sum_k \mu_k \phi(P_k')h, h \rangle.$$

Since this holds true for every $h \in H$, we obtain that ψ is well-defined.

As $F(H)$ is the linear span of $P_f(H)$, the definition (3.5.3) clearly implies that ψ is a linear transformation on $F(H)$. Since ψ sends projections to idempotents, it follows from the argument given in the proof of Theorem A.4 that ψ is a Jordan homomorphism of $F(H)$.

As $F(H)$ is a locally matrix algebra, it follows from the classical result Theorem A.5 of Jacobson and Rickart that ψ can be written as $\psi = \psi_1 + \psi_2$, where ψ_1 is a homomorphism and ψ_2 is an antihomomorphism. Let $P \in P(H)$ be a rank-one projection. Because the idempotent $\psi(P) = \phi(P)$ is also rank-one, we obtain that one of the idempotents $\psi_1(P), \psi_2(P)$ is zero. Since $F(H)$ is a simple ring, it is easy to see that this implies that either ψ_1 or ψ_2 is identically zero, that is, ψ is either a homomorphism or an antihomomorphism of $F(H)$. In what follows we can assume without loss of generality that ψ is a homomorphism.

We show that ψ preserves the rank. Let $A \in F(H)$ be a rank-n operator. Then there is a rank-n projection P such that $PA = A$. The rank of $\phi(P)$ is also n. We have $\psi(A) = \psi(P)\psi(A) = \phi(P)\psi(A)$ which proves that $\psi(A)$ is of rank at most n. If Q is any rank-n projection, then there are finite rank operators U, V such that $Q = UAV$. Since $\phi(Q) = \psi(Q) = \psi(U)\psi(A)\psi(V)$ and the rank of $\phi(Q)$ is n, it follows that the rank of $\psi(A)$ is at least n. Therefore, ψ is rank preserving and hence its form can be described. It follows from Theorem A.1 that there are linear operators T, S on H such that ψ is of the form

$$\psi(x \otimes y) = (Tx) \otimes (Sy) \qquad (x, y \in H). \tag{3.5.4}$$

(Recall that in Theorem A.1 the form $\psi(x \otimes y) = (S'y) \otimes (T'x)$ also appears but as ψ is a homomorphism, this possibility can be easily excluded.)

We claim that the operators T, S are bounded. This follows, for example, from [167, Lemma 1] which states that if T, S are linear operators on H with the property that the map $x \mapsto (Tx) \otimes (Sx)$ is continuous on the unit ball of H, then T, S are bounded. Since ψ equals ϕ on the set of all rank-one projections, we have that continuity property, so we obtain that our operators T and S are really bounded.

Next, we infer from (3.5.4) that $\langle Tx, Sx \rangle = \langle x, x \rangle$ for every unit vector $x \in H$. This is because ϕ sends rank-one projections to rank-one idempotents. By polarization it follows that $\langle Tx, Sy \rangle = \langle x, y \rangle$ $(x, y \in H)$. Consequently, we have $S^*T = I$. From (3.5.4) we deduce that $\phi(P) = TPS^*$ for every rank-one projection P. By the finite orthoadditivity property of ϕ (see (3.5.1)) we have

$$\phi(P) = TPS^* \qquad (P \in P_f(H)). \tag{3.5.5}$$

Since $S^*T = I$, it follows that $Q = TS^*$ is an idempotent. We can write

$$\phi(P) = Q\phi(P)Q + Q\phi(P)(I - Q) + (I - Q)\phi(P)Q + (I - Q)\phi(P)(I - Q)$$

for every $P \in I(H)$. We claim that the two middle terms on the right hand side of this equality are in fact zero. Denote

$$\phi_{11}(P) = Q\phi(P)Q, \quad \phi_{12}(P) = Q\phi(P)(I - Q),$$

$$\phi_{21}(P) = (I - Q)\phi(P)Q, \quad \phi_{22}(P) = (I - Q)\phi(P)(I - Q).$$

Let $P \in P(H)$ be fixed and let P' be an arbitrary finite rank projection with $P' \leq P$. We know that $\phi(P') \leq \phi(P)$. Since $\phi(P') = TP'S^*$, we obtain that $\phi(P')Q = Q\phi(P') = \phi(P')$. Hence,

$$\phi(P')\phi_{11}(P) = (\phi(P')Q)\phi(P)Q = \phi(P')\phi(P)Q = \phi(P')Q = \phi(P').$$

Therefore, $TP'S^*\phi_{11}(P) = TP'S^*$. Similarly, we have $\phi_{11}(P)TP'S^* = TP'S^*$. By the arbitrariness of P' it follows that

$$\phi_{11}(P)TPS^* = TPS^*\phi_{11}(P) = TPS^*.$$

This means that

$$TPS^* \leq \phi_{11}(P). \tag{3.5.6}$$

The local form of ϕ implies that $\phi(P) + \phi(I - P) = I$. Therefore, $\phi(I - P) = I - \phi(P)$. Writing $I - P$ in the place of P in (3.5.6), we have

$$Q - TPS^* = T(I - P)S^* \leq \phi_{11}(I - P) = Q - \phi_{11}(P).$$

We deduce that $\phi_{11}(P) \leq TPS^*$ and conclude that

$$TPS^* = \phi_{11}(P).$$

We next compute

$$TP'S^*\phi_{12}(P) = \phi(P')Q\phi(P)(I - Q) = \phi(P')\phi(P)(I - Q) =$$

$$\phi(P')(I - Q) = \phi(P')Q(I - Q) = 0.$$

Since this holds for every finite rank projection P' for which $P' \leq P$, we infer that

$$TPS^*\phi_{12}(P) = 0. \tag{3.5.7}$$

Since $\phi(I - P) = I - \phi(P)$, we have $\phi_{12}(I - P) = -\phi_{12}(P)$. Therefore, using (3.5.7) for P and $I - P$ as well, it follows that

$$0 = T(I - P)S^*\phi_{12}(I - P) = TS^*(-\phi_{12}(P)) - TPS^*(-\phi_{12}(P)) =$$

$$-Q\phi_{12}(P) + TPS^*\phi_{12}(P) = -\phi_{12}(P) + TPS^*\phi_{12}(P) = -\phi_{12}(P).$$

Hence, we obtain that

$$\phi_{12}(P) = 0 \quad (P \in P(H)).$$

Similarly, one can verify that $\phi_{21}(P) = 0$ holds for every $P \in P(H)$. Consequently, our map ϕ is of the form

$$\phi(P) = TPS^* + \phi_{22}(P) \quad (P \in P(H)). \tag{3.5.8}$$

It follows from (3.5.5) that $\phi_{22}(P) = 0$ for every finite rank projection P. We claim that $\phi_{22}(P) = 0$ holds for every $P \in I(H)$ as well.

Assume for a moment that $\phi_{22}(P) \neq 0$ for every projection P of infinite rank and infinite corank. We can choose uncountably many projections P_α of infinite rank and infinite corank such that $P_\alpha P_\beta$ is a finite rank projection for every $\alpha \neq \beta$ (see the proof of Theorem 3.1.1). Using the local form of ϕ we see that the rank of the idempotent $\phi(P_\alpha)\phi(P_\beta)$ is equal to the rank of $P_\alpha P_\beta$. On the other hand, referring to the injectivity of T and to the surjectivity of S^* (these follow from $S^*T = I$), we find that the rank of $TP_\alpha S^* T P_\beta S^* = T P_\alpha P_\beta S^*$ is the same as that of $P_\alpha P_\beta$. By (3.5.8) this gives us that the rank of $\phi_{22}(P_\alpha)\phi_{22}(P_\beta)$ is 0, that is, we have

$$\phi_{22}(P_\alpha)\phi_{22}(P_\beta) = 0$$

for every $\alpha \neq \beta$. This means that the range of ϕ_{22} contains uncountably many nonzero pairwise orthogonal idempotents which contradicts the separability of H (once again, see the proof of Theorem 3.1.1). Therefore, $\phi_{22}(P) = 0$ holds for at least one projection P with infinite rank and infinite corank. The projections P and $I - P$ can be connected by a continuous curve inside the set of projections (this is an easy consequence of the arcwise connectedness of the unitary group of $B(H)$). If $R, R' \in P(H)$ are lying on the same arc in $P(H)$, then by the continuity of ϕ, the idempotents $\phi_{22}(R), \phi_{22}(R')$ are close enough to each other if $\|R - R'\|$ is sufficiently small. Taking into account that the norm of a nonzero idempotent is not less than 1, this, together with $\phi_{22}(P) = 0$, yields that $\phi_{22}(I - P) = 0$. Hence, we have

$$I - Q = \phi_{22}(I) = \phi_{22}(P) + \phi_{22}(I - P) = 0.$$

This implies that $Q = I$ and $\phi_{22} = 0$.

Since $TS^* = Q = I = S^*T$, we obtain $S^* = T^{-1}$. Therefore, ϕ is of the form

$$\phi(P) = TPT^{-1} \qquad (P \in P(H)) \tag{3.5.9}$$

where T is an invertible bounded linear operator on H. We show that our map ϕ is of this form on the whole set $I(H)$. To verify this, we can obviously suppose that $T = I$. So, assume that ϕ is the identity on the set of all projections.

Let P be any idempotent. Pick an arbitrary unit vector $x \in H$ and consider the operator $\phi(x \otimes x)\phi(P)\phi(x \otimes x)$. Taking into account the local form of ϕ and that ϕ is the identity on the set of all projections, we have either

$$\langle \phi(P)x, x \rangle x \otimes x = \phi(x \otimes x)\phi(P)\phi(x \otimes x) =$$

$$A \cdot x \otimes x \cdot A^{-1} A P A^{-1} A \cdot x \otimes x \cdot A^{-1} = A \cdot (\langle Px, x \rangle x \otimes x) \cdot A^{-1} =$$

$$\langle Px, x \rangle \phi(x \otimes x) = \langle Px, x \rangle x \otimes x,$$

(here A is an invertible bounded linear operator such that

$$\phi(x \otimes x) = A \cdot x \otimes x \cdot A^{-1}, \quad \phi(P) = APA^{-1}$$

hold) or, using similar computation,

$$\langle \phi(P)x, x \rangle x \otimes x = \overline{\langle Px, x \rangle} x \otimes x,$$

or

$$\langle \phi(P)x, x \rangle x \otimes x = \langle P^*x, x \rangle x \otimes x,$$

or

$$\langle \phi(P)x, x \rangle x \otimes x = \overline{\langle P^*x, x \rangle} x \otimes x.$$

This gives us that for every $x \in H$ we have either $\langle \phi(P)x, x \rangle = \langle Px, x \rangle$ or $\langle \phi(P)x, x \rangle = \langle P^*x, x \rangle$. By Lemma 3.5.4 we obtain that for any $P \in I(H)$ we have either $\phi(P) = P$ or $\phi(P) = P^*$. We assert that this results in either $\phi(P) = P$ for all $P \in I(H)$ or $\phi(P) = P^*$ for all $P \in I(H)$.

By (3.5.9) we know that this holds for every $P \in P(H)$. Now, let $P \in I(H)$ be a non-self-adjoint finite rank idempotent for which we have, say, $\phi(P) = P$. Consider any non-self-adjoint finite rank idempotent P' which is orthogonal to P. If $\phi(P') = P'^*$, then by $\phi(P) + \phi(P') = \phi(P + P')$ we would arrive at a contradiction. So, $\phi(P') = P'$ for every finite rank idempotent which is orthogonal to P. This implies that $\phi(R) = R$ for every finite rank idempotent with $P \leq R$. Let P' be any non-self-adjoint finite rank idempotent. Suppose that $\phi(P') = P'^*$. If R is a non-self-adjoint finite rank idempotent for which $P' \leq R$, then one can verify similarly as above that $\phi(R) = R^*$. Now, if we choose a non-self-adjoint finite rank idempotent R such that $P, P' \leq R$, then we obtain on the one hand that $\phi(R) = R$ and on the other hand that $\phi(R) = R^*$. But this is a contradiction. Therefore, we have $\phi(P') = P'$ for every finite rank idempotent P'. By the order preserving property of ϕ it now follows that $P \leq \phi(P)$ for every $P \in I(H)$. Putting $I - P$ in the place of P we finally obtain that $\phi(P) = P$ $(P \in I(H))$.

In case ψ is an antihomomorphism, one can follow a similar argument. The proof is complete. $\square$

Proof of Theorem 3.5.2. One can simply follow the proof of Theorem 3.5.1. The only thing which deserves attention is that here we do not need the continuity of ϕ. To see this, we go through those parts of the proof of Theorem 3.5.1 where we have used continuity.

The first such place was where we applied Gleason's theorem. But as in the present case ϕ sends projections to projections, it follows that ϕ is bounded, so we do not need continuity here.

We next used the continuity when showing that the operators T, S in (3.5.4) are continuous. Since in the present case ψ sends projections to projections, it follows that for every unit vector $x \in H$, the operator $Tx \otimes Sx$ is a projection. This implies that $Tx = Sx$ and $\|Tx\| = \|Sx\| = 1$. So, $T = S$ is an isometry.

The third appearance of the continuity of ϕ was where we proved that if $\phi_{22}(P) = 0$ holds for at least one projection P of infinite rank and infinite corank, then $\phi_{22} = 0$. Observe that in our present case the terms in the decomposition $\phi(R) = \phi_{11}(R) + \phi_{22}(R)$ $(R \subset P(H))$ are projections. Let $P' \in P(H)$ be such that $\|P - P'\| < 1$. By the local form of ϕ we see that $\|\phi(P) - \phi(P')\| = \|P - P'\| < 1$. Since

$$\|\phi_{22}(P')\| = \|\phi_{22}(P) - \phi_{22}(P')\| \leq \|\phi(P) - \phi(P')\| < 1,$$

we deduce that $\phi_{22}(P') = 0$. As we can get from P to $I - P$ in finitely many steps $P = P_0, P_1, \ldots, P_n = I - P$ such that $\|P_{k-1} - P_k\| < 1$, we obtain that $\phi_{22}(I - P) = 0$. This gives us that $\phi_{22}(I) = 0$ which implies $\phi_{22} = 0$. The proof is complete.   □

Proof of Theorem 3.5.3. Let $\phi : B_s(H) \to B_s(H)$ be a 2-local automorphism. By the forms of the automorphisms of $P(H)$ and $B_s(H)$ it follows that $\phi_{|P(H)}$ is a 2-local automorphism of $P(H)$. Using Theorem 3.5.2 we obtain that there exists a unitary or antiunitary operator U on H such that

$$\phi(P) = UPU^* \qquad (P \in P(H)).$$

We can assume without loss of generality that $\phi(P) = P$ for every $P \in P(H)$. Now, similarly as in the proof of Theorem 3.5.1, picking any operator $A \in B_s(H)$ and unit vector $x \in H$ and considering the operator $\phi(x \otimes x)\phi(A)\phi(x \otimes x)$, we find that

$$\langle \phi(A)x, x \rangle x \otimes x = \langle Ax, x \rangle x \otimes x$$

which implies that

$$\langle \phi(A)x, x \rangle = \langle Ax, x \rangle.$$

Therefore, we have $\phi(A) = A$ for every $A \in B_s(H)$ and this completes the proof.   □

3.5.4 Remarks

In conclusion we make the following remarks. First of all, we repeat our convincement from the Introduction (see Section 0.5) that the concept of 2-local automorphisms is more fruitful than that of the linear local automorphisms. As we have already mentioned, the main reason is that by the help of this former notion one can study the local actions of automorphism groups of algebraic structures of any kind. We believe that such investigations can provide noteworthy new information on the automorphism groups appearing in many different parts of mathematics. Therefore, we hope that the intensity of research in this direction will increase in the near future and we can see nice new results. As the definition of 2-local automorphisms in its full generality is quite new, there are only few results available on them in non-linear structures. Here we mention the paper [13] of Barczy and Tóth, and our work [193].

In [13] it was proved that for infinite dimensional separable H, every 2-local automorphism of the set of all states on H is an automorphism and a similar result was obtained for the 2-local automorphisms of the effect algebra $E(H)$. For the finite dimensional case see the paper [125] of Kim. In our paper [193] we gained results of the same spirit concerning the 2-local automorphisms of the unitary group and the general linear group of $B(H)$.

Appendix

Below we collect some basic results that we apply several times in our book.

We begin with a description of rank-one preserving linear transformations. If X is a real or complex Banach space, then let $X^\sharp$ denote its dual. If $x \in X$ and $f \in X^\sharp$, then $x \otimes f$ stands for the operator (of rank at most 1) defined by

$$(x \otimes f)(z) = f(z)x \qquad (z \in X).$$

Denote by $F(X)$ the algebra of all bounded linear finite rank operators on X. The following result could be derived from the arguments used in the first half of Section I in the paper [104] by Hou.

Theorem A.1. *Let X be a real or complex Banach space and $\phi : F(X) \to F(X)$ be a linear transformation which maps rank-one operators to rank-one operators. Suppose that the range of ϕ contains an operator with rank greater than 1. Then either there are linear operators $S : X \to X$ and $R : X^\sharp \to X^\sharp$ such that*

$$\phi(x \otimes f) = Sx \otimes Rf \qquad (x \in X, f \in X^\sharp)$$

or there are linear operators $R' : X \to X^\sharp$ and $S' : X^\sharp \to X$ such that

$$\phi(x \otimes f) = S'f \otimes R'x \qquad (x \in X, f \in X^\sharp).$$

The next result on the structure of linear maps between C^*-algebras which preserve the unitary group is due to Russo and Dye. See [219, Corollary 2].

Theorem A.2. *Let $\mathcal{A}, \mathcal{B}$ be C^*-algebras. Suppose that $\phi : \mathcal{A} \to \mathcal{B}$ is a linear map sending unitaries to unitaries. Then ϕ can be written in the form*

$$\phi(A) = U\psi(A) \qquad (A \in \mathcal{A})$$

*where $U \in \mathcal{B}$ is a fixed unitary element and $\psi : \mathcal{A} \to \mathcal{B}$ is a unital Jordan *-homomorphism.*

The following description of order-preserving linear bijections between C^*-algebras was obtained by Kadison. See [118, Corollary 5].

Theorem A.3. *Any bijective linear transformation between C^*-algebras which sends the unit to the unit and preserves the order in both directions is necessarily a Jordan *-isomorphism.*

The next result gives a very useful characterization of Jordan homomorphisms and Jordan *-homomorphisms. We present it with proof as we refer to the argument below several times in the previous parts of the book.

Theorem A.4. *Let $\mathcal{A}$ be a von Neumann algebra and $\mathcal{B}$ be a Banach algebra. If $\phi : \mathcal{A} \to \mathcal{B}$ is a continuous linear transformation which sends projections to idempotents, then ϕ is a Jordan homomorphism. If $\mathcal{B}$ is a C^*-algebra and $\phi : \mathcal{A} \to \mathcal{B}$ is a continuous linear transformation which maps projections to projections, then ϕ is a Jordan *-homomorphism.*

Proof. The argument is borrowed from the proof of [29, Proposition 3.7]. If $P, Q \in \mathcal{A}$ are orthogonal projections, then $P + Q$ is also a projection. It follows that $\phi(P), \phi(Q)$ and $\phi(P) + \phi(Q) = \phi(P + Q)$ are idempotents. From $(\phi(P) + \phi(Q))^2 = \phi(P) + \phi(Q)$ we infer that $\phi(P)\phi(Q) + \phi(Q)\phi(P) = 0$. One can verify that this implies the orthogonality (in the algebraic sense) of the idempotents $\phi(P)$ and $\phi(Q)$, i.e., that $\phi(P)\phi(Q) = \phi(Q)\phi(P) = 0$. Let $P_1, ..., P_n \in \mathcal{A}$ be pairwise orthogonal projections and $\lambda_1, ..., \lambda_n \in \mathbb{R}$. We compute

$$(\phi(\sum_{k=1}^{n} \lambda_k P_k))^2 = (\sum_{k=1}^{n} \lambda_k \phi(P_k))^2 = \sum_{k=1}^{n} \lambda_k^2 \phi(P_k) =$$

$$\phi(\sum_{k=1}^{n} \lambda_k^2 P_k) = \phi((\sum_{k=1}^{n} \lambda_k P_k)^2).$$

Using the continuity of ϕ and the spectral theorem of self-adjoint operators, we obtain that $\phi(A^2) = \phi(A)^2$ holds for every self-adjoint element $A \in \mathcal{A}$. Linearizing this equality, i.e., replacing A by $A + B$ ($B \in \mathcal{A}$ is also self-adjoint), we obtain $\phi(AB + BA) = \phi(A)\phi(B) + \phi(B)\phi(A)$. Next, it follows that

$$\phi((A + iB)^2) = \phi(A^2) - \Phi(B^2) + i\phi(AB + BA) =$$
$$\phi(A)^2 - \phi(B)^2 + i(\phi(A)\phi(B) + \phi(B)\phi(A)) = (\phi(A) + i\phi(B))^2$$

which implies that ϕ is a Jordan homomorphism. If ϕ sends projections to projections, then we see that ϕ maps self-adjoint elements to self-adjoint elements and one can easily verify that this gives us that ϕ is a Jordan *-homomorphism. $\square$

The following theorem is due to Jacobson and Rickart [106]. It states that every Jordan homomorphism from a locally matrix algebra can be written as the sum of a homomorphism and an antihomomorphism. We point out that in [106, Theorem 8] the authors presented their result in the setting of rings. However, for our purpose we need the following version of the statement (cf. [206, 6.3.12 Theorem]).

Theorem A.5. *Let $\mathcal{A}, \mathcal{B}$ be algebras over a field $\mathbb{F}$. Suppose that $\mathcal{A}$ is a locally matrix algebra, i.e., every finite subset of $\mathcal{A}$ can be included in a subalgebra of $\mathcal{A}$ which is isomorphic to a full matrix algebra $M_n(\mathbb{F})$ with $n \geq 2$. If $\phi : \mathcal{A} \to \mathcal{B}$ is a Jordan homomorphism, then we have a homomorphism $\phi_1 : \mathcal{A} \to \mathcal{B}$ and an antihomomorphism $\phi_2 : \mathcal{A} \to \mathcal{B}$ such that $\phi = \phi_1 + \phi_2$.*

The following analogue of the theorem of Jacobson and Rickart concerning Jordan *-homomorphisms of C^*-algebras was obtained by Størmer in [235] (also see pp. 773–776 in [120]).

Theorem A.6. *Let $\mathcal{A}$ be a C^*-algebra and H be a Hilbert space. Suppose that $\phi : \mathcal{A} \to B(H)$ is a Jordan *-homomorphism. Then there exist two central projections E, F in the von Neumann algebra generated by the range of ϕ such that the map $\phi_1 : A \mapsto \phi(A)E$ is a *-homomorphism, the map $\phi_2 : A \mapsto \phi(A)F$ is a *-antihomomorphism, $E + F = I$, and $\phi = \phi_1 + \phi_2$.*

The next theorem is due Herstein [100]. Recall that a ring $\mathcal{R}$ is prime if $a\mathcal{R}b = \{0\}$ implies $a = 0$ or $b = 0$ $(a, b \in \mathcal{R})$.

Theorem A.7. *If ϕ is a Jordan homomorphism of a ring onto a prime ring of characteristic different from 2 and 3, then either ϕ is a homomorphism or an antihomomorphism.*

It is easy to see that every standard operator algebra on an arbitrary Banach space is prime. Therefore, any Jordan homomorphism onto such an algebra is either a homomorphism or an antihomomorphism. The next folklore result describes the forms of the algebra automorphisms, algebra antiautomorphisms, algebra *-automorphisms and algebra *-antiautomorphisms of standard operator algebras. One can easily demonstrate it after having a look at the proof of [223, Theorem] (also see [51]).

Theorem A.8. *Let H be a (real or complex) Hilbert space and $\mathcal{A}$ be a standard operator algebra on H. Then the following assertions hold.*

(i)   *Every algebra automorphism of $\mathcal{A}$ is of the form $A \mapsto TAT^{-1}$ for some invertible bounded linear operator T on H.*

(ii) *Every algebra antiautomorphism of $\mathcal{A}$ is of the form $A \mapsto TA^{tr}T^{-1}$ for some invertible bounded linear operator T on H.*

(iii) *If $\mathcal{A}$ is a *-subalgebra of $B(H)$, then every algebra *-automorphism of $\mathcal{A}$ is of the form $A \mapsto UAU^*$ for some unitary operator U on H.*

(iv) *If $\mathcal{A}$ is a *-subalgebra of $B(H)$, then every algebra *-antiautomorphism of $\mathcal{A}$ is of the form $A \mapsto UA^{tr}U^*$ for some unitary operator U on H.*

It is easy to verify that the statement (ii) above can be reformulated in the following way. Every algebra antiautomorphism of $\mathcal{A}$ is of the form $A \mapsto TA^*T^{-1}$ for some invertible bounded conjugate-linear operator T on H. Of course, similar reformulation applies also for the statement (iv).

According to an important result of Kadison [117], every surjective linear isometry between C^*-algebras can be written as a Jordan *-isomorphism multiplied by a fixed unitary element. This together with the previous two theorems provides the structure of all surjective linear isometries of $B(H)$.

Theorem A.9. *Let H be a Hilbert space. If $\phi : B(H) \to B(H)$ is a surjective linear isometry, then there are unitary operators U, V on H such that ϕ is either of the form*

$$\phi(A) = UAV \qquad (A \in B(H))$$

or of the form

$$\phi(A) = UA^{tr}V \qquad (A \in B(H)).$$

The following statement which is called Banach-Stone theorem describes the form of all surjective linear isometries of the function spaces $C_0(L)$. If L is a locally compact Hausdorff space, then let $C_0(L)$ denote the space of all continuous complex valued functions on L which vanish at infinity. We endow $C_0(L)$ with the usual sup-norm.

Theorem A.10. *Let X and Y be locally compact Hausdorff spaces, and $\phi : C_0(X) \to C_0(Y)$ be a surjective linear isometry. Then ϕ is of the form*

$$\phi(f) = \tau \cdot f \circ \varphi \qquad (f \in C_0(X)),$$

where $\tau : X \to \mathbb{C}$ is a continuous function of modulus 1 and $\varphi : Y \to X$ is a homeomorphism.

It is an important consequence of the representation theory of $B(H)$ that in the case of a separable space, the *-endomorphisms of $B(H)$ can be written in a regular form. The following result can be found in [121, 10.4.14. Corollary].

Theorem A.11. *Let H be a separable infinite dimensional Hilbert space. If $\phi : B(H) \to B(H)$ is an algebra *-homomorphism, then there exists a collection $\{U_\alpha\}$ of isometries on H with pairwise orthogonal ranges such that ϕ is of the form*

$$\phi(A) = \sum_\alpha U_\alpha A U_\alpha^* \qquad (A \in B(H)).$$

The next result is due to Bunce and Wright [35]. It gives the solution of the famous Mackey-Gleason problem.

Theorem A.12. *Let $\mathcal{A}$ be a von Neumann algebra without type I_2 direct summand and let $P(\mathcal{A})$ be its lattice of projections. Let X be a Banach space. Suppose that $m : P(\mathcal{A}) \to X$ is a bounded function such that $m(P + Q) = m(P) + m(Q)$ holds whenever $P, Q \in \mathcal{A}$ are orthogonal projections. Then m has a unique extension to a bounded linear transformation from $\mathcal{A}$ to X.*

The following important result was obtained by Ovchinnikov [204]. It describes the general form of the order automorphisms of the poset of all idempotents on a Hilbert space.

Theorem A.13. *Let H be a Hilbert space with $\dim H \geq 3$. Denote $I(H)$ the set of all idempotents in $B(H)$ equipped with the partial order $\leq$ defined as $P \leq Q$ iff $PQ = QP = P$. Then every automorphism of $I(H)$ as a poset (i.e., every bijective map of $I(H)$ which preserves the order in both directions) is either of the form $P \mapsto APA^{-1}$ or of the form $P \mapsto AP^*A^{-1}$ with some $A \in \mathfrak{S}$. Here $\mathfrak{S}$ is the set of all semilinear bijections $A : H \to H$ if H is finite dimensional and the set of all invertible bounded linear or conjugate-linear operators on H if H is infinite dimensional.*

The content of the last result which is well-known for real or complex vector spaces is that under mild conditions locally linearly dependent linear operators are necessarily linearly dependent. The presented version of the result can be found in [30, Theorem 2.3].

Theorem A.14. *Let X be a linear space over an infinite field. Let $A, B : X \to X$ be linear operators. Suppose that for every $x \in X$ the vectors Ax, Bx are linearly dependent. Then either the operators A, B are linearly dependent or the ranges of A and B are included in the same one-dimensional subspace of X.*

Recent Results Added in Revision

This short chapter is devoted to overview some of the very recent papers relating the topics discussed in our work which have come out or come to our knowledge after the submission of the manuscript.

In Section 1.4 we mentioned the problem of additivity of bijective multiplicative or multiplicative-like maps between general rings or operator algebras. For recent results in this direction we first refer to the paper [255] by An and Hou. There they considered the products $(A, B) \mapsto ABA$, $(A, B) \mapsto AB + BA$ and $(A, B) \mapsto \frac{1}{2}(AB + BA)$ on the set of all self-adjoint operators on a Hilbert space H and also on nest algebras. They proved that if $\dim H > 1$, then every bijective map on the set of all self-adjoint operators which is multiplicative with respect to any one of the above listed products is automatically additive. In fact, they determined the general forms of those transformations. Furthermore, they presented a similar result concerning the second product on standard subalgebras of nest algebras.

Next we mention our paper [282] where we considered bijective maps on the set of all invertible positive (respectively, all invertible self-adjoint) operators which are multiplicative with respect to the Jordan triple product $(A, B) \mapsto ABA$. Under the assumption of continuity we determined the general forms of those transformations. It turned out that in the infinite dimensional case any such map defined on the set of all invertible positive operators is equal, up to composition by a linear *-automorphism or a linear *-antiautomorphism of the full operator algebra, either to the identity map $A \mapsto A$ or to the inverse operation $A \mapsto A^{-1}$ which is obviously non-additive. In the finite dimensional case the function of the determinant also shows up in the general forms of our transformations. What concerns our maps on the set of all invertible self-adjoint operators, the situation with them is a little bit more complicated. In the proofs we used some of our former results from the paper [283]. There we described the general forms of all continuous (non-linear) functionals on the sets of positive definite, positive semi-definite and Hermitian matrices which are multiplicative on the commuting elements. That investigation was motivated by the famous Kochen-Specker theorem concerning the problem of hidden variables in quantum mechanics. As applications we obtained new characterizations of the determinant on the above mentioned classes of matrices by means of its multiplicativity property with respect to the Jordan triple product.

In relation with the same product we refer to the paper [271] by Lešnjak and Sze where they described all injective maps (surjectivity was not assumed) of the full matrix algebra over a field which are multiplicative with respect to the Jordan triple product.

There are some recent results on automatic additivity also for the so-called elementary maps and Jordan elementary maps. These transformations are abstract generalizations of some algebraic transformations like double centralizers and isomorphisms. For corresponding results we refer to the papers [274, 275, 276, 277] by Li, Jing and Lu and to the paper [296] by Timmermann. In our paper [285] we obtained the additivity of surjective Jordan elementary maps defined on self-adjoint operators. Namely, we proved the following statement. Assume that $M : \mathcal{A}_s \rightarrow \mathcal{B}_s$ and $M^* : \mathcal{B}_s \rightarrow \mathcal{A}_s$ are surjective maps between the self-adjoint parts of standard *-operator algebras $\mathcal{A}$ and $\mathcal{B}$ on an infinite dimensional Hilbert space which satisfy the equations $M(AM^*(B)A) = M(A)BM(A)$ and $M^*(BM(A)B) = M^*(B)AM^*(B)$ for every pair $A \in \mathcal{A}_s$, $B \in \mathcal{B}_s$. Then there exist an invertible bounded linear or conjugate-linear operator $T : H \rightarrow H$ and a constant $c \in \{-1, 1\}$ such that $M(A) = cTAT^*$ $(A \in \mathcal{A}_s)$ and $M^*(B) = cT^*BT$ $(B \in \mathcal{B}_s)$.

Finally, we draw the reader's attention to the interesting papers [262, 263, 264] of Chebotar et al. where they discussed linear or additive maps on operator algebras, matrix algebras or general rings with certain product preserving properties. Their approaches were based on deep algebraic tools like the modern theory of functional identities. This idea was utilized already in the basic paper [256] by Beidar, Brešar, Chebotar and Fong where functional identities were applied to certain linear preserver problems, for example, to the problem of commutativity preservers. For important new achievements in this direction we refer to the papers [257, 258] of Brešar and Šemrl.

In Section 2.2 we presented a generalization of Uhlhorn's theorem concerning orthogonality preserving maps for the case of indefinite inner product spaces. Our approach was based on the description of all bijective maps on the set of rank-one idempotents which preserve zero product in both directions (see Theorem 2.2.1). In relation with this result we refer to a series of recent papers by Šemrl in which he presented a number of deep results concerning various transformations on the set of idempotents. In fact, in those papers he continued and extended his previous work that we already mentioned in 2.2.4. Remarks. First of all, we refer to his substantial work [291] where he gave a complete and fine analysis of transformations on the set of all $n \times n$ $(n \geq 3)$ idempotent matrices over a division ring which preserve either commutativity, or order, or orthogonality. This analysis extends to the study of non-bijective maps as well as maps which preserve any of the mentioned relations only in one direction. As an application, he obtained a quaternionic analogue of Ovchinnikov's theorem (see Theorem A.13). Other applications were also given on automorphisms of operator and matrix semigroups, local automorphisms, linear preserver problems and the geometry of matrices and Grassmannians. The main results in [291] appear also in the expository paper [292]. There he gave a short proof of the well-known statement that every automorphism of the full matrix algebra is inner and discussed several extensions of this theorem. Those include, among others, the descriptions of multiplicative maps on matrix algebras and some non-linear preserver results. In the paper [293], beside structural results concerning non-linear maps on full matrix algebras preserving commutativity and maps on idempotent matrices and operators preserving orthogonality, the general form of all bijective non-linear maps which preserve the Lie product of operators was completely determined both in the finite and in the infinite dimensional cases. In [294] it was shown that in infinite dimensional Banach spaces the structural results concerning bijective maps on idempotent operators which preserve the order or the orthogonality in both directions can be derived from the corresponding result concerning commutativity preserving maps. Moreover, the form of bijective maps which preserve comparability in both directions was also described. Next we mention another expository paper of Šemrl [295] where beside order or orthogonality preserving maps on idempotent matrices or operators, he surveyed some recent results concerning adjacency preserving maps on linear matrix spaces, maps on the set of space-time events preserving coherency, maps on bounded observables preserving compatibility, Wigner's unitary-antiunitary theorem, and the geometry of Grassmann spaces. Moreover, in that paper he pointed out the interrelations between those, seemingly rather unrelated areas of research.

In the paper [290] Rodman and Šemrl presented an Uhlhorn-type result for orthogonality preserving maps between projective spaces corresponding to different indefinite inner products on $\mathbb{K}^n$ ($n \geq 3$ and $\mathbb{K}$ denotes the field of real or complex numbers) which arise from invertible $n \times n$ matrices. The authors obtained their result for bijective maps which preserve orthogonality only in one direction.

Next we refer to the papers [266, 267] of Fošner. It turns out from [266] that the structure of the order automorphisms of the set of all upper triangular idempotent matrices is not so nice as in the case of the set of all idempotent matrices. In [267] Fošner described the general form of all bijective maps on the set of upper triangular idempotent matrices which preserve both the order and the orthogonality in one direction.

Finally, we mention the papers [265] and [297] for interesting results on additive or linear maps defined on operator algebras which preserve certain kinds of orthogonality relations between operators.

Section 2.3 was devoted to determine the fidelity preserving bijective maps on the sets of states and density operators. In our recent paper [286] we obtained a generalization of the corresponding result. Namely, among others, we determined the so-called Θ-fidelity preserving bijective maps on the same sets.

In Section 2.5 we described the form of all bijective maps (no linearity was assumed) on the set $B_s(H)$ of all bounded observables (i.e., self-adjoint operators) which preserve the usual order $\leq$ in both directions. The order is a rather important relation between observables but there is another even more important correspondence which is called compatibility. Mathematically this means simply commutativity. In our paper [284] we characterized those bijective maps on $B_s(H)$ (once again without assuming linearity) which preserve commutativity in both directions. In particular, we proved that for a complex separable Hilbert space H with dim $H \geq 3$, if $\phi : B_s(H) \to B_s(H)$ is a bijective map preserving commutativity in both directions, then there exists an either unitary or antiunitary operator U on H, and for every operator $A \in B_s(H)$ there is a real valued bounded Borel function f_A on the spectrum $\sigma(A)$ of A such that $\phi(A) = U f_A(A) U^*$. In other words, this means that, up to composition by either a linear *-isomorphism or a linear *-antiisomorphism of the operator algebra $B(H)$, for every $A \in B_s(H)$ the operator $\phi(A)$ is a real valued bounded Borel function of A. Concerning the same problem for full matrix algebras we refer to the papers [293] by Šemrl and [268] by Fošner. There they described all the continuous bijective maps on the algebra of $n \times n$ ($n \geq 3$) complex, respectively real matrices which preserve commutativity in both directions.

In Section 2.8 we studied the structure of sequential isomorphisms between the sets of von Neumann algebra effects. In 2.8.4. Remarks we mentioned a result of Marovt on the form of sequential isomorphisms between the sets of commutative C^*-algebra effects. In fact, in [280] he described all the bijective multiplicative maps on the set $C(X, I)$ of continuous functions from a first countable compact Hausdorff space X into the unit interval I. Furthermore, in his recent paper [279] he presented a similar structural result for the bijective maps of $C(X, I)$ preserving the order in both directions, while in [281] he gave the complete description of the affine bijections of $C(X, I)$.

In Chapter 3 we considered certain local transformations on structures of linear operators and continuous functions.

As for recent results in this direction concerning function spaces we mention the paper [259] of Cabello Sánchez. In Section 3.2 we proved the algebraic reflexivity of the automorphism and isometry groups of the complex function space $C(X)$ on a first countable compact Hausdorff space X. In our proof we used the theorem of Russo and Dye (see Theorem A.2) concerning the structure of linear maps on C^*-algebras preserving the unitary group which result do require that the underlying field is $\mathbb{C}$. In the paper [259] Cabello Sánchez proved that for various classes of locally compact spaces L, the space $C_0^{\mathbb{R}}(L)$ of all real-valued continuous functions on L vanishing at infinity has the property that its local isometries are all surjective. This means the algebraic reflexivity of the isometry group of $C_0^{\mathbb{R}}(L)$. The classes in question include totally disconnected locally compact spaces whose one-point compactification is metrizable and manifolds with and without boundary.

In the paper [287] Pop studied bounded linear maps from a von Neumann algebra into $B(H)$ which are (1-)local or 2-local representations (the definition should be self-explanatory). He obtained the interesting result that every bounded 2-local representation is necessarily a representation but, in contrast, 1-local representations may fail to be multiplicative even at the 2 by 2 matrix algebra level.

In a series of papers Zhang and his coauthors studied 1-local and 2-local transformations on nest subalgebras of von Neumann algebras. In [299] they proved that every norm-continuous linear local derivation of a nest subalgebra of a factor von Neumann algebra is a derivation and that every linear 2-local derivation (not necessarily norm-continuous) of the same structure is a derivation. In [300] it was shown that every surjective weakly continuous linear local automorphism of a nest subalgebra corresponding to a non-trivial nest in a factor von Neumann algebra is an automorphism and that every surjective linear 2-local automorphism (no continuity is assumed) of the same structure is an automorphism. In the paper [298] they considered local 2-cocyles of operator algebras. It was proved there that every weakly continuous local 2-cocycle of a nest subalgebra of a factor von Neumann algebra M with coefficients in M is a 2-cocycle. In their paper [278] Liu and Wong presented results on the 2-local automorphisms of standard operator algebras over locally convex spaces, especially, over Fréchet spaces with Schauder basis and also on the 2-local automorphisms of some function algebras.

Finally, we recall Theorem 3.4.1 which was our key tool in the proof of a theorem on the 2-local automorphisms of standard operator algebras over Banach spaces. This result gives a non-linear characterization of the automorphisms of matrix algebras by means of their property that they leave the set of eigenvalues (counted according to multiplicity) of matrix products invariant. In the papers [260, 261, 272, 273] by Chan, Li, Poon, Sze and [270] by Hou and Di one can find several recent results of similar spirit concerning the invariance of certain functional values of matrix or operator products. Corresponding results for function algebras relating maps which preserve the range of products were presented in the papers [269, 288, 289].

References

1. J. Aczél, *Lectures on Functional Equations and their Applications*, Academic Press, New York–London, 1966.
2. P.M. Alberti, *Playing with fidelities*, Rep. Math. Phys. **51** (2003), 87–125.
3. P.M. Alberti and A. Uhlmann, *On Bures distance and *-algebraic transition probability between inner derived positive linear forms over W^*-algebras*, Acta Appl. Math. **60** (2000), 1-37.
4. G. An and J. Hou, *Rank-preserving multiplicative maps on $B(X)$*, Linear Algebra Appl. **342** (2002), 59–78.
5. J. Araujo, E. Beckenstein and L. Narici, *Biseparating maps and homeomorphic realcompactifications*, J. Math. Anal. Appl. **192** (1995), 258–265.
6. J. Araujo and K. Jarosz, *Biseparating maps between operator algebras*, J. Math. Anal. Appl. **282** (2003), 48–55.
7. J. Arazy, *The isometries of C_p*, Israel J. Math. **22** (1975), 247–256.
8. B. Aupetit, *A Primer on Spectral Theory*, Springer-Verlag, 1991.
9. B. Aupetit, *Sur les transformations qui conservent la spectre*, in Banach Algebras'97, Walter de Gruyter, 1998, pp. 55–78.
10. B. Aupetit, *Spectrum-preserving linear mappings between Banach algebras or Banach-Jordan algebras*, J. London Math. Soc. **62** (2000), 917–924.
11. D. Bakić and B. Guljaš, *Wigner's theorem in Hilbert C^*-modules over C^*-algebras of compact operators*, Proc. Amer. Math. Soc. **130** (2002), 2343–2349.
12. D. Bakić and B. Guljaš, *Wigner's theorem in a class of Hilbert C^*-modules*, J. Math. Phys. **44** (2003), 2186–2191.
13. M. Barczy and M. Tóth, *Local automorphisms of the sets of states and effect on a Hilbert space*, Rep. Math. Phys. **48** (2001), 289–298.
14. V. Bargmann, *Note on Wigner's theorem on symmetry operations*, J. Math. Phys. **5** (1964), 862–868.
15. B.A. Barnes and A.K. Roy, *Diameter preserving maps on various classes of functions spaces*, Studia Math. **153** (2002), 127–145.
16. C.J.K. Batty and L. Molnár, *On topological reflexivity of the groups of *-automorphisms and surjective isometries of $\mathcal{B}(H)$*, Arch. Math. **67** (1996), 415–421.
17. L.B. Beasley, *Linear operators on matrices: the invariance of rank-k matrices*, Linear Algebra Appl. **107** (1988), 161–167.

18. L.B. Beasley, A.H. Kim and W.Y. Lee, *On a positive linear map preserving absolute values,* Linear Algebra Appl. **260** (1997), 311–318.
19. R. Bhatia, *Matrix Analysis,* Springer-Verlag, 1997.
20. R. Bhatia and P. Šemrl, *Approximate isometries on Euclidean spaces,* Amer. Math. Monthly **104** (1997), 497-504.
21. P. Botta, *Linear maps preserving rank less than or equal to one,* Linear Multilinear Algebra **20** (1987), 197–201.
22. L. Bracci, G. Morchio, and F. Strocchi, *Wigner's theorem on symmetries in indefinite metric spaces,* Commun. Math. Phys. **41** (1975), 289–299.
23. M. Brešar, *Characterizations of derivations on some normed algebras with involution,* J. Algebra **152** (1992), 454–462.
24. M. Brešar, W.S. Martindale and C.R. Miers, *Maps preserving nth powers,* Commun. Algebra **26** (1998), 117-138.
25. M. Brešar and P. Šemrl, *Mappings which preserve idempotents, local automorphisms, and local derivations,* Canad. J. Math. **45** (1993), 483–496.
26. M. Brešar and P. Šemrl, *Normal-preserving linear mappings,* Canad. Math. Bull. **37** (1994), 306-309.
27. M. Brešar and P. Šemrl, *On local automorphisms and mappings that preserve idempotents,* Studia Math. **113** (1995), 101–108.
28. M. Brešar and P. Šemrl, *Linear maps preserving the spectral radius,* J. Funct. Anal. **142** (1996), 360–368.
29. M. Brešar and P. Šemrl, *Linear preservers on $B(X)$,* Banach Cent. Publ. **38** (1997), 49–58.
30. M. Brešar and P. Šemrl, *On locally linearly dependent linear operators and derivations,* Trans. Amer. Math. Soc. **351** (1999), 1257–1275.
31. P.M. Van den Broek, *Twistor space, Minkowski space and the conformal group,* Physica A **122** (1983), 587–592.
32. P.M. Van den Broek, *Symmetry transformations in indefinite metric spaces: A generalization of Wigner's theorem,* Physica A **127** (1984), 599-612.
33. P.M. Van den Broek, *Group representations in indefinite metric spaces,* J. Math. Phys. **25** (1984), 1205-1210.
34. L.G. Brown, R.G. Douglas and P.A. Fillmore, *Unitary equivalence modulo the compact operators and extensions of C^*-algebras,* Lect. Notes in Math. **345** (1973) pp. 58–128.
35. L.J. Bunce and D.M. Wright, *The Mackey-Gleason problem,* Bull. Amer. Math. Soc. **26** (1992), 288–293.
36. P. Busch, *Stochastic isometries in quantum mechanics,* Math. Phys. Anal. Geom. **2** (1999), 83–106.
37. P. Busch and S.P. Gudder, *Effects as functions on projective Hilbert spaces,* Lett. Math. Phys. **47** (1999), 329–337.
38. P. Busch, P.J. Lahti and P. Mittelstaedt, *The Quantum Theory of Measurement,* Springer-Verlag, 1991.
39. F. Cabello Sánchez, *Diameter preserving linear maps and isometries,* Arch. Math. **73** (1999), 373–379.
40. F. Cabello Sánchez, *Diameter preserving linear maps and isometries II,* Proc. Indian Acad. Sci. **110** (2000), 205–211.
41. F. Cabello Sánchez, *The group of automorphisms of L_∞ is algebraically reflexive,* Studia Math. **161** (2004), 19–32.
42. F. Cabello Sánchez, *Convex transitive norms on spaces of continuous functions,* Bull. London. Math. Soc. **37** (2005), 107–118.

43. A. Cabello Sánchez and F. Cabello Sánchez, *Maximal norms on Banach spaces of continous functions*, Math. Proc. Camb. Philos. Soc. **129** (2000), 325–330.

44. F. Cabello Sánchez and L. Molnár, *Reflexivity of the isometry group of some classical spaces*, Rev. Mat. Iberoam. **18** (2002), 409–430.

45. G. Cassinelli, E. De Vito, P. Lahti and A. Levrero, *Symmetry groups in quantum mechanics and the theorem of Wigner on the symmetry transformations*, Rev. Math. Phys. **8** (1997), 921–941.

46. G. Cassinelli, E. De Vito, P. Lahti and A. Levrero, *A theorem of Ludwig revisited*, Found. Phys. **30** (2000), 1755–1761.

47. G. Cassinelli, E. De Vito, P.J. Lahti and A. Levrero, *The Theory of Symmetry Actions in Quantum Mechanics*, Lecture Notes in Physics 654, Springer, 2004.

48. G.H. Chan and M.H. Lim, *Linear preservers on powers of matrices*, Linear Algebra Appl. **162-164** (1992), 615–626.

49. Z. Charzyński, *Sur les transformations isomtriques des espaces du type (F)*, Studia Math. **13** (1953), 94–121.

50. M.A. Chebotar, W.F. Ke and P.H. Lee, *Maps characterized by action on zero products*, Pacific J. Math. **216** (2004), 217–228.

51. P.R. Chernoff, *Representations, automorphisms and derivations of some operator algebras*, J. Funct. Anal. **12** (1973), 275–289.

52. W.S. Cheung, S. Fallat and C.K. Li, *Multiplicative preservers on semigroups of matrices*, Linear Algebra Appl. **355** (2002), 173–186.

53. P. Civin and B. Yood, *Lie and Jordan structures in Banach algebras*, Pacific J. Math. **15** (1965), 775–797.

54. J.B. Conway, *A Course in Functional Analysis*, Springer-Verlag, 1985.

55. J.B. Conway, *A Course in Operator Theory*, American Mathematical Society, 2000.

56. R.L. Crist, *Local derivations on operator algebras*, J. Funct. Anal. **135** (1996), 76–92.

57. R. Crist, *Local automorphisms*, Proc. Amer. Math. Soc. **128** (2000), 1409–1415.

58. J. Cui and J. Hou, *Linear maps on von Neumann algebras preserving zero products or tr-rank*, Bull. Austral. Math. Soc. **65** (2002), 79–91.

59. J. Cui and J. Hou, *Linear maps preserving ideals of C^*-algebras*, Proc. Amer. Math. Soc. **131** (2003), 3441–3446.

60. J. Cui and J. Hou, *A characterization of homomorphisms between Banach algebras*, Acta Math. Sin. (Engl. Ser.) **20** (2004), 761–768.

61. T. Dang, Y. Friedman and B. Russo, *Affine geometric proofs of the Banach Stone theorems of Kadison and Kaup*, Rocky Mountain J. Math. **20** (1990), 409–428.

62. K.R. Davidson, *C^*-Algebras by Example*, American Mathematical Society, 1996.

63. E.B. Davies, *Quantum Theory of Open Systems*, Academic Press, 1976.

64. C. Davis, *Separation of two linear subspaces*, Acta Sci. Math. (Szeged) **19** (1958), 172–187.

65. J. Dieudonné, *Sur une généralisation du groupe orthogonal à quatre variables*, Arch. Math. **1** (1949), 282–287.

66. D.Z. Djokovič, *Linear transformations of tensor products preserving a fixed rank*, Pacific J. Math. **30** (1969), 411–414.

67. R.G. Douglas, *On majorization, factorization, and range inclusion of operators on Hilbert space*, Proc. Amer. Math. Soc. **17** (1966), 413–415.

68. A. Dvurečenskij, *Gleason's Theorem and Its Applications*, Kluwer Academic Publishers, 1993.

69. A. Dvurečenskij and S. Pulmannová, *Recent Trends in Quantum Structures*, Kluwer Academic Publisher, 2000.

70. P.A. Fillmore, *Sums of operators with square zero*, Acta Sci. Math. (Szeged) **28** (1967), 285–288.

71. P.A. Fillmore and W.E. Longstaff, *On isomorphisms of lattices of closed subspaces*, Canad. J. Math. **36** (1984), 820–829.

72. C.K. Fong, C.R. Miers and A.R. Sourour, *Lie and Jordan ideals of operators on Hilbert space*, Proc. Amer. Math. Soc. **84** (1982), 516–520.

73. J.J. Font and S. Hernandez, *On separating maps between locally compact spaces*, Arch. Math. **63** (1994), 158–165.

74. J.J. Font and M. Sanchis, *A characterization of locally compact spaces with homeomorphic one-point compactifications*, Topology Appl., **121** (2002), 91–104.

75. J.J. Font and M. Sanchis, *Extreme points and the diameter norm*, Rocky Mountain J. Math. **34** (2004), 1325–1332.

76. G. Frobenius, *Über die Darstellung der endlichen Gruppen durch lineare Substitutionen*, Sitzungsber. Königl. Preuss. Akad. Wiss. Berlin (1897), 994–1015.

77. I.C. Gohberg and M.G. Krein, *Introduction to The Theory of Linear Nonselfadjoint Operators*, American Mathematical Society, 1969.

78. González and V.V. Uspenskij, *On homomorphisms of groups of integer-valued functions*, Extracta Math. **14** (1999), 19–29.

79. S. Gudder and R. Greechie, *Sequential products on effect algebras*, Rep. Math. Phys. **49** (2002), 87–111.

80. S. Gudder and R. Greechie, *Uniqueness and order in sequential effect algebras*, Int. J. Theor. Phys. **44** (2005), 755–770.

81. S. Gudder and G. Nagy, *Sequential quantum measurements*, J. Math. Phys. **42** (2001), 5212–5222.

82. S. Gudder and G. Nagy, *Sequentially independent effects*, Proc. Amer. Math. Soc. **130** (2002), 1125–1130.

83. R.M. Guralnick, C.K. Li and L. Rodman, *Multiplicative maps on invertible matrices that preserve matricial properties*, Electron. J. Linear Algebra **10** (2003) 291–319.

84. M. Győry, *Diameter preserving linear bijections of $C_0(X)$*, Publ. Math. (Debrecen) **54** (1999), 207–215.

85. M. Győry, *2-local isometries of $C_0(X)$*, Acta Sci. Math. (Szeged) **67** (2001), 735–746.

86. M. Győry, *Preserver Problems and Reflexivity Problems on Operator Algebras and on Function Algebras*, PhD dissertation, University of Debrecen, 2003.

87. M. Győry, *Transformations on the set of all n-dimensional subspaces of a Hilbert space preserving orthogonality*, Publ. Math. (Debrecen) **65** (2004), 233–242.

88. M. Győry, *On the topological reflexivity of the isometry group of the suspension of $B(H)$*, Studia Math. **166** (2005), 287–303.

89. M. Győry and L. Molnár, *Diameter preserving linear bijections of C(X)*, Arch. Math. **71** (1998), 301–310.

90. M. Győry, L. Molnár and P. Šemrl, *Linear rank and corank preserving maps on B(H) and an application to *-semigroup isomorphisms of operator ideals*, Linear Algebra Appl. **280** (1998), 253–266.

91. D. Hadwin, *A general view of reflexivity*, Trans. Amer. Math. Soc. **344** (1994), 325–360.

92. D. Hadwin and J.K. Li, *Local derivations and local automorphisms*, J. Math. Anal. Appl. **290** (2004), 702–714.

93. N. Hadjisavvas, *Metrics on the set of states of a W*-algebra*, Linear Algebra Appl. **84** (1986), 281–287.

94. J. Hakeda, *Additivity of *-semigroup isomorphisms among *-algebras*, Bull. London Math. Soc. **18** (1986), 51–56.

95. J. Hakeda, *Additivity of Jordan *-maps on AW*-algebras*, Proc. Amer. Math. Soc. **96** (1986), 413–420.

96. J. Hakeda and K. Saitô, *Additivity of Jordan *-maps between operator algebras*, J. Math. Soc. Japan **38** (1986), 403–408.

97. P.R. Halmos, *A Hilbert Space Problem Book*, Van Nostrand, 1967.

98. P.R. Halmos, *Two subspaces*, Trans. Amer. Math. Soc. **144** (1969), 381–389.

99. S. Hernandez, E. Beckenstein and L. Narici, *Banach-Stone theorems and separating maps*, Manuscripta Math. **86** (1995), 409–416.

100. I.N. Herstein, *Jordan homomorphisms*, Trans. Amer. Math. Soc. **81** (1956), 331–341.

101. S.H. Hochwald, *Multiplicative maps on matrices that preserve the spectrum*, Linear Algebra Appl. **212/213** (1994), 339–351.

102. J.R. Holub, *On the metric geometry of ideals of operators on Hilbert space*, Math. Ann. **201** (1973), 157–163.

103. H. Hotelling, *Relations between two sets of variates*, Biometrika **28** (1935), 321-377.

104. J. Hou, *Rank-preserving linear maps on B(X)*, Sci. China Ser. A **32** (1989), 929–940.

105. J. Hou and J. Cui, *Additive maps on standard operator algebras preserving invertibilities or zero divisors*, Linear Algebra Appl. **359** (2003), 219–233.

106. N. Jacobson and C. Rickart, *Jordan homomorphisms of rings*, Trans. Amer. Math. Soc. **69** (1950), 479–502.

107. A.A. Jafarian and A.R. Sourour, *Spectrum-preserving linear maps*, J. Funct. Anal. **66** (1986), 255–261.

108. K. Jarosz, *Automatic continuity of separating linear isomorphisms*, Canad. Math. Bull. **33** (1990), 139-144.

109. K. Jarosz and T.S.S.R.K. Rao, *Local isometries of function spaces*, Math. Z. **243** (2003), 449–469.

110. J.S. Jeang and N.C. Wong, *Weighted composition operators of $C_0(X)$'s*, J. Math. Anal. Appl. **201** (1996), 981–996.

111. W. Jing and S.J. Lu, *Topological reflexivity of the spaces of (α, β)-derivations on operator algebras*, Studia Math. **156** (2003), 121–131.

112. W. Jing, S. Lu and G. Han, *On topological reflexivity of the spaces of derivations on operator algebras*, Appl. Math. Ser. B. **17** (2002), 75–79.

113. M. Jodeit Jr. and T.Y. Lam, *Multiplicative maps of matrix semi-groups*, Arch. Math. **20** (1969), 10–16.

114. B.E. Johnson, *Local derivations on C^*-algebras are derivations,* Trans. Amer. Math. Soc. **353** (2001), 313–325.
115. C. Jordan, *Essai sur la géométrie á n dimensions,* Bull. Soc. Math. France **3** (1875), 103–174.
116. R. Jozsa, *Fidelity for mixed quantum states,* J. Modern Opt. **41** (1994), 2315–2323.
117. R.V. Kadison, *Isometries of operator algebras,* Ann. Math. **54** (1951) 325–338.
118. R.V. Kadison, *A generalized Schwarz inequality and algebraic invariants for operator algebras,* Ann. of Math. **56** (1952), 494–503.
119. R.V. Kadison, *Local derivations,* J. Algebra **130** (1990), 494–509.
120. R.V. Kadison and J.R. Ringrose, *Fundamentals of the Theory of Operator Algebras, Vol I.,* Academic Press, 1983.
121. R.V. Kadison and J.R. Ringrose, *Fundamentals of the Theory of Operator Algebras, Vol II.,* Academic Press, 1986.
122. I. Kaplansky, *Algebraic and Analytical Aspects of Operator Algebras,* CBMS Regional Conf. Ser. in Math. 1, Amer. Math. Soc., Providence, 1970.
123. W.F. Ke, B.R. Li and N.C. Wong, *Zero product preserving maps of continuous operator valued functions,* Proc. Amer. Math. Soc. **132** (2004), 1979-1985.
124. S.O. Kim, *Linear maps preserving ideals of C^*-algebras,* Proc. Amer. Math. Soc. **129** (2001), 1665–1668.
125. S.O. Kim, *Automorphisms of Hilbert space effect algebras,* Linear Algebra Appl. **402** (2005), 193–198.
126. S.O. Kim and J.S. Kim, *Local automorphisms and derivations on M_n,* Proc. Amer. Math. Soc. **132** (2004), 1389–1392.
127. S.O. Kim and J.S. Kim, *Local automorphisms and derivations on certain C^*-algebras,* Proc. Amer. Math. Soc. **133** (2005), 3303-3307.
128. A.A. Kirillov and A.D. Gvishiani, *Theorems and Problems in Functional Analysis,* Springer-Verlag, 1982.
129. S. Kowalski and Z. Slodkowski, *A characterization of multiplicative linear functionals in Banach algebras,* Studia Math. **67** (1980), 215–223.
130. K. Kraus, *States, Effects and Operations,* Lecture Notes in Physics, Vol. 190, Springer-Verlag, 1983.
131. M. Kuczma, *An Introduction to the Theory of Functional Equations and Inequalities,* Państwowe Wydawnictwo Naukowe, Warszawa–Kraków–Katowice, 1985.
132. L.E. Labuschagne and V. Mascioni, *Linear maps between C^*-algebras whose adjoint preserve extreme points of the unit ball,* Adv. Math. **138** (1998), 15–45.
133. D.R. Larson, *Reflexivity, algebraic reflexivity and linear interpolation,* Amer. J. Math. **110** (1988), 283–299.
134. D.R. Larson and A.R. Sourour, *Local derivations and local automorphisms of $B(X)$,* Proc. Sympos. Pure Math. 51, Providence, Rhode Island 1990, Part 2, pp. 187–194.
135. J.S. Lemont and P. Mendelson, *The Wigner unitary-antiunitary theorem,* Ann. Math. **78** (1963), 548–559.
136. C.K. Li and N.K. Tsing, *Linear preserver problems: A brief introduction and some special techniques,* Linear Algebra Appl. **162-164** (1992), 217–235.

137. C.K. Li and S. Pierce, *Linear preserver problems,* Amer. Math. Monthly **108** (2001), 591–605.

138. Z. Ling and F. Lu, *Jordan maps of nest algebras,* Linear Algebra Appl. **387** (2004), 361–368.

139. J. Lindenstrauss and L. Tzafriri, *Classical Banach Spaces I,* Springer-Verlag, 1977.

140. R. Loewy, *Linear mappings which are rank-k nonincreasing,* Linear and Multilinear Algebra **34** (1993), 21–32.

141. A.J. Loginov and V.S. Shulman, *Hereditary and intermediate reflexivity of W^*-algebras,* Izv. Akad. Nauk SSSR **39** (1975), 1260–1273. English transl. in USSR-Isv. **9** (1975), 1189–1201.

142. F. Lu, *Multiplicative mappings of operator algebras,* Linear Algebra Appl. **347** (2002), 283–291.

143. F. Lu, *Additivity of Jordan maps on standard operator algebras,* Linear Algebra Appl. **357** (2002), 121–131.

144. F. Lu, *Jordan maps on associative algebras,* Commun. Algebra **31** (2003), 2273–2286.

145. F. Lu, *Jordan triple maps,* Linear Algebra Appl. **375** (2003), 311–317.

146. G. Ludwig, *Foundations of Quantum Mechanics, Vol. I.,* Springer Verlag, 1983.

147. B. Magajna, *Hilbert C^*-modules in which all closed submodules are complemented,* Proc. Amer. Math. Soc. **125** (1997), 849–852.

148. P. Mankiewicz, *On extension of isometries in normed linear spaces,* Bull. Acad. Pol. Sci., Sér. Sci. Math. Astron. Phys. **20** (1972), 367–371.

149. M. Marcus, *All linear operators leaving the unitary group invariant,* Duke Math. J. **26** (1959), 155–163.

150. M. Marcus and B.N. Moyls, *Transformations on tensor product spaces,* Pacific J. Math. **9** (1959), 1215–1221.

151. M. Marcus and R. Purves, *Linear transformations on algebras of matrices: the invariance of elementary symmetric functions,* Canad. J. Math. **11** (1959), 383–396.

152. J. Marovt and T. Petek, *Automorphisms of Hilbert space effect algebras equipped with Jordan triple product, the two-dimensional case,* Publ. Math. (Debrecen) **66** (2005), 245-250.

153. W.S. Martindale III, *Jordan homomorphisms of the symmetric elements of a ring with involution,* J. Algebra **5** (1967), 232–249.

154. W.S. Martindale III, *When are multiplicative mappings additive?,* Proc. Amer. Math. Soc. **21** (1969), 695–698.

155. V. Mascioni and L. Molnár, *Linear maps on factors which preserve the extreme points of the unit ball,* Canad. Math. Bull. **41** (1998), 434–441.

156. M. Matvejchuk, *Gleason's theorem in W^*J-algebras in spaces with indefinite metric,* Internat. J. Theoret. Phys. **38** (1999), 2065–2093.

157. S. Mazur and S.M. Ulam, *Sur les transformations isométriques des espaces vectoriels normés* C.R. Acad. Sci. Paris **194** (1932), 946–948.

158. J. Miao and A. Ben-Israel, *On principal angles between subspaces in $\mathbb{R}^n$,* Linear Algebra Appl. **171** (1992), 81–98.

159. L. Molnár, *Two characterizations of additive *-automorphisms of $\mathcal{B}(H)$,* Bull. Austral. Math. Soc. **53** (1996), 391–400.

160. L. Molnár, *Wigner's unitary-antiunitary theorem via Herstein's theorem on Jordan homomorphisms,* J. Nat. Geom. **10** (1996), 137–148.

224   References

161. L. Molnár, *The set of automorphisms of B(H) is topologically reflexive in B(B(H))*, Studia Math. **122** (1997), 183–193.
162. L. Molnár, *A proper standard C*-algebra whose automorphism and isometry groups are topologically reflexive*, Publ. Math. (Debrecen) **52** (1998), 563–574.
163. L. Molnár, *The automorphism and isometry groups of $\ell_\infty(\mathbb{N}, \mathcal{B}(\mathcal{H}))$ are topologically reflexive*, Acta Sci. Math. (Szeged) **64** (1998), 671–680.
164. L. Molnár, *An algebraic approach to Wigner's unitary-antiunitary theorem*, J. Austral. Math. Soc. **65** (1998), 354–369.
165. L. Molnár, *Some linear preserver problems on B(H) concerning rank and corank*, Linear Algebra Appl. **286** (1999), 311-321.
166. L. Molnár, *A generalization of Wigner's unitary-antiunitary theorem to Hilbert modules*, J. Math. Phys. **40** (1999), 5544–5554.
167. L. Molnár, *Some multiplicative preservers on B(H)*, Linear Algebra Appl. **301** (1999), 1–13.
168. L. Molnár, *Reflexivity of the automorphism and isometry groups of C*-algebras in BDF theory*, Arch. Math. **74** (2000), 120–128.
169. L. Molnár, *Generalization of Wigner's unitary-antiunitary theorem for indefinite inner product spaces*, Commun. Math. Phys. **210** (2000), 785–791.
170. L. Molnár, *On some automorphisms of the set of effects on Hilbert space*, Lett. Math. Phys. **51** (2000), 37–45.
171. L. Molnár, *A Wigner-type theorem on symmetry transformations in type II factors*, Int. J. Theor. Phys. **39** (2000), 1463–1466.
172. L. Molnár, *A Wigner-type theorem on symmetry transformations in Banach spaces*, Publ. Math. (Debrecen) **58** (2000), 231–239.
173. L Molnár, *A reflexivity problem concerning the C*-algebra $C(X) \otimes B(H)$*, Proc. Amer. Math. Soc. **129** (2001), 531–537.
174. L. Molnár, *Transformations on the set of all n-dimensional subspaces of a Hilbert space preserving principal angles*, Commun. Math. Phys. **217** (2001), 409–421.
175. L. Molnár, *Characterizations of the automorphisms of Hilbert space effect algebras*, Commun. Math. Phys. **223** (2001), 437–450.
176. L. Molnár, **-semigroup endomorphisms of B(H)*, in I. Gohberg et al. (Edt.), Operator Theory: Advances and Applications, Vol. 127, pp. 465–472, Birkhäuser, 2001.
177. L. Molnár, *Order-automorphisms of the set of bounded observables*, J. Math. Phys. **42** (2001), 5904–5909.
178. L. Molnár, *Fidelity preserving maps on density operators*, Rep. Math. Phys. **48** (2001), 299–303.
179. L. Molnár, *Local automorphisms of some quantum mechanical structures*, Lett. Math. Phys. **58** (2001), 91–100.
180. L. Molnár, *Some characterizations of the automorphisms of B(H) and C(X)*, Proc. Amer. Math. Soc. **130** (2002), 111-120.
181. L. Molnár, *On certain automorphisms of sets of partial isometries*, Arch. Math. **78** (2002), 43–50.
182. L. Molnár, *2-local isometries of some operator algebras*, Proc. Edinb. Math. Soc. **45** (2002), 349–352.
183. L. Molnár, *Orthogonality preserving transformations on indefinite inner product spaces: generalization of Uhlhorn's version of Wigner's theorem*, J. Funct. Anal., **194** (2002), 248–262.

184. L. Molnár, *Local automorphisms of operator algebras on Banach spaces*, Proc. Amer. Math. Soc. **131** (2003), 1867–1874.

185. L. Molnár, *Preservers on Hilbert space effects,* Linear Algebra Appl. **370** (2003), 287–300.

186. L. Molnár, *Sequential isomorphisms between the sets of von Neumann algebra effects,* Acta Sci. Math. (Szeged) **69** (2003), 755–772.

187. L. Molnár and M. Barczy, *Linear maps on the space of all bounded observables preserving maximal deviation,* J. Funct. Anal. **205** (2003), 380–400.

188. L. Molnár and M. Győry, *Reflexivity of the automorphism and isometry groups of the suspension of* $\mathcal{B}(\mathcal{H})$, J. Funct. Anal. **159** (1998), 568–586.

189. L. Molnár and E. Kovács, *An extension of a characterization of the automorphisms of Hilbert space effect algebras,* Rep. Math. Phys. **52** (2003), 141–149.

190. L. Molnár and Zs. Páles, $^{\perp}$*-order automorphisms of Hilbert space effect algebras: The two-dimensional case,* J. Math. Phys. **42** (2001), 1907–1912.

191. L. Molnár and P. Šemrl, *Local Jordan *-derivations of standard operator algebras,* Proc. Amer. Math. Soc. **125** (1997), 447–454.

192. L. Molnár and P. Šemrl, *Order isomorphisms and triple isomorphisms of operator ideals and their reflexivity,* Arch. Math. **69** (1997), 497–506.

193. L. Molnár and P. Šemrl, *Local automorphisms of the unitary group and the general linear group on a Hilbert space,* Expo. Math. **18** (2000), 231–238.

194. L. Molnár and W. Timmermann, *Isometries of quantum states,* J. Phys. A: Math. Gen., **36** (2003), 267–273.

195. L. Molnár and W. Timmermann, *Preserving the measure of compatibility between quantum states,* J. Math. Phys. **44** (2003), 969–973.

196. L. Molnár and B. Zalar, *Reflexivity of the group of surjective isometries on some Banach spaces,* Proc. Edinb. Math. Soc. **42** (1999), 17–36.

197. L. Molnár and B. Zalar, *On local automorphism of group algebras of compact groups,* Proc. Amer. Math. Soc. **128** (2000), 93–99.

198. G.J. Murphy, *C*-algebras and Operator Theory,* Academic Press, 1990.

199. M.A. Naimark, *Normed Algebras,* Wolters-Noordhoff Publishing, 1972.

200. A. Nowicki, *On local derivations in the Kadison sense,* Colloq. Math. **89** (2001), 193–198.

201. M. Omladič, *On operators preserving commutativity,* J. Funct. Anal., **66** (1986), 105–122.

202. M. Omladič, H. Radjavi and P. Šemrl, *Preserving commutativity,* J. Pure Appl. Algebra **156** (2001), 309-328.

203. M. Omladič and P. Šemrl, *Additive mappings preserving operators of rank one,* Linear Algebra Appl. **182** (1993), 239-256.

204. P.G. Ovchinnikov, *Automorphisms of the poset of skew projections,* J. Funct. Anal. **115** (1993), 184–189.

205. C.C. Paige and M. Wei, *History and generality of the CS decomposition,* Linear Algebra Appl. **208/209** (1994), 303-326.

206. T.W. Palmer, *Banach Algebras and The General Theory of *-Algebras, Vol. I.,* Encyclopedia Math. Appl. 49, Cambridge University Press, 1994.

207. R. S. Pierce, *Associative Algebras,* Springer-Verlag, 1982.

208. S. Pierce et al., *A survey of linear preserver problems,* Linear and Multilinear Algebra **33** (1992), 1-129.

209. H. Porta and J.T. Schwartz, *Representations of the algebra of all operators in Hilbert space, and related analytic function algebras,* Commun. Pure Appl. Math. **20** (1967), 457–492.

210. M. Radjabalipour, *Additive mappings on von Neumann algebras preserving absolute values,* Linear Algebra Appl. **368** (2003), 229–241.

211. M. Radjabalipour, K. Seddighi and Y. Taghavi, *Additive mappings on operator algebras preserving absolute values,* Linear Algebra Appl. **327** (2001), 197–206.

212. J. Rätz, *On Wigner's theorem: remarks, complements, comments, and corollaries,* Aequationes Math. **52** (1996), 1–9.

213. M. Rais, *The unitary group preserving maps (the infinite dimensional case),* Linear Multilinear Algebra **20** (1987), 337–345.

214. T.S.S.R.K. Rao, *Local surjective isometries of function spaces,* Expositiones Math. **18** (2000), 285–296.

215. T.S.S.R.K. Rao, *Some generalizations of Kadison's theorem: A survey,* Extracta Math. **19** (2004), 319–334.

216. T.S.S.R.K. Rao, *Local isometries of $\mathcal{L}(X, C(K))$,* Proc. Amer. Math. Soc. **133** (2005), 2729-2732.

217. T.S.S.R.K. Rao and A.K. Roy, *Diameter preserving linear bijections of functions spaces,* J. Austral. Math. Soc. **70** (2001), 323–335.

218. O.S. Rothaus, *Order isomorphisms of cones,* Proc. Amer. Math. Soc. **17** (1966), 1284–1288.

219. B. Russo and H.A. Dye, *A note on unitary operators in C^*-algebras,* Duke Math. J. **33** (1966), 413–416.

220. S. Sakai, *C^*-Algebras and W^*-Algebras,* Springer-Verlag, 1971.

221. E. Scholz and W. Timmermann, *Local derivations, automorphismsm and commutativity preserving maps on $L^+(D)$,* Publ. Res. Inst. Math. Sci. **29** (1993), 977–995.

222. O. Schreier and B.L. van der Waerden, *Die Automorphismen der projektiven Gruppen,* Abh. Math. Sem. Hamburgischen Univ. **6** (1928), 303–322.

223. P. Šemrl, *Isomorphisms of standard operator algebras,* Proc. Amer. Math. Soc. **123** (1995), 1851–1855.

224. P. Šemrl, *Local automorphisms and derivations on $B(H)$,* Proc. Amer. Math. Soc. **125** (1997), 2677–2680.

225. P. Šemrl, *Generalized symmetry transformations on quaternionic indefinite inner product spaces: An extension of quaternionic version of Wigner's theorem,* Commun. Math. Phys. **242** (2003), 579–584.

226. P. Šemrl, *Order-preserving maps on the poset of idempotent matrices,* Acta Sci. Math. (Szeged) **69** (2003), 481–490.

227. P. Šemrl, *Applying projective geometry to transformations on rank one idempotents,* J. Funct. Anal. **210** (2004), 248–257.

228. P. Šemrl, *Orthogonality preserving transformations on the set of n-dimensional subspaces of a Hilbert space,* Illinois J. Math. **48** (2004), 567–573.

229. P. Šemrl, *Maps on idempotents,* Studia Math. **169** (2005), 21–44.

230. C.S. Sharma and D.F. Almeida, *A direct proof of Wigner's theorem on maps which preserve transition probabilities between pure states of quantum systems,* Ann. Phys. **197** (1990), 300–309.

231. V.S. Shulman, *Operators preserving ideals in C^*-algebras,* Studia Math. **109** (1994), 67–72.

232. B. Simon, *Quantum dynamics: from automorphism to hamiltonian,* in *Studies in Mathematical Physics. Essays in Honor of Valentine Bargmann,* eds. E.H. Lieb, B. Simon, A.S. Wightman, Princeton Series in Physics, Princeton University Press, 1976, pp. 327–349.

233. B. Simon, *Trace ideals and their applications,* London Math. Soc., Lecture Notes Series 35, Cambridge University Press, 1979.

234. A.R. Sourour, *Invertibility preserving linear maps on $\mathcal{L}(X)$,* Trans. Amer. Math. Soc. **348** (1996), 13–30.

235. E. Størmer, *On the Jordan structure of C^*-algebras,* Trans. Amer. Math. Soc. **120** (1965), 438–447.

236. S. Strătilă and L. Zsidó, *Lectures on von Neumann Algebras,* Abacus Press, 1979.

237. U. Uhlhorn, *Representation of symmetry transformations in quantum mechanics,* Ark. Fysik **23** (1963), 307–340.

238. A. Uhlmann, *The "transition probability" in the state space of a *-algebra,* Rep. Math. Phys. **9** (1976), 273–279.

239. A. Uhlmann, *Geometric phases and related structures,* Rep. Math. Phys. **36** (1995), 461–481.

240. A. Uhlmann, *Spheres and hemispheres as quantum state spaces,* J. Geom. Phys. **18** (1996), 76–92.

241. A. Uhlmann, *On "partial" fidelities,* Rep. Math. Phys. **45** (2000), 407–418.

242. A. Uhlmann, *Simultaneous decomposition of two states,* Rep. Math. Phys. **46** (2000), 319–324.

243. V.S. Varadarajan, *Geometry of Quantum Theory, Vol. I.,* D Van Nostrand Company, Inc., 1968.

244. R. C. Walker, *The Stone-Čech Compactification,* Springer, 1974

245. R. Wang, *Linear isometric operators on the $C_0^{(n)}(X)$ type spaces,* Kodai Math. J. **19** (1996), 259–281.

246. W.C. Waterhouse, *On linear transformations preserving rank one matrices over commutative rings,* Linear Multilinear Algebra **17** (1985), 101–106.

247. W.C. Waterhouse, *Linear transformations on self-adjoint matrices: The preservation of rank-one-plus-scalar,* Linear Algebra Appl. **74** (1986), 73–85.

248. N. Weaver, *Isometries of noncompact Lipschitz spaces,* Canad. Math. Bull. **38** (1995), 242–249.

249. M. Wiehl, *Local derivations on the Weyl algebra with one pair of generators,* Acta Math. Hung. **92** (2001), 51–59.

250. E.P. Wigner, *Gruppentheorie und ihre Anwendung auf die Quantenmechanik der Atomspektrum,* Fredrik Vieweg und Sohn, 1931.

251. J. Wu, *Local derivations of reflexive algebras,* Proc. Amer. Math. Soc. **125** (1997), 869–873.

252. J. Wu, *Local derivations of reflexive algebras II.,* Proc. Amer. Math. Soc. **129** (2001), 1733–1737.

253. J. Xie and F. Lu, *A note on 2-local automorphisms of digraph algebras* Linear Algebra Appl. **378** (2004), 93–98.

254. J. Zielinski, *Local derivations in polynomial and power series rings,* Colloq. Math. **92** (2002), 295–305.

References Added in Revision

255. R.L. An and J.C. Hou, *Additivity of Jordan multiplicative maps on Jordan operator algebras,* Taiwanese J. Math. **10** (2006), 45–64.
256. K.I. Beidar, M. Brešar, M.A. Chebotar, Y. Fong, *Applying functional identities to some linear preserver problems,* Pacific J. Math. **204** (2002), 257–271.
257. M. Brešar, P. Šemrl, *Commutativity preserving maps on central simple algebras,* J. Algebra **284** (2005), 102–110.
258. M. Brešar and P. Šemrl, *On bilinear maps on matrices with applications to commutativity preservers,* J. Algebra (to appear)
259. F. Cabello Sánchez, *Local isometries on spaces of continuous functions,* Math. Z. **251** (2005), 735–749.
260. J.T. Chan, C.K. Li and N.S. Sze, *Mappings on matrices: invariance of functional values of matrix products,* J. Austral. Math. Soc. (to appear)
261. J.T. Chan, C.K. Li and N.S. Sze, *Mappings preserving spectra of product of matrices,* Proc. Amer. Math. Soc. (to appear)
262. M.A. Chebotar, Y. Fong and P.H. Lee, *On maps preserving zeros of the polynomial $xy - yx^*$,* Linear Algebra Appl. **408** (2005), 230–243.
263. M.A. Chebotar, W. Ke and P. Lee, *Maps preserving zero Jordan products on Hermitian operators,* Illinois J. Math. **49** (2005), 445–452.
264. M.A. Chebotar, W.F. Ke, P.H. Lee and L.S. Shiao, *On maps preserving certain algebraic properties,* Canad. Math. Bull. **48** (2005), 355–369.
265. J. Cui, J. Hou and C-G. Park, *Indefinite orthogonality preserving additive maps,* Arch. Math. **83** (2004), 548–557.
266. A. Fošner, *Automorphisms of the poset of upper triangular idempotent matrices,* Linear Multilinear Algebra **53** (2005), 27–44.
267. A. Fošner, *Order preserving maps on the poset of upper triangular idempotent matrices,* Linear Algebra Appl. **403** (2005), 248–262.
268. A. Fošner, *Non-linear commutativity preserving maps on $M_n(\mathbb{R})$,* Linear Multilinear Algebra **53** (2005), 323–344.
269. O. Hatori, T. Miura and H. Takagi, *Characterizations of isometric isomorphisms between uniform algebras via non-linear range-preserving properties,* Proc. Amer. Math. Soc. (to appear)
270. J. Hou and Q. Di, *Maps preserving numerical ranges of operator products,* Proc. Amer. Math. Soc. **134** (2006), 1435–1446.
271. G. Lešnjak and N.S. Sze, *On injective Jordan semi-triple maps of matrix algebras,* Linear Algebra Appl. **414** (2006), 383–388.
272. C.K. Li and E. Poon, *Schur product of matrices and numerical radius (range) preserving maps,* Linear Algebra Appl. (to appear)
273. C.K. Li and N.S. Sze, *Product of operators and numerical range preserving maps,* Studia Math. (to appear)
274. P. Li, *Elementary maps on nest algebras,* J. Math. Anal. Appl. (to appear)
275. P. Li and W. Jing, *Jordan elementary maps on rings,* Linear Algebra Appl. **382** (2004), 237–245.
276. P. Li and F. Lu, *Additivity of elementary maps on rings,* Commun. Alg. **32** (2004), 3725–3737.
277. P. Li and F. Lu, *Additivity of Jordan elementary maps on nest algebras,* Linear Algebra Appl. **400** (2005), 327–338.
278. J.H. Liu and N.C. Wong, *2-local automorphisms of operator algebras,* J. Math. Anal. Appl. **321** (2006), 741–750.

279. J. Marovt, *Order preserving bijections of* $C(\mathcal{X}, I)$, J. Math. Anal. Appl. **311** (2005), 567–581.

280. J. Marovt, *Multiplicative bijections of* $C(\mathcal{X}, I)$, Proc. Amer. Math. Soc. **134** (2006), 1065–1075.

281. J. Marovt, *Affine bijections of* $C(\mathcal{X}, I)$, Studia Math. **173** (2006), 295–309.

282. L. Molnár, *Non-linear Jordan triple automorphisms of sets of self-adjoint matrices and operators*, Studia Math., **173** (2006), 39–48.

283. L. Molnár, *A remark to the Kochen-Specker theorem and some characterizations of the determinant on sets of Hermitian matrices*, Proc. Amer. Math. Soc. **134** (2006), 2839–2848.

284. L. Molnár and P. Šemrl, *Non-linear commutativity preserving maps on self-adjoint operators*, Quart. J. Math., **56** (2005), 589–595.

285. L. Molnár and P. Šemrl, *Elementary operators on self-adjoint operators*, J. Math. Anal. Appl. (to appear)

286. L. Molnár and W. Timmermann, *Transformations on the sets of states and density operators*, Linear Algebra Appl. (to appear)

287. F. Pop, *On local representations of von Neumann algebras*, Proc. Amer. Math. Soc. **132** (2004), 3569–3576.

288. N.V. Rao and A.K. Roy, *Multiplicatively spectrum-preserving maps of function algebras*, Proc. Amer. Math. Soc. **133** (2005), 1135–1142.

289. N.V. Rao and A.K. Roy, *Multiplicatively spectrum-preserving maps of function algebras II*, Proc. Edinb. Math. Soc. **48** (2005), 219–229.

290. L. Rodman and P. Šemrl, *Orthogonality preserving bijective maps on real and complex projective spaces*, Linear and Multilinear Algebra (to appear)

291. P. Šemrl, *Maps on idempotent matrices over division rings*, J. Algebra **298** (2006), 142–187.

292. P. Šemrl, *Maps on matrix spaces*, Linear Algebra Appl. **413** (2006), 364–393.

293. P. Šemrl, *Non-linear commutativity preserving maps*, Acta Sci. Math. (Szeged) **71** (2005), 781–819.

294. P. Šemrl, *Maps on idempotent operators*, Banach Cent. Publ. (to appear)

295. P. Šemrl, *Maps on matrix and operator algebras*, Jahresber. Deutsch. Math.-Verein. (to appear)

296. W. Timmermann, *Elementary operators on algebras of unbounded operators*, Acta Math. Hung. **107** (2005) 149–160.

297. A. Turnšek, *On operators preserving James' orthogonality*, Linear Algebra Appl. **407** (2005), 189–195.

298. J.H. Zhang, S. Feng and R.H. Wu, *Local 2-cocycles of nest subalgebras of factor von Neumann algebras*, Linear Algebra Appl. **416** (2006), 908–916.

299. J.H. Zhang, G.X. Ji and H.X. Cao, *Local derivations of nest subalgebras of von Neumann algebras*, Linear Algebra Appl. **392** (2004), 61–69.

300. J.H. Zhang, A.L. Yang and F.F. Pan, *Local automorphisms on nest subalgebras of factor von Neumann algebras*, Linear Algebra Appl. **402** (2005), 335–344.

Index

Lecture Notes in Mathematics

For information about earlier volumes
please contact your bookseller or Springer
LNM Online archive: springerlink.com

sanone/Brixen, Italy, 2003. Editors: M. Fritelli, W. Runggaldier (2004)

Vol. 1857: M. Émery, M. Ledoux, M. Yor (Eds.), Séminaire de Probabilités XXXVIII (2005)

Vol. 1858: A.S. Cherny, H.-J. Engelbert, Singular Stochastic Differential Equations (2005)

Vol. 1859: E. Letellier, Fourier Transforms of Invariant Functions on Finite Reductive Lie Algebras (2005)

Vol. 1860: A. Borisyuk, G.B. Ermentrout, A. Friedman, D. Terman, Tutorials in Mathematical Biosciences I. Mathematical Neurosciences (2005)

Vol. 1861: G. Benettin, J. Henrard, S. Kuksin, Hamiltonian Dynamics – Theory and Applications, Cetraro, Italy, 1999. Editor: A. Giorgilli (2005)

Vol. 1862: B. Helffer, F. Nier, Hypoelliptic Estimates and Spectral Theory for Fokker-Planck Operators and Witten Laplacians (2005)

Vol. 1863: H. Fürh, Abstract Harmonic Analysis of Continuous Wavelet Transforms (2005)

Vol. 1864: K. Efstathiou, Metamorphoses of Hamiltonian Systems with Symmetries (2005)

Vol. 1865: D. Applebaum, B.V. R. Bhat, J. Kustermans, J. M. Lindsay, Quantum Independent Increment Processes I. From Classical Probability to Quantum Stochastic Calculus. Editors: M. Schürmann, U. Franz (2005)

Vol. 1866: O.E. Barndorff-Nielsen, U. Franz, R. Gohm, B. Kümmerer, S. Thorbjønsen, Quantum Independent Increment Processes II. Structure of Quantum Levy Processes, Classical Probability, and Physics. Editors: M. Schürmann, U. Franz, (2005)

Vol. 1867: J. Sneyd (Ed.), Tutorials in Mathematical Biosciences II. Mathematical Modeling of Calcium Dynamics and Signal Transduction. (2005)

Vol. 1868: J. Jorgenson, S. Lang, $Pos_n(R)$ and Eisenstein Sereies. (2005)

Vol. 1869: A. Dembo, T. Funaki, Lectures on Probability Theory and Statistics. Ecole d'Eté de Probabilités de Saint-Flour XXXIII-2003. Editor: J. Picard (2005)

Vol. 1870: V.I. Gurariy, W. Lusky, Geometry of Müntz Spaces and Related Questions. (2005)

Vol. 1871: P. Constantin, G. Gallavotti, A.V. Kazhikhov, Y. Meyer, S. Ukai, Mathematical Foundation of Turbulent Viscous Flows, Martina Franca, Italy, 2003. Editors: M. Cannone, T. Miyakawa (2006)

Vol. 1872: A. Friedman (Ed.), Tutorials in Mathematical Biosciences III. Cell Cycle, Proliferation, and Cancer (2006)

Vol. 1873: R. Mansuy, M. Yor, Random Times and Enlargements of Filtrations in a Brownian Setting (2006)

Vol. 1874: M. Yor, M. Émery (Eds.), In Memoriam Paul-André Meyer - Séminaire de Probabilités XXXIX (2006)

Vol. 1875: J. Pitman, Combinatorial Stochastic Processes. Ecole d'Eté de Probabilités de Saint-Flour XXXII-2002. Editor: J. Picard (2006)

Vol. 1876: H. Herrlich, Axiom of Choice (2006)

Vol. 1877: J. Steuding, Value Distributions of L-Functions(2006)

Vol. 1878: R. Cerf, The Wulff Crystal in Ising and Percolation Models, Ecole d'Eté de Probabilits de Saint-Flour XXXIV-2004. Editor: Jean Picard (2006)

Vol. 1879: G. Slade, The Lace Expansion and its Appli- cations, Ecole d'Eté de Probabilités de Saint-Flour XXXIV-2004. Editor: Jean Picard (2006)

Vol. 1880: S. Attal, A. Joye, C.-A. Pillet, Open Quantum Systems I, The Hamiltonian Approach (2006)

Vol. 1881: S. Attal, A. Joye, C.-A. Pillet, Open Quantum Systems II, The Markovian Approach (2006)

Vol. 1882: S. Attal, A. Joye, C.-A. Pillet, Open Quantum Systems III, Recent Developments (2006)

Vol. 1883: W. Van Assche, F. Marcellàn (Eds.), Orthogonal Polynomials and Special Functions, Computation and Application (2006)

Vol. 1884: N. Hayashi, E.I. Kaikina, P.I. Naumkin, I.A. Shishmarev, Asymptotics for Dissipative Nonlinear Equations (2006)

Vol. 1885: A. Telcs, The Art of Random Walks (2006)

Vol. 1886: S. Takamura, Splitting Deformations of Degenerations of Complex Curves (2006)

Vol. 1887: K. Habermann, L. Habermann, Introduction to Symplectic Dirac Operators (2006)

Vol. 1888: J. van der Hoeven, Transseries and Real Differential Algebra (2006)

Vol. 1889: G. Osipenko, Dynamical Systems, Graphs, and Algorithms (2006)

Vol. 1890: M. Bunge, J. Funk, Singular Coverings of Toposes (2006)

Vol. 1891: J.B. Friedlander, D.R. Heath-Brown, H. Iwaniec, J. Kaczorowski, Analytic Number Theory, Cetraro, Italy, 2002. Editors: A. Perelli, C. Viola (2006)

Vol. 1892: A. Baddeley, I. Bárány, R. Schneider, W. Weil, Stochastic Geometry, Martina Franca, Italy, 2004. Editor: W. Weil (2007)

Vol. 1893: H. Hanßmann, Local and Semi-Local Bifurcations in Hamiltonian Dynamical Systems, Results and Examples (2007)

Vol. 1894: C.W. Groetsch, Stable Approximate Evaluation of Unbounded Operators (2007)

Vol. 1895: L. Molnár, Selected Preserver Problems on Algebraic Structures of Linear Operators and on Function Spaces (2007)

Recent Reprints and New Editions

Vol. 1618: G. Pisier, Similarity Problems and Completely Bounded Maps. 1995 – Second, Expanded Edition (2001)

Vol. 1629: J.D. Moore, Lectures on Seiberg-Witten Invariants. 1997 – Second Edition (2001)

Vol. 1638: P. Vanhaecke, Integrable Systems in the realm of Algebraic Geometry. 1996 – Second Edition (2001)

Vol. 1702: J. Ma, J. Yong, Forward-Backward Stochastic Differential Equations and their Applications. 1999. – Corrected 3rd printing (2005)